Earth and Environmental Sciences Library

Series Editor

Abdelazim M. Negm, Faculty of Engineering, Zagazig University, Zagazig, Egypt

Earth and Environmental Sciences Library (EESL) is a multidisciplinary book series focusing on innovative approaches and solid reviews to strengthen the role of the Earth and Environmental Sciences communities, while also providing sound guidance for stakeholders, decision-makers, policymakers, international organizations, and NGOs.

Topics of interest include oceanography, the marine environment, atmospheric sciences, hydrology and soil sciences, geophysics and geology, agriculture, environmental pollution, remote sensing, climate change, water resources, and natural resources management. In pursuit of these topics, the Earth Sciences and Environmental Sciences communities are invited to share their knowledge and expertise in the form of edited books, monographs, and conference proceedings.

Abdelazim M. Negm · Mustafa El-Rawy
Editors

Sustainability of Groundwater in the Nile Valley, Egypt

Springer

Editors
Abdelazim M. Negm
Faculty of Engineering
Zagazig University
Zagazig, Egypt

Mustafa El-Rawy
Faculty of Engineering
Minia University
Minia, Egypt

ISSN 2730-6674 ISSN 2730-6682 (electronic)
Earth and Environmental Sciences Library
ISBN 978-3-031-12678-9 ISBN 978-3-031-12676-5 (eBook)
https://doi.org/10.1007/978-3-031-12676-5

This Springer imprint is published by the registered company Springer Nature Switzerland AG
The registered company address is: Gewerbestrasse 11, 6330 Cham, Switzerland

Preface

Groundwater is the world's largest source of freshwater, but its safe and sustainable exploitation remains a challenge. Proper knowledge of the current state of the groundwater quantity and quality in the Nile Valley is vital for the development and management of groundwater resources in Egypt. Due to Egypt's water scarcity, the projected decline in Nile River flow due to climate change, and the development and possible improper operation of Grand Ethiopian Renaissance Dam, the water resources in Egypt would be greatly reduced. Furthermore, Egypt's growing population puts significant strain on groundwater, which is the second most important freshwater supply after the Nile River's surface water supply, which primarily comes from outside the country. Several books on the various aspects of Egypt's water resources have been published, but there is insufficient recent information on groundwater in Nile Valley aquifer, which is essential for Egyptians populations for domestic and irrigation purposes. As a result, the idea of publishing a book on the groundwater in the Nile Valley aquifer emerges to complete the picture of the Egypt's water resources in Egypt. It is a good example of an arid country of the almost arid MENA region. Consequently, the lessons learned from this book could be extremely beneficial to other countries in MENA regions, particularly those in North Africa. Egypt's Nile Valley aquifer is the most important renewable aquifer, accounting for approximately 3.5 million cubic meters (MCM) per year, or nearly half of all groundwater abstracted in Egypt. Egypt's long-term development and socioeconomic growth are dependent on the Nile Valley aquifer. Concerns about groundwater assessment, quality, management, and sustainability frame the current status of Nile Valley groundwater supplies.

This book is divided into 14 chapters, written by more than 25 researchers, scientists, and hydrologist experts from various institutions and countries with good experience in the field of water resources, particularly groundwater. The book's goal is to assess the groundwater quantity, quality, management, and sustainability in the Nile Valley, Egypt, which serves as a model for an arid country with limited water resources. In addition, the book highlights some meaningful experiences and successful case studies that reflect the current challenges and potential solutions.

In addition to the Introduction and Conclusions chapters, the remaining 12 chapters are presented in three parts. Part II discusses groundwater modeling, and Part III discusses the groundwater quality assessment. Part IV focuses on the long-term sustainability of groundwater. The introduction chapter summarizes the technical elements of each chapter, whereas the conclusions chapter summarizes the most important conclusions and recommendations of the book while also providing an update on the literature on groundwater in the Nile Valley. It is worth mentioning that Chapter 1 will provide more details about the objectives of the book chapters.

The editors would like to express their heartfelt gratitude to everyone who contributed to making this one-of-a-kind, high-quality book a real source of knowledge and most recent findings in the field of groundwater in Egypt's Nile Valley. They value contributions of all authors too much without their patience and efforts in writing and revising the various versions of the manuscripts to meet Springer's high-quality standards. Many thanks also go to the series' editors, publishing editor, and all of Springer's team for their collaboration and support throughout all stages of the book production.

Acknowledgments must be extended to include all Springer team members who have worked long and hard to produce this book and make it a reality for the researchers, graduate students, and scientists in Egypt and worldwide, particularly the countries with similar aridity conditions.

Any feedback will be greatly appreciated by the book's editors in order to improve the future editions. Comments, reviews, enhancement recommendations, and new chapters for future editions are all welcome and should be directed to the volume editors. The editors' emails can be found in several chapters of the book.

Zagazig, Egypt — Abdelazim M. Negm
Minia, Egypt — Mustafa El-Rawy
February 2022

Contents

Introducing the Book

Introduction to "Sustainability of Groundwater in the Nile Valley, Egypt"

Mustafa El-Rawy, El-Sayed Ewis Omran, and Abdelazim M. Negm

Abstract In this chapter, the book titled "Groundwater in the Nile River Valley" is introduced. In addition to one chapter in the section Conclusions and two chapters in the introduction, the book consisted of 11 core chapters, which entirely included 14 chapters. The 11 chapters fall under three themes. The themes include (a) Groundwater Modeling, which is covered in one chapter, (b) Assessment of Groundwater Quality, which is covered in five chapters (c) Sustainability of Groundwater, which is covered in five chapters too.

Keywords Water resources · Nile Valley · Groundwater modeling · Agriculture · Wastewater · Water quality · GIS · RS · Egypt

1 Introduction

Egypt is one of the world's driest nations (Ragab and Abdelrady 2020). As a result, Egypt is classified as a water-scarce country (Luo et al. 2020). According to the Nile Water Agreement of 1959, Egypt's water supplies are regulated by the Nile River, which has a fixed share of 55.5 BCM/year. Egypt's water supplies are classified into two categories: renewable and nonrenewable. Renewable water supplies are reflected by the Nile River inflow and rainfall. Egypt's second water supply, groundwater, is a critical source for domestic and other uses, as well as irrigation, especially in

M. El-Rawy
Civil Engineering Department, Faculty of Engineering, Minia University, Minia 61111, Egypt
e-mail: mustafa.elrawy@mu.edu.eg; mustafa.elrawy@su.edu.sa

Civil Engineering Department, College of Engineering, Shaqra University, Dawadmi, Ar Riyadh 11911, Saudi Arabia

E.-S. E. Omran
Soil and Water Department, Faculty of Agriculture, Suez Canal University, Ismailia 41522, Egypt

A. M. Negm (✉)
Water and Water Structures Department Engineering, Faculty of Engineering, Zagazig University, Zagazig 44519, Egypt
e-mail: amnegm@zu.edu.eg; amnegm85@yahoo.com

A. M. Negm and M. El-Rawy (eds.), *Sustainability of Groundwater in the Nile Valley, Egypt*, Earth and Environmental Sciences Library,
https://doi.org/10.1007/978-3-031-12676-5_1

desert areas (Abdel-Shafy and Kamel 2016; El-Rawy et al. 2020). The abstractions to groundwater account for around 7–8 BCM/year (El-Rawy et al. 2020; Salman et al. 2019).

Groundwater, seawater desalination, and wastewater usage, on the other hand, are non-renewable water supplies (Djuma et al. 2016). Since it is replenished from the Nile River by seepage from canals and deep percolation as a result of flood irrigation, the shallow aquifer of the Nile River Valley is considered a sustainable water supply with extraction mainly from shallow wells (El-Rawy et al. 2020, 2021). The Nile River and groundwater systems in desert areas are considered Egypt's largest water systems. Other water sources, such as rainfall and flash flooding, are often considered, but in smaller and varying amounts. Groundwater is not regarded a nonrenewable resource in the same way as minerals and petroleum reserves are, nor is it entirely renewable in the same way that solar energy is, but it is dependent on many factors such as recharge, discharge, and the amount of water used (Alley et al. 2020). The Nile River replenishes groundwater aquifers in the Nile Valley, Nile Delta, and Sinai Peninsula by seepage from canals and deep percolation from irrigation (El Arabi 2012).

The groundwater tank below the Nile Valley is the second largest renewable underground reservoir in Egypt and North Africa and it composes Delta, Giza, El-Minia, Assuit, Sohag, Qena, Luxor, and Esna Aquifers (Omran 2017).

The Nile system's groundwater quality is adequate, but it is increasingly declining in many aquifers around the world, not just in Egypt (Mirzavand et al. 2020). Pollution has harmed some shallow groundwater bodies, with nearly 20% of groundwater in the Nile aquifer failing to meet drinking water requirements, especially near the aquifer's edges, where there is little or no protective clay cap. The Nile Delta's groundwater is typically of higher quality than that of the Nile Valley.

The book's intention is therefore to improve and tackle the following main theme.

- Groundwater Modeling.
- Assessment of Groundwater Quality.
- Sustainability of Groundwater.

The summary of the chapters under each theme will be presented in the following sections.

2 Summaries of the Book Chapters

2.1 Groundwater Modeling

One chapter on groundwater modeling methods and a based analysis on groundwater models of the Nile Valley Aquifer were displayed in this section. The use of groundwater is critical to the overall management of water supplies. Groundwater modeling can be used to evaluate the quantity and quality of groundwater in aquifers and how aquifers' conditions change under climate change, different groundwater extraction rates, changes in human activities, and under a wide variety of environmental conditions. To investigate the system response under certain phenomena or forecast the system's behavior, groundwater modeling represents reality in a condensed form without making invalid assumptions. The function of the model is defined as the first step in the modeling process, and data collection is a significant issue in the modeling process. The model's conceptualization is the fundamental step in modeling, followed by setting up the numerical model. Model calibration and validation, and sensitivity analysis is performed after the model completion. Several numerical groundwater flow models have been developed for different parts of the Nile Valley aquifer to assess the interaction between the surface water and the aquifer, to apply various management recharge and discharge scenarios. Studies have shown that groundwater modeling can be used successfully to help understand the Nile Valley aquifer's behavior.

2.2 *Assessment of Groundwater Quality*

The second part of this book, consisting of five approaches, which are used under the title "**Assessment of Groundwater Quality**." The first method is concerned with pollution sources in Egypt's Nile Valley and their effects on groundwater. In Egypt from Aswan to Cairo, the Nile receives discharge of wastewater from over 124 point sources along the banks of the Nile; more than 72 points are agricultural, while the rest are industrial drains. These disposed waters contain untreated or partially treated wastewater; accordingly, different contaminants might be identified in the water of the River. In the water treatment process, all chemical disinfectants are toxic to some extent, with oxidizers like ozone and chlorine being the most common. Many authors have reported on the potential effects of anthropogenic activities such as domestic activities, particularly sewage disposal practises, industrial, and agricultural activities in the Nile valley region. An evaluation of the hydrochemical and bacteriological characteristics of the surface and groundwater resources are necessary to assess the adverse impacts of wastewater disposal in the area. More Advanced methods for monitoring and detecting the source of contamination using remote sensing and GIS techniques are required.

The second chapter linked the role of environmental impacts of treated wastewater contaminates on groundwater quality in the Nile River Valley, Egypt. The discharge of untreated or partially treated wastewater into open streams increases the risk of pollution of surface and groundwater. In Egypt, non-traditional solutions are desperately required to prevent groundwater quality deterioration caused by the long-term effects of treated wastewater pollutants. Thus, the aquifer vulnerability zones should be identified for the planning and implementation of the sustainability strategy pursued by all world countries to conserve groundwater resources. This chapter presented a summary of the effects of treated wastewater disposals on groundwater in the Nile River valley, as well as a review of the results of key related studies and the most up-to-date methodologies for assessing quaternary aquifer hazard zones. Satellite remote sensing (RS) and GIS techniques have been internationally developed to collect information needed to observe different groundwater in several regions across the world. Accordingly, these tools are lower cost and more useful instruments for measuring water quality parameters in groundwater than in situ investigations where measurements are limited to select sampling points. Additionally, in this chapter, a case study was selected in North–East of Cairo due to the presence of disposal of the largest two conventional wastewater treatment plants in Egypt. The results indicated that the studied area was highly polluted due to the discharge of the effluent from wastewater treatment plants. This chapter's outcomes may allow developers and decision-makers to decide the regions need critical and special treatment.

The third approach is the hydrochemistry and hydrogeology aspects of alluvial aquifer in Aswan City, Egypt. The groundwater characteristics of the Aswan aquifer are highlighted in this chapter. Samples were taken from surface water sources and groundwater at various locations to determine the water quality. According to World Health Organization regulations, the quality of the groundwater was found to be acceptable for drinking water after chemical analyses. But a noticeable increase in the organic matter concentration was observed. It can deteriorate the quality of the water and limit its uses. To help further studying the system's dynamic, the groundwater system was modelled as well using MODFLOW. In addition to the water quality, some areas experienced concerns with the groundwater table getting worse as a result of some of the extraction wells being shut off, which can lead to damage in the infrastructure. Several solutions have been proposed to tackle down both problems. Further studies are needed to maintain the stability of the soil during the water table lowering process. A coupled analysis of groundwater models and soil models need to be conducted to estimate the amount of subsidence that might occurs.

The fourth approach is related to the hydrochemical analysis of groundwater in the area Northwest of El-Sadat City, West Nile Delta, Egypt. The Nile River is Egypt's largest surface water source, and the area along its banks includes some of the country's most fertile ground. El Sadat is one of Egypt's most important modern industrial cities. Due to the existence of agricultural soil potentialities, gentle relief, and groundwater resources, a lot of effort has gone into developing this promising field. The main purpose for the present study is to evaluate the hydrological settings of the groundwater and overcome the groundwater shortage problem in this area. Results obtained from the hydrochemical interpretation shows a clear harmony deduced from EC, TDS, and TH distribution maps. In the northwest parts, it appears to have high values of EC, TDS, and TH, moreover cations distribution maps present high values at the same area. Therefore, this part is little bit not suitable for domestic purposes and could be used for agriculture utilities with some restrictions. While the rest parts of the study area have moderate and low values of EC, TDS, TH, cations, and anions levels, hence at these area waters are good for the different uses.

The final approach is related to groundwater quality for irrigation "As a Water Resources Management Tool, Groundwater Quality Assessment for Irrigation in the Young Alluvial Plain of Western Nile Delta, Egypt." Water quality is determined by a variety of factors, including physical, chemical, and biological characteristics.

Conclusions about water quality must be drawn from analyses of a wide number of variables, which is a challenging task (Mukate et al. 2019). To make this difficult task easier, a large number of parameters must be incorporated into a single composite value or model that accurately describes water quality without losing sight of its scientific foundation (Ponsadailakshmia et al. 2018; Wu et al. 2018). As a result, the water quality index for irrigation, rather than conventional methods of water quality measurement, serves as an important instrument for tracking and assessing the overall quality of water. Different types of water quality index models have been produced by different researchers to assess water quality (Khangembam and Kshetrimayum 2019; Mukate et al. 2019; Solangi et al. 2019).

2.3 *Sustainability of Groundwater*

The third part of this book introduces also five potential practices for groundwater sustainable. The first potential practice is gives an overview of agricultural drainage strategies in Egypt as a protection tool against groundwater contamination by fertilizers. Due to the improper application of fertilizers in waterlogged lands, the quality of groundwater started to deteriorate because of the contamination by these fertilizers. This raised the issue of "poor drainage" problem in agricultural lands in the Nile River valley and the need to practice proper drainage strategies to enhance groundwater quality in these lands and ensure better moisture conditions for crops. As a result, Egypt has implemented a strategy for drainage technology at the farm level, which is characterised by gravity subsurface drainage systems, and has begun to introduce a number of drainage projects in the Nile River valley. We investigated existing drainage schemes in the El-Minia governorate, which is located in Upper Egypt, as a case study. In addition, the governorate has a set of open drains with a total length exceeds 1200 km. In Egypt, agricultural drainage strategies are necessary to ensure a better quality of aquifers that underlain agricultural lands along with the protection and sustainability of agricultural productivity.

The second potential practice is the Pliocene aquifer characterization using transient electromagnetic (TEM) and vertical electrical sounding (VES) geophysical techniques: Case study at the area to the East of Wadi El-Natrun City, West Nile Delta, Egypt. Using an integrated analysis of vertical electrical sounding (VES) and transient electromagnetic (TEM) geophysical data, the current study aims to characterise the Pliocene groundwater aquifer east of Wadi El-Natrun city and northwest of El-Sadat city in the Western Nile Delta region. The results revealed the aquifer geometries and its spatial extents and determined the most promising sites for drilling new productive boreholes. They also show that the Pliocene aquifer is the main groundwater aquifer in the survey area, where it is divided into two sub-aquifers. The upper Pliocene aquifer is encountered at a depth ranging from 30 to 70 m and has a thickness ranges from 20 to 60 m and a resistivity varies from 28 to 70 Ωm. The lower Pliocene aquifer is recorded at a depth range between 95 and 135 m with resistivity values of 33–57 Ωm. The findings of this research can serve as a valuable groundwater information base to help establishing new cities and land reclamation projects supported by the Egyptian government plan of the sustainable development strategy for the year 2030.

The third potential practice is concerned with the evaluation of groundwater potential zones using electrical resistivity and hydrogeochemistry in West Tahta Region, Upper Egypt. In this chapter, the geoelectrical and hydrogeochemistry studies were carried out in the western Tahta region to identify the subsurface hydrostratigraphic aquifer units and obtain a proposed potential groundwater development. From the isopach map of the saturated layer, it has been found that the Pleistocene aquifer becoming thicker to the northeast. The resistivity contour map of the Quaternary aquifer showed that average values of about 40 Ω min the region of cultivated land (near to the Nile) increased and reached about 90 Ωm in the west. The groundwater table above sea level contour map showed that there are high values in the southwest and low values in the northeast. The majority of the surface water appears to be suitable for drinking and irrigation, while the majority of the collected groundwater samples are unsuitable for drinking, and half of them are unsuitable for irrigation due to their high salinity.

The fourth potential practice gives an overview of the mapping groundwater recharge potential in the Nile Basin using remotely sensed data and GIS techniques. Remotely sensed data were used followed by applying GIS techniques. After applying weight factors, several evidential maps were extracted from radar/optical remotely sensed data, as well as geologic and precipitation data, which monitor the incidence, movements, and infiltration capacities. The results revealed that about 3.56 and 10.36% of the surface areas of Wadi Qena, and Wadi Matula basins, respectively, were characterized by excellent recharging potential and represent the most promising zones for future groundwater extraction. Field data confirmed that the detected zones are cultivated and groundwater is suitable for irrigation of several strategic crops such as wheat, tomato and alfalfa, by pumping wells. It is concluded that storing flood water during rainy storms in man-made reservoirs would provide additional new water resources to help sustaining the life of the people in these remote areas.

The fifth potential practice is the sustainability of groundwater in the Nile River Valley: is it possible after the construction of GERD? Groundwater is the second largest water resources in Egypt after the Nile River. The shallow aquifer of the Nile Valley and Delta is known to be a renewable water source, extracted from shallow wells, as it is replenished from the Nile River by seepage from canals and excess of irrigation water. Decreasing the flow of the Nile to Egypt due to the construction of the Grand Ethiopian Renaissance Dam (GERD) might be a major problem for Egypt's groundwater resources. This chapter investigates the possibility of the correctness of the assumption that the improper operation of GERD will negatively affect the groundwater along the Nile valley aquifer and reduce or might stop its sustainable nature. The results of GERD filling in 10 years with the normal flow (NF), minimum of average flow (MAF), and minimum flow (MF) conditions, showed that the average groundwater levels reduced by 0.15 m, 0.573 m, and 1.25 m for NF, MAF, and MF with respect to the base case, respectively. It can be concluded that the filling of GERD's reservoir in 10 years could be an optimal option, especially in cases of normal flow or minimum of average flow conditions; otherwise, the aquifer's sustainability will not be maintained.

The book comes to a close with a one-chapter update, conclusion, and recommendation section.

References

Abdel-Shafy HI, Kamel AH (2016) Groundwater in Egypt issue: resources, location, amount, contamination, protection, renewal, future overview. Egypt J Chem 59:321–362

Alley WM, Reilly TE, Franke OL (2020) Sustainability of ground-water resources. The United States Geological Survey (USGS), Report available at https://pubs.usgs.gov/circ/circ1186/html/intro.html

Djuma H, Bruggeman A, Eliades M, Lange M (2016) Non-conventional water resources research in semi-arid countries of the Middle East. Desalin Water Treat 57(5):2290–2303. https://doi.org/10.1080/19443994.2014.984930

El Arabi N (2012) Environmental management of groundwater in Egypt via artificial recharge extending the practice to soil aquifer treatment (SAT). Int J Environ Sustain 1(3). https://doi.org/10.24102/ijes.v1i3.91

El-Rawy M, Abdalla F, El Alfy M (2020) Water resources in Egypt. In: Hamimi Z, El-Barkooky A, Martínez Frías J, Fritz H, Abd El-Rahman Y (eds) The geology of Egypt. Regional geology reviews. Springer Nature, Cham, Switzerland, pp 687–711

El-Rawy M, Moghazy HE, Eltarabily MG (2021) Impacts of decreasing Nile flow on the Nile Valley aquifer in El-Minia Governorate, Egypt. Alexandria Eng J 60:2179–2192

Khangembam S, Kshetrimayum KS (2019) Evaluation of hydrogeochemical controlling factors and water quality index of water resources of the Barak valley of Assam, Northeast India. Groundwater Sustain Dev

Luo P, Sun Y, Wang S, Wang S, Lyu J, Zhou M, Nakagami K, Takara K, Nover D (2020) Historical assessment and future sustainability challenges of Egyptian water resources management. J Cleaner Prod 121154

Mirzavand M, Ghasemieh H, Sadatinejad SJ, Bagheri R (2020) An overview on source, mechanism and investigation approaches in groundwater salinization studies. Int J Environ Sci Technol. https://doi.org/10.1007/s13762-020-02647-7

MukateaS WV, Panaskara D, Jacobsb JA, Sawantc A (2019) Development of new integrated water quality index (IWQI) model to evaluate the drinking suitability of water. Ecol Ind 101:348–354

Omran E-SE (2017) Land and groundwater resources in the Egypt's Nile Valley, Delta, and its fringes. In: The handbook of environmental chemistry. https://doi.org/10.1007/698_2017_64

Ponsadailakshmi S, Ganapathy Sankari S, Mythili Prasanna S, Madhurambal G (2018) Evaluation of water quality suitability for drinking using drinking water quality index in Nagapattinam District, Tamil Nadu in Southern India. Groundwater Sustain Dev 6:43–49

Ragab A, Abdelrady A (2020) Impact of green roofs on energy demand for cooling in Egyptian buildings. Sustainability 12:5729

Salman SA, Arauzo M, Elnazer AA (2019) Groundwater quality and vulnerability assessment in west Luxor Governorate, Egypt. Groundwater Sustain Dev 8:271–280. https://doi.org/10.1016/j.gsd.2018.11.009

Solangi GS, Siyal AA, Babar MM, Siyal P (2019) Groundwater quality evaluation using the water quality index (WQI), the synthetic pollution index (SPI), and geospatial tools: a case study of Sujawal District, Pakistan. Human Ecol Risk Assess. https://doi.org/10.1080/10807039.2019.1588099

Wu Z, Wang Z, Chen Y, Cai Y, Deng J (2018) Assessing river water quality using water quality index in Lake Taihu Basin, China. Sci Total Environ 612:914–922

Evaluation of the Groundwater Resources in the Nile Valley, Egypt

Menna Haggag, Hosni H. Ghazala, and Ismael M. Ibraheem

Abstract The deficit of freshwater represents a global challenge as there are become many countries suffer from the water poverty. Egypt is regarded one of these countries as most of its land are deserts due to its location within the barren zone of North Africa besides the lack of fresh water sources. The River Nile is the main source of fresh water in Egypt and constitutes the lifeline of many human activities. The river is fed by heavy rains falling on the highlands of Ethiopia and provides around 85% of the water supply in Egypt. Several reasons such as; population growth, climate change, pollution of industrial waste and agricultural drainage that ask questions about whether the amount of fresh water and the efficiency of available sources can meet the growing needs of different scales. Accordingly, it must be critical alternatives for providing water, especially after the construction of the Grand Ethiopian Dam (GERD) as it would be an important reason of aggravating the crisis of fresh water in Egypt by 2025. Therefore, it is expected that the demand for groundwater resources will increase as they are considered the backbone of the development plan along with seawater desalination. The Nile Valley, Delta and Nubia aquifers represent the main aquifers in Egypt. Due to the importance of The Nile Valley, and Delta aquifers, many efforts and studies have been directed during the last decades to evaluate the potentialities of these aquifers. These efforts included attempts to simulate groundwater conditions in some areas in and around the Nile Valley basin through digital and Electric-analog computer models. Based on these results, several proposals have been made to exploit the efficiency of groundwater in the Nile basin.

M. Haggag (✉) · H. H. Ghazala
Department of Geology, Faculty of Science, Mansoura University, Mansoura 35116, Egypt
e-mail: mennahaggag216@gmail.com

H. H. Ghazala
e-mail: Ghazala59@yahoo.com; Ghazala@mans.edu.eg

I. M. Ibraheem
Institute of Geophysics and Meteorology, University of Cologne, Pohligstr. 3, 50969 Cologne, Germany
e-mail: ismail.geo@gmail.com; ismael.ibraheem@geo.uni-koeln.de

A. M. Negm and M. El-Rawy (eds.), *Sustainability of Groundwater in the Nile Valley, Egypt*, Earth and Environmental Sciences Library,
https://doi.org/10.1007/978-3-031-12676-5_2

Keywords Groundwater · Water resources · Nile Valley aquifer · Nile Delta · Salinity · Total dissolved solid (TDS) · Desalination

1 Introduction

Recently, it has become evident that water is in greater demand, as access to clean water is one of the fundamental requirements to manufacture sustainable development and has become a major challenge for many countries because of many reasons such as the increase in population, human activity, and extreme climate change. Egypt is one of the most drought countries due to its location in the arid belt of North Africa so, it severely needs developing clean water resources (Fig. 1).

Around 96% of the total area of Egypt is desert land depends on the groundwater that leaks through springs and seepage zones which makes most of population density confined in the remaining 4% is dissected by the river and its valley and delta called the green zone. Hence, the Nile River is the essential source of freshwater in this area as it is used for all modalities of human activities since ancient times. Most of the Nile River originates from water precipitation over mountainous areas in the Ethiopian highland (Elbeih 2015).

Many factors negatively affect the future Nile River revenues such as climate change, which will provide the inability of the Nile River to provide water, which will lead to a shortage of water per capita in Egypt. In the last decades, the amount of water from the Nile River is no longer adequate to meet the needs of different activities, which created a gap between the requirements and the quantity of available freshwater that reached to 20 BCM annually (Nofal et al. 2018). This affected many water-based

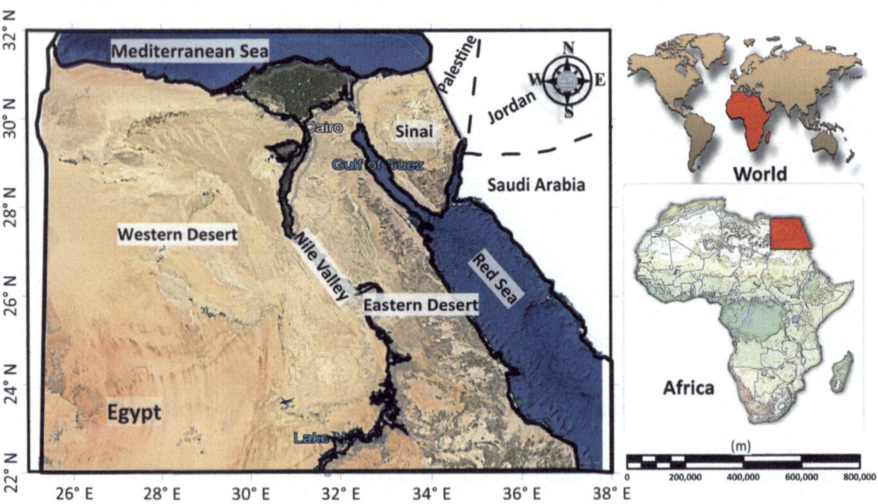

Fig. 1 Location map of Egypt

services and production that essential for sustainable development. So that, there is an importance to wisely manage water resources and identify possibilities of the recent recharge of aquifers (Hamza et al. 1999).

2 Water Resources

In Egypt, the water resources are divided into renewable and non-renewable systems. The Nile River inflow and rainfall represent the renewable resources. However, groundwater, seawater desalination and treated water are non-renewable (Djuma et al. 2016). The Nile River and groundwater are considered the main water systems in Egypt (Fig. 2). Other systems such as rainfall and flash floods are also considered as water sources but with small and varying proportions. They have restrictions related to their amount, character, location, time, and cost of water production (Mohamed 2007).

2.1 Renewable Water Resources

2.1.1 The Nile River

The structure of the Nile system consists of; the Nile River and its branches, irrigation canals, agricultural drains, and aquifers of the valley and its delta. All these resources are interconnected and are renewed annually with 58.5 BCM of freshwater (MWRI 2005). The Nile River is regarded as the longest river in the world because of its

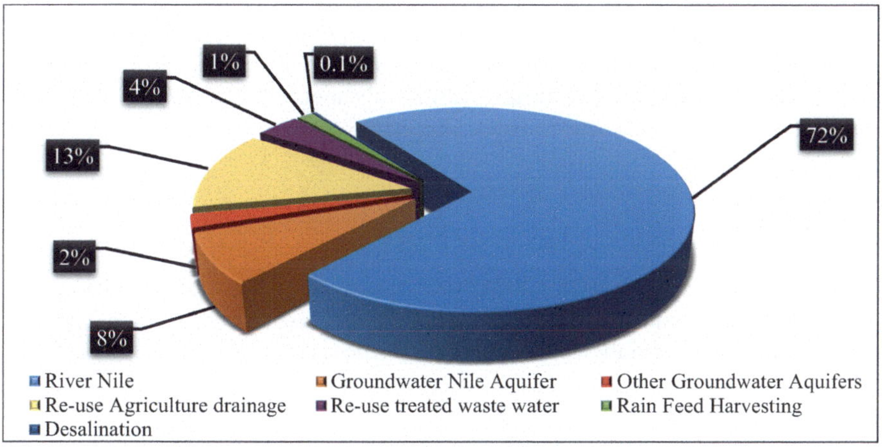

Fig. 2 Water supply in Egypt (BCM) (MWRI 2005; CAPMAS 2017)

approximate 6800 km length. It extends over a wide range of latitudes (from 4°S to 31°N) and ecosystems flowing through highland regions with abundant moisture to lowlands with arid conditions. The Nile Basin covers a surface area of around 3.1 million km^2 shared by eleven countries (MWRI 2017) and has two main tributaries; the White Nile originates from Lake Victoria basin and the Blue Nile from the Ethiopian plateau (MWRI 2005) (Fig. 3). About 85% of water released from the High Aswan Dam is used for irrigation with 15% remaining for other purposes (e.g. industry, domestic, and water supply). As a result of the scarcity of rainfall, the Egyptian populations rely on the Nile River for all their water needs and the vast majority of the population accumulates around the Nile (Ibrahim et al. 2018).

2.2 Non-renewable Water Resources

2.2.1 Groundwater

Groundwater is not considered as a nonrenewable resource, such as minerals and petroleum deposits, nor a completely renewable in the same manner and timeframe as solar energy but, it depends on several aspects like recharge, discharge, and the quantity of water used (Alley et al. 2020). Groundwater aquifers in the Nile Valley, Nile Delta and Sinai Peninsula are renewed from the Nile River through seepage even from canals or deep percolation resulting from irrigation application (El Arabi 2012). In Western Desert, groundwater is the only available resource for interdisciplinary development and is considered as a non-renewable resource. The contribution of groundwater to the total water supply in Egypt has become very moderate over the years. However, it remains an important freshwater resource and the sole source for people living in Egyptian deserts (CEDARE 2014). The main groundwater systems in Egypt are Nile valley and Delta aquifer system, Fissured carbonate aquifer, Coastal aquifer, Moghra aquifer, Nubia aquifer, and the Hardrock aquifer system (Fig. 4). Deep aquifers in the Western Desert and Sinai regions are considered as non-renewable with current total abstraction only about 1.65 BCM/year (CEDARE 2011).

2.2.2 Rainfall

According to El-Tantawi et al. (2021) the system of the rainfall has changed in all levels as the amount decreased parallel to the increase in temperature. The total authentic amount of rainfall in use for living activities is about 1.3 BCM/year (MWRI 2005).

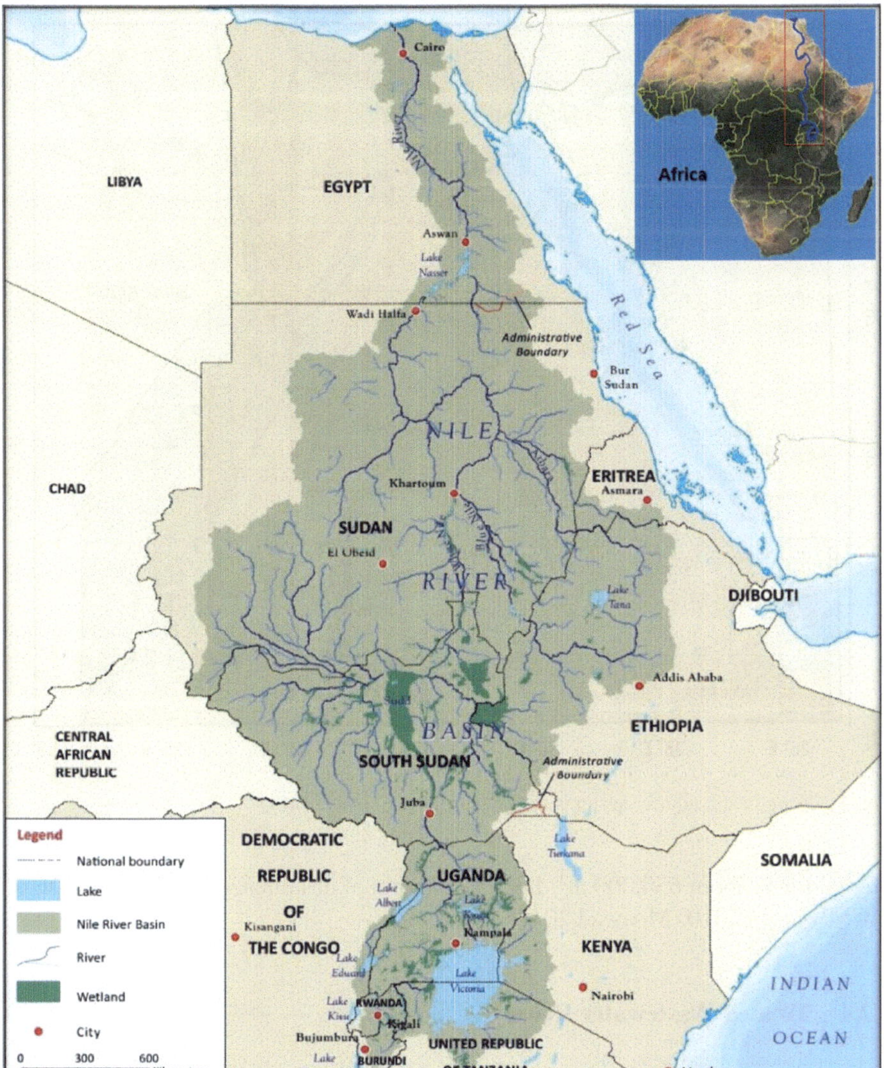

Fig. 3 The drainage system of the Nile River (after The World Bank 2000)

2.2.3 Desalination

The desalination plans for drinking water and industrial purposes are developed in conjunction with growing demands (MWRI 2010). According to MWRI (2020), until the mid-1980s, the capacity produced from the desalinated sea water was only 20,000 m^3/day, and beyond 2000–2011, production was 145 m^3/day, while the current

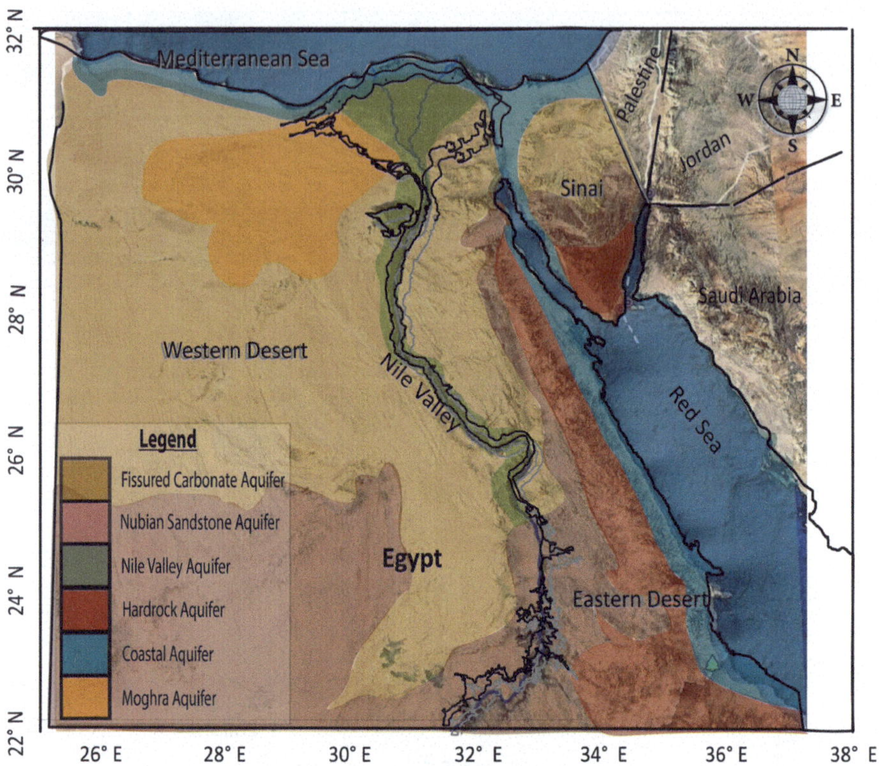

Fig. 4 Groundwater aquifer systems of Egypt (after CEDARE 2014)

production is about 650,000 m^3/day. The amount of desalinated water reached to 60 MCM/year in 2010 (Moawad 2012).

2.2.4 Treated Wastewater Reuse

Wastewater recently has been treated, reclaimed, and reused in irrigating the agricultural lands in order to alleviate the burdens of increased demands for freshwater resources. In year 2010, Egypt treated about 2.97 BCM/year, but only 0.7 BCM/year was utilized for agriculture (0.26 BCM/year was undergoing secondary treatment and 0.44 BCM/year undergoing primary treatment) (Abdelwahab and Omar 2011; MWRI 2010). According to MWRI (2017), the amount of the treated wastewater production has reached to 21 BCM/year recently.

Although, the multitude sources of freshwater in Egypt, these sources suffer many challenges and face several constraints. The most significant challenges are presented in:

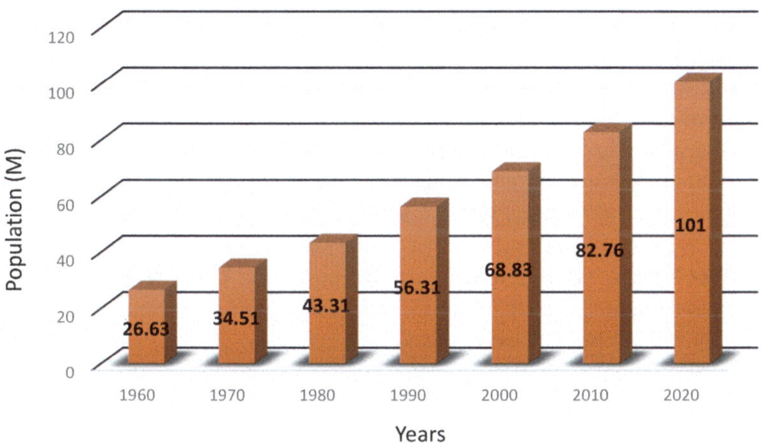

Fig. 5 Population of Egypt according to CAPMAS (2020)

- **Population Increase**: According to FAOSTAT (2013) and CAPMAS (2020) Egypt's population was expected to reach 98.7 million by 2025, but it reached > 100 million by 2020 (Fig. 5). The average availability of freshwater in Egypt has steadily decreased from 1893 m^3/year in 1959 to about 570 m^3 in 2020. This decreasing is expected to continue yearly until reaching to 500 m^3 by 2025 (Arab Water Council 2012; MWRI 2017). This amount is less than the international water poverty limit. This limit equals 1000 m^3 of renewable water available per person/year according to the World Bank and this will put the country below the threshold of water scarcity (Drainage Research Institute 2010).
- **Climate Change**: The Middle East and North Africa suffer from a critical situation because of rising the earth's temperature about 1.2 °C above the pre-industrial levels. The high temperature reduces the amount of rainfall and increases evaporation of water that leads to increase in dry seasons (Badawy 2009). Increasing evaporation decreases the Nile flow in a way cannot be neglected and reduces soil moisture levels that in turn increase the frequency of drought in the region. In addition, the soil moisture and the filtration ratios will also decrease leading to a decrease in the rate of groundwater recharge (Radwan 1997).
- **Rise of Sea Level**: The Egyptian coast stretches for 3500 km including 1150 km on the Mediterranean coast and 1500 km on the Red Sea coast. This extension makes it more vulnerable to sea level rising and soil salinity. The sea levels rise by about 19–58 cm and is expected to increase nearly 100 cm by year 2100 with considering a drop in the Delta land (Ministry of Environment 2016). This rising in sea level may cause imbalance in nature and decreasing the abundance of the groundwater in the coastal regions. The big effect of reducing freshwater flow is polluting soil, causing poor crop quality and loss productivity (IDSC 2013; Abu Hatab et al. 2019).

- **Economic Growth**: It threatens the quality and quantity of water resources and exacerbates the current problem of polluting the shallow groundwater by overusing industrial chemicals, fertilizers, and pesticides. In addition, farmers continue to practice ineffective submerged irrigation methods which results in water losses due to evaporation and over-irrigation and causes soil damage (Abdel-Dayem 2011). According to the Ministry of Water Resources and Irrigation, Egypt consumes about 147% of its water resources and will need about 20% additional water by 2025. This means that 47% of the water is imported.
- **The effect of the Grand Ethiopian Renaissance Dam (GERD)**: This dam is expected to reduce Egypt's amount of water, which is currently 55 BCM/year. The amount of shortage is expected to range between 5 and 15 BCM. It can also lead to a permanent reduction in water levels in Lake Nasser if floods are stored in Ethiopia instead (Sallam 2018).

Consequently, according to the international studies and statistics, since the Nile is Egypt's largest freshwater resource, it is going to be one of the most affected one among other resources by the above-mentioned hazards. Therefore, using groundwater is considered the one of the best available solutions and alternative resource for freshwater as it is the second largest resource of freshwater in accordance with the sustainable development plans.

Groundwater seems to be one of the most promising solutions to face the water shortage problem. Although, groundwater is not a resource, even in the Nile aquifer or desert fringes as it is renewed obliquely from the Nile water. Furthermore, there are serious limitations about the intensive use of groundwater resources as some of them are non-renewable.

3 Groundwater Usage

The idea of using the groundwater started from long time ago; however, it did not receive any explicit attention until 1996. Since then, the government intended to develop it, especially the Nile aquifer system (Elnashar 2014).

Groundwater and surface water are linked through the underground leaked water (Winter et al. 1998). This water is collected by many ways such as drainage, tunnels and wells. It also may flow naturally to the earth's surface through seeps or springs and deep percolation from irrigation applications (Abdelhalim et al. 2020).

Groundwater is a vital resource that aged more than 31,000 years (Münnich and Vogel 1962). It is used for many purposes such as public and domestic water supplies, irrigation, industry, mining, and thermo-electrical power production. The quantity of groundwater in Egypt was about 4.80 BCM/year in 2005 according to Ashour et al. (2009) but, it has become about 2.54 BCM/year recently, according to MWRI (2017).

4 Groundwater Quality

Groundwater quality in the Nile system is reasonable and is gradually being reduced not only in the Nile system of Egypt but also in many other aquifers around the world (Mirzavand et al. 2020). Almost 20% of groundwater in the Nile aquifer no longer meets the drinking water standards because of pollution, especially at fringes where there is little or no protective clay cap. Generally, the quality of groundwater in the Nile Delta and Western Desert was considered better than that in the Nile Valley while in the Eastern Desert and Sinai was highly saline. The carbonate aquifer is generally brackish but has some freshwater in recharge areas (MWRI 2013).

5 The Nile Valley and Delta Aquifers

The groundwater tank below the Nile Valley is the largest renewable underground aquifer in Egypt and North Africa and it composes Delta, Giza, El-Minia, Assuit, Sohag, Qena, Luxor, and Esna Aquifers (Omran 2017) (Fig. 6). It extends between Giza and Aswan for about 900 km and about 14 km wide, with a maximum width about 20 km at Minia and minimum width about 2 km at Aswan. The total area of the Nile Valley basin between Cairo and Aswan is about 100 km^2. The edges of the valley on both the east and west flanks are marked by steep erosional scarps that rise greatly onto adjacent desert plateaus (Abdel-Lah and Shamrukh 2001). The Nile Delta and its fringes occupy an area equals 22,000 km^2 and have land levels range from 17 m southward to less than one m above sea level to the north (Farid 1985).

6 Geomorphology and Main Characteristics of the Nile Valley

The latest differential pattern of the Nile Valley that has been made by Abdelkareem and El-Baz (2016) shows uplifts that caused by Ma'aza and Western Desert plateaus. The uplift has led to compose a variety of drainage systems, basins, and structural features. In addition, they mentioned that there is a tilting in the eastern Ma'aza plateau formed a distortion in the topography differs from that surrounding region. This tilt controls the drainage direction and incisions along the Nile's main channel. The eastern plateau is composed of a series of fractures and normal faults that formed several small blocks. The boundaries of these blocks separate the watersheds into valleys and sets of tributaries (Youssef 2003) (Fig. 7).

The Nile Valley has a relative uneven hot and dry climate. The temperature reaches up to 40 °C in summer and 12 °C with some precipitation in winter. The Nile Delta has climate specifications as it is considered less arid compare with the rest of the valley (Nofal et al. 2018). The annual average precipitation is less than 20 mm, which

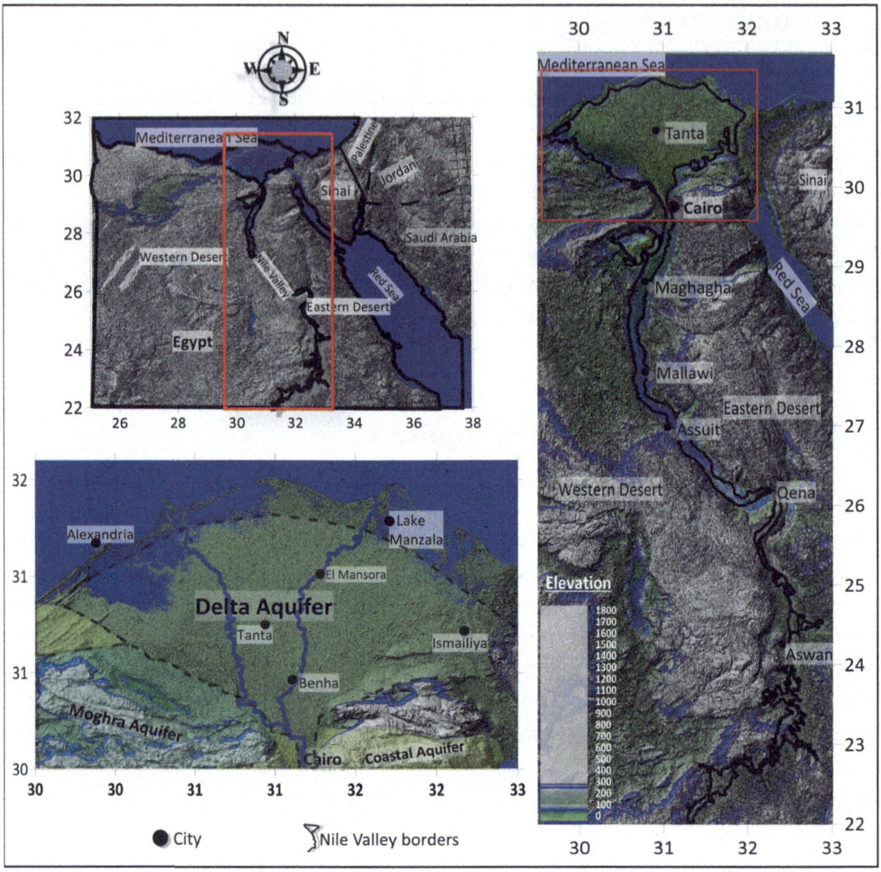

Fig. 6 Location map of the Nile Valley and its Delta

is insignificant. The relative of humidity in winter is higher than that in summer and varies between 25 and 55% (Abdel-Lah and Shamrukh 2001). The rainfall exceeds around 100 mm per year along the coastal zone and rapidly decreases southward along the upper Nile Valley to less than 5 mm per year at Aswan (MWRI 2013). Table 1 shows the hydrogeological characteristics of the Nile Valley and Delta aquifer in 2002.

7 Hydrological Settings

The Nile Valley and Delta aquifers are located at the bottom of the Nile River mainly in water-carrying sediments layers that belong to the Quaternary Period (Fig. 8). The Quaternary aquifer system can be divided into three main hydrogeological units:

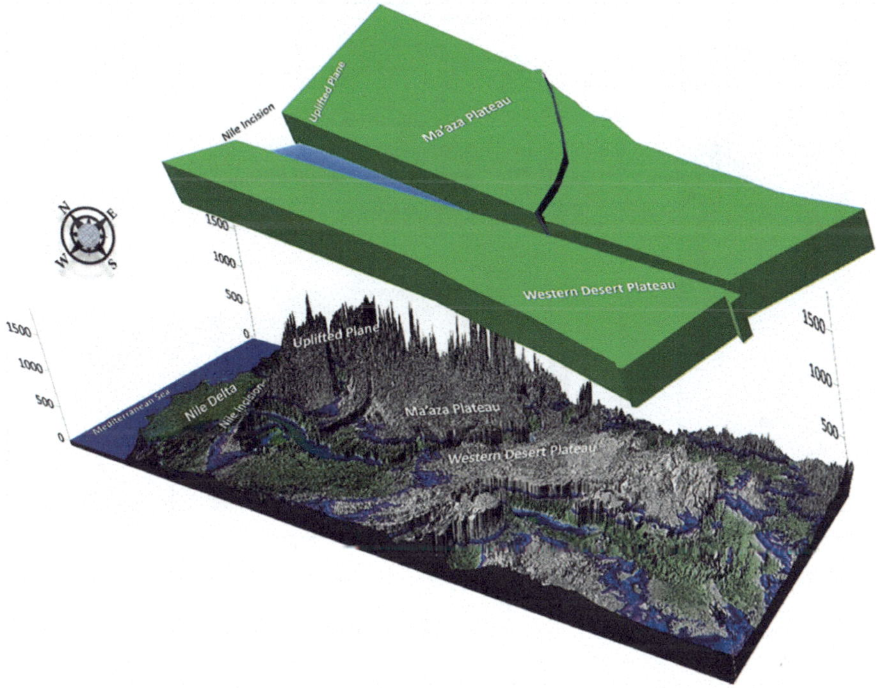

Fig. 7 Sketch of the relation between the dissimilar block boundaries (Youssef 2003)

Table 1 Hydrogeological characteristics of the Nile Valley and Delta aquifer year 2002 (after Allam et al. 2003; Abdel Latif and El Kashouty 2010)

Location	Top aquifer (m)	Standard thickness (m)	Depth to groundwater level (m)	Salinity
Along Nile Valley	0–20	10–260	0–5	< 1500
South Nile Delta	0–20	100–500	0–5	< 1500
North Nile Delta	0–200	500–1000	0–3	< 5000

The upper Holocene unit (Aquitard): Generally, it covers 70% of the Nile Valley land formed of clay and silty clay deposits of low lateral and vertical permeability. It also functions as a semi-confined layer. The average hydraulic conductivity of the unit is about 0.6 m per day (Barber and Carr 1980).

This unit extends laterally having a greater thickness near the river channel about 40 m in the North and reaches up to 95 m easting the Delta (Fig. 9) (Nofal et al. 2015). However, it vanishes near the desert fringes where its thickness ranges between 6 and 22 m (Selim et al. 2000).

The lower Pleistocene unit: This unit consists of graded sand and gravel sediments. Less permeable sediments, the Pliocene clays, and the Holocene silty clays, bound it downward and upward, respectively. Therefore, the aquifer is under semi-confined

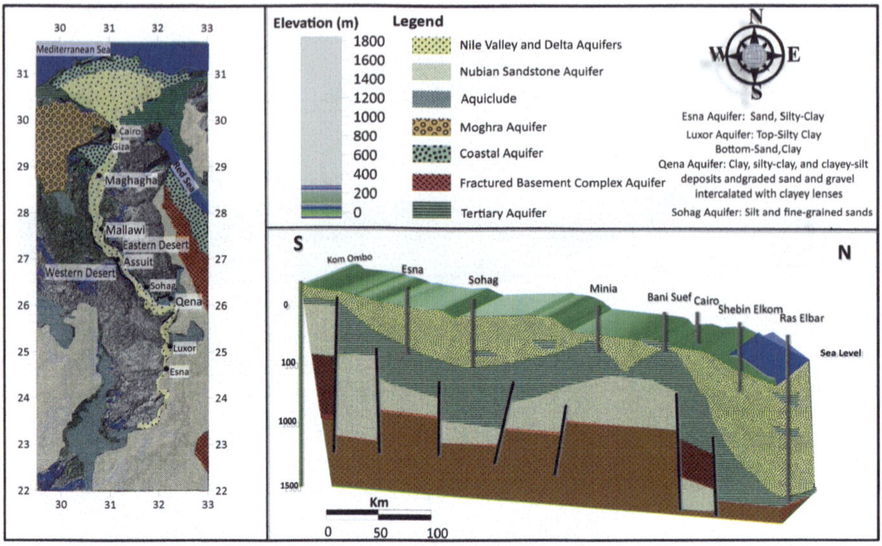

Fig. 8 Hydrogeological cross-section along Egypt shows the average thickness of the Nile Valley aquifer from south–north modified (after Hefny and Shata 2004; El Tahlawi et al. 2008; Mahmod and Watanabe 2014)

conditions (Selim et al. 2000). Selim (1995) and Farid (1980) mentioned that the thickness of this aquifer varies from 150 m northward to more than 1000 m near the coast and from 18 to 35 m downward Upper Egypt.

The average hydraulic conductivity increases to more than 100 m per day, northward at the Delta and about 50 m per day in the area between Idfu and Giza, then reaches its minimum to be 36 m per day around Aswan area (Todini and O'Connell 1981; Farid 1980). The Basal unit of the Pliocene (Aquiclude): This layer is the base of the deltaic deposits and consists of a dense thick clay section (Farid 1980).

8 Discharge and Recharge

The Nile valley aquifer consists of almost 3000 main basins as the pop-up map from the raster image shows (Fig. 10). The main source of the Nile Valley reservoir's water supply is the deep penetration of irrigation water and canals (Abdel-Shafy and Mansour 2013). However, Delta drains are mainly discharged from poorly treated domestic and industrial wastewater (Brown et al. 2003). According to RIGW (2006) and Abdel-Shafy and Mansour (2013), the total amount to recharge the Nile Valley basin's reservoir (river sediments) was about 5.43 BCM/year while it was about 6.78 BCM in the Nile Delta. The discharge amount to the Nile River was around 2.287 BCM while the amount was about 7 BCM in the Delta (Molle et al. 2016).

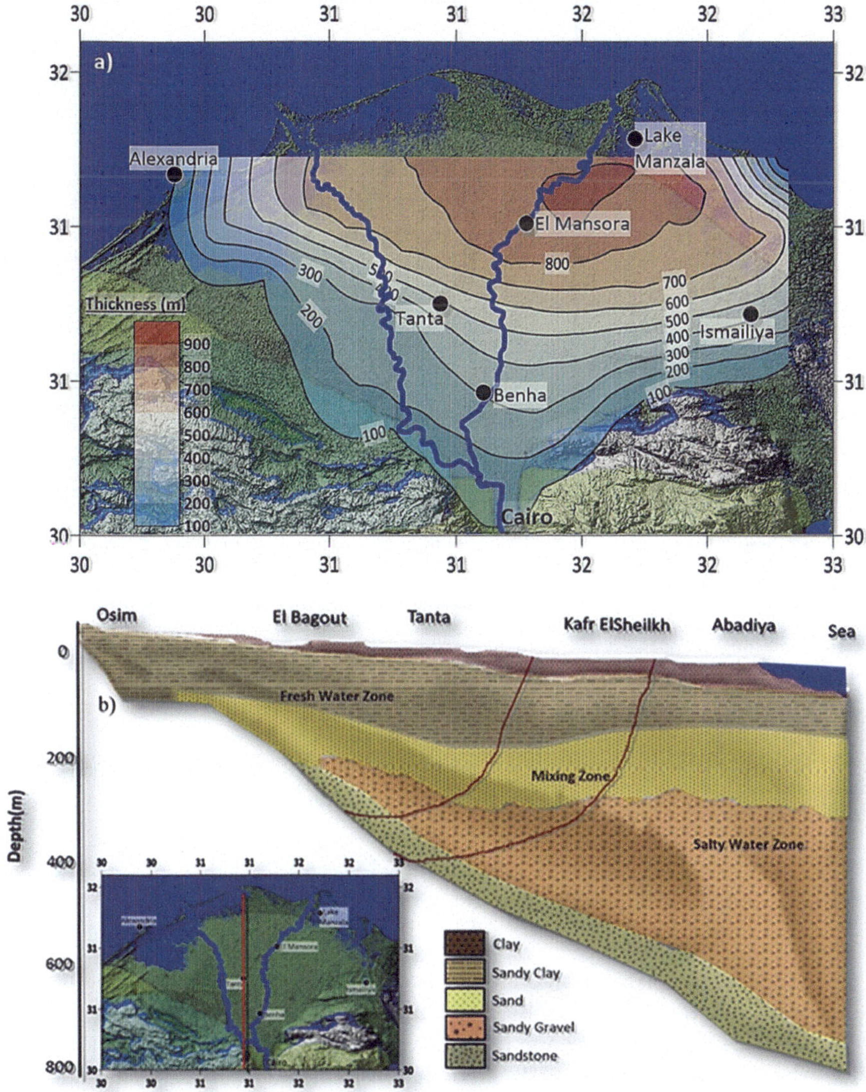

Fig. 9 **a** Thickness contour map of the Delta aquifer and **b** hydrogeological cross-section along the Nile Delta from south–north (modified after Farid 1985; Sherif 1999; Zeidan 2017)

The salinity ranges from 1500 ppm along the Nile Valley and South Delta to 5000 ppm in North Delta (Dawoud and Ismail 2013). The Nile valley and Delta aquifers are, by far, the largest aquifers followed by the Nubian Sandstone then Moghra aquifer. Sinai aquifer is relatively the smallest one (Elnashar 2014) (Fig. 11). The annual groundwater abstraction in the Nile Valley and Delta aquifer systems have

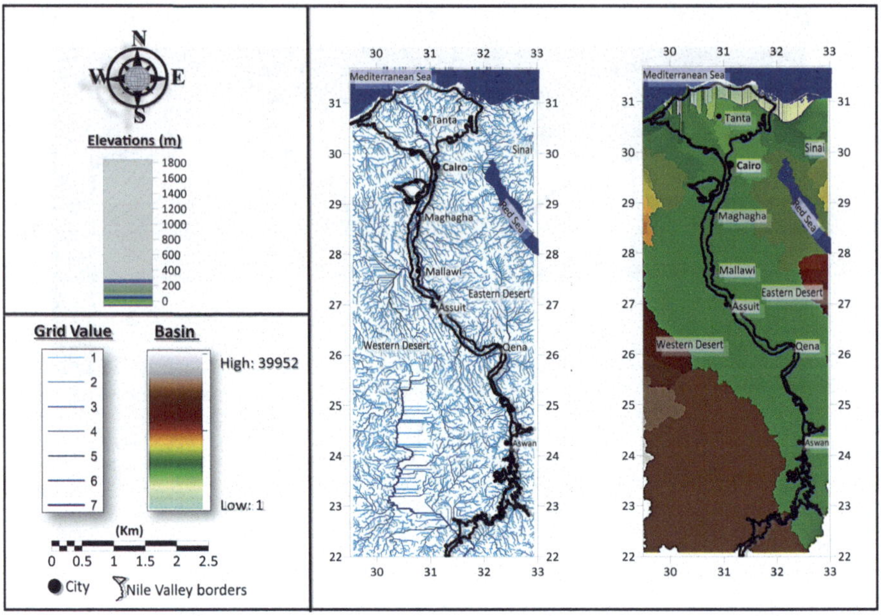

Fig. 10 Identification of the stream network and main basins in the Nile Valley as extracted from the Raster image (made by the authors)

reached up to 7 BCM in 2020 which exceeds the amount of production from all other aquifers in the rest of Egypt (Abd Ellah 2020) (Fig. 12).

9 Vulnerability Assessment

In the recent decades, several studies have been conducted on; the nature of the Nile Valley aquifer, its quality, suitability, and water levels. Some of these studies (Goode 1982; Sherif 1999; Thorweihe and Heinl 2002; Morsy 2009; Mahmoud and Tawfik 2015; Abdalla and Moubark 2018; Abdelhalim et al. 2020; El-Rawy et al. 2020, 2021) have tracked rising and droping in the groundwater levels for the last 50 years. The results of those studies showed that the groundwater level was ranging from 100 to 200 m around the Nile Valley region and decreased periodically. Nevertheless, from 20 years, the level of groundwater has been rising significantly, especially in the middle and north of Egypt (Fig. 13). In the Nile Delta region, groundwater levels were decreasing with a slight rate. However, they have got to rise again, recently (Fig. 14).

Many studies contributed to simulate the percentage of salinity and the total dissolved solid (TDS) in the Nile Valley aquifer. NWRC (2003), EcoconServ (2005), FAOSTAT (2013), El-Aassar et al. (2016), Elbeih (2016), Ismail and El-Rawy (2018),

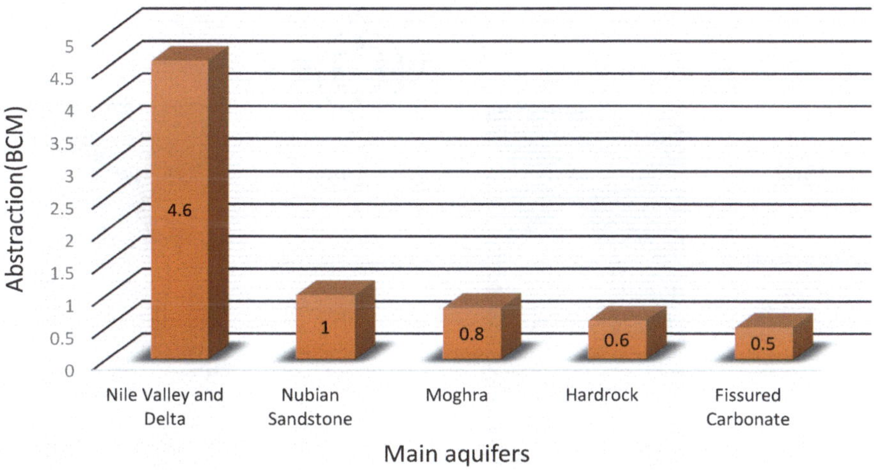

Fig. 11 Distribution of groundwater abstraction in the main aquifers after (Abdel-Shafy and Kamel 2016)

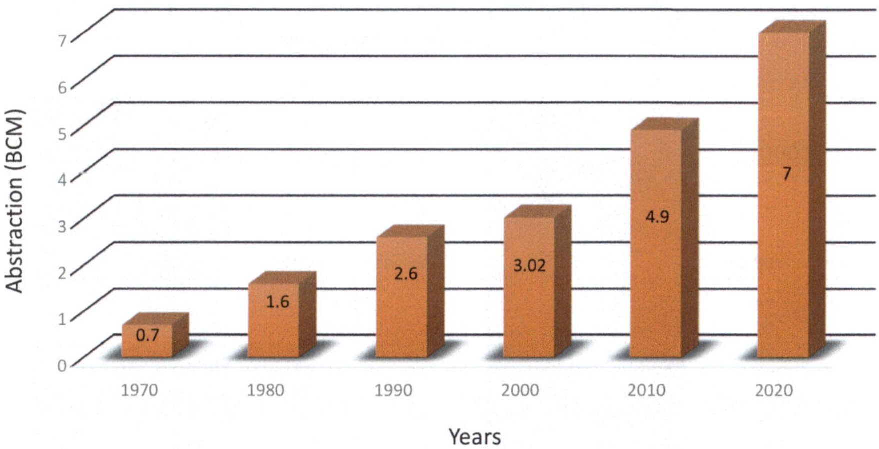

Fig. 12 Distribution of groundwater abstraction in the Nile Valley and Delta aquifer after (Sallam 2018; Abd Ellah 2020)

El-Rawy et al. (2019), and Gedamy et al. (2019) showed that the groundwater of Delta aquifer has been significantly affected by intrusion of sea water. The worst hit region is the north of the Delta, as the proportion of TDS has increased from 5000 ppm in 1970 to more than 40,000 ppm nowadays, while the effect decreases southward (Fig. 15).

The effect of seawater intrusion with the aquifer along the Nile Valley was decreasing downward, the proportion of the TDS was between 500 and 1550 ppm, which was the ratio allowed by the World Health Organization (WHO). However,

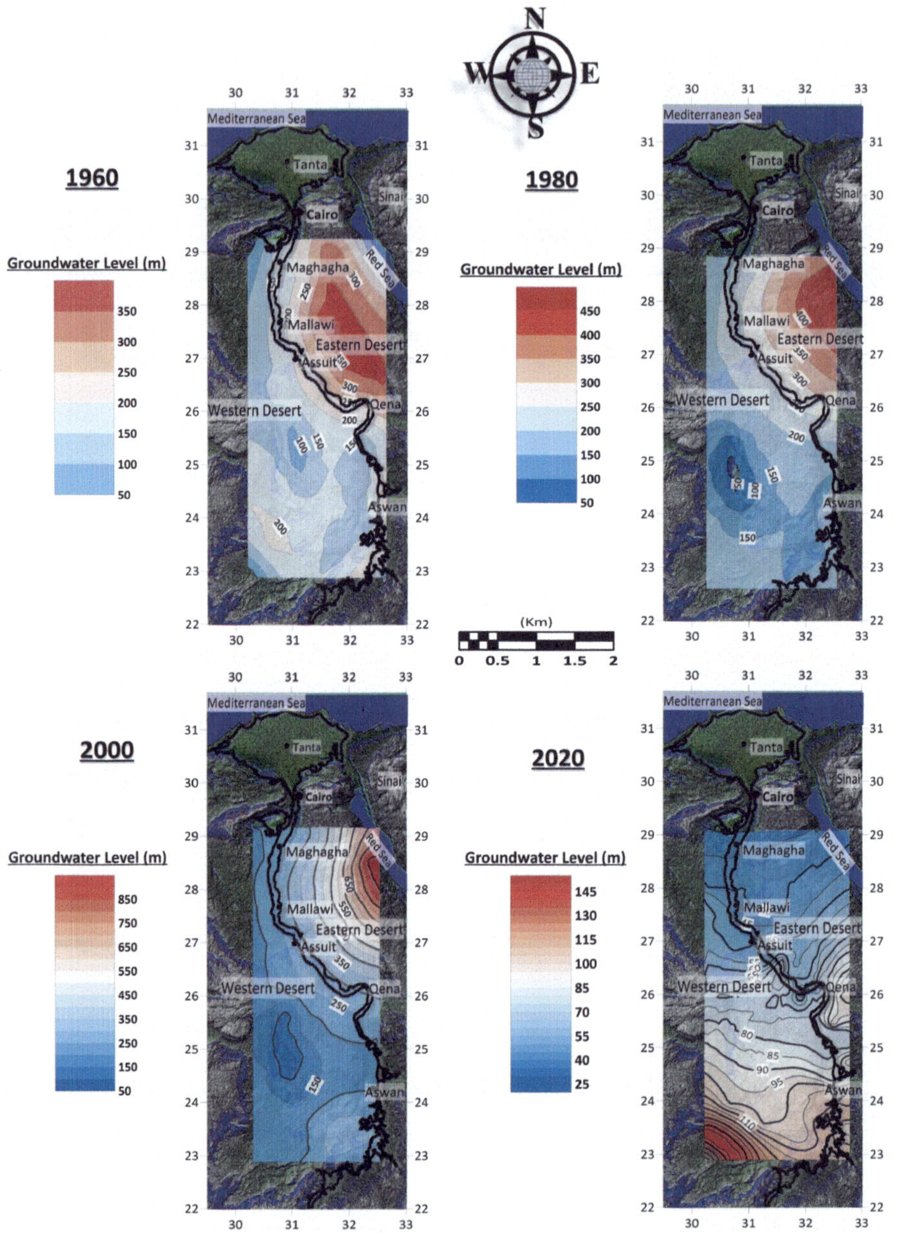

Fig. 13 Groundwater contour lines from year 1960 to recent of the Nile Valley aquifer from different sources

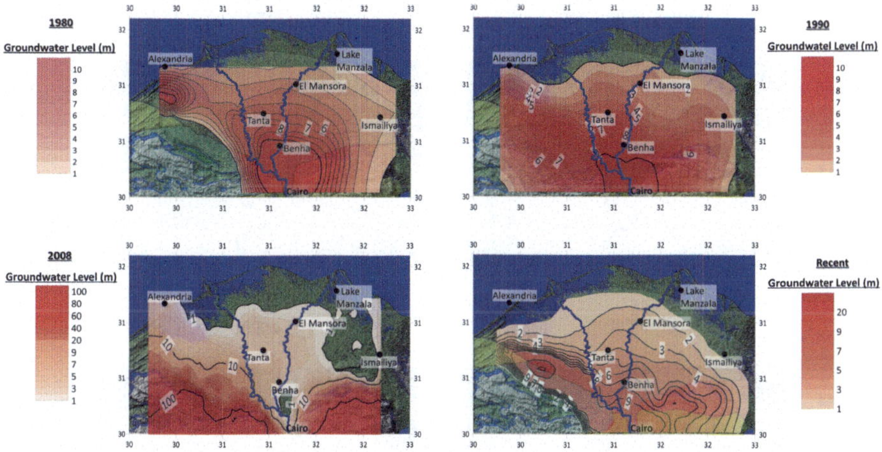

Fig. 14 Groundwater contour lines from year 1980 to recent of the Nile Delta aquifer from different sources

in the last few years, there has been an unprecedented rising in values specifically southwest of Aswan and northward to Fayoum as the value up to between 2500 and 7000 ppm, respectively (Fig. 16).

10 Proposed Scenarios and Solutions

Several researchers have represented lots of scenarios, solutions, and recommendation through many studies in the last decades in order to protect and desalinate groundwater. Abdelhalim et al. (2020) have presented a study discussing the various impacts that the groundwater in the Nile Valley aquifer will be affected with by year 2050. They suggested three different scenarios: the first scenario was evaluating the decreasing in the surface water standard levels by 0.5, 1 and 1.5 m due to the different challenges that mentioned before. The second scenario, which has the least effects on the groundwater levels, was about studying the effect of increasing extraction by 25 and 50% from the normal extraction rate. This will be an inevitable result for increasing demand of groundwater. The third scenario was studying the effect of increasing extraction from groundwater with decreasing surface water standard levels by 0.5, 1 and 1.5 m on the aquifer. The last one showed the maximum change in groundwater budget and levels that will decrease from 23.31 to 43.07 m.

One of the best solutions that has been represented by Salim (2012) was desalination of groundwater by using solar energy, which is a particular benefit to poor communities suffering from polluted water, taking in consideration some criteria such as aquifer depth, aquifer salinity, incidence of flash floods, sand dunes, and faults (Fig. 17). The results show that groundwater solar desalination is suitable in remote regions on the Northwestern Coast, and in the South toward Aswan (Fig. 18).

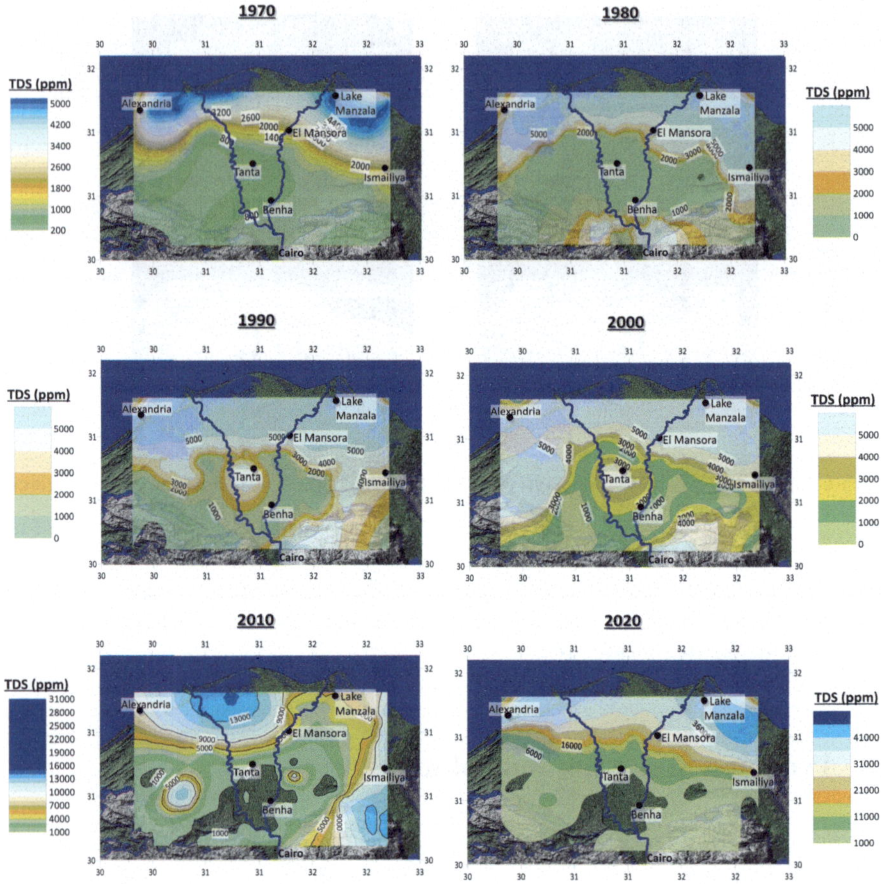

Fig. 15 TDS values from year 1970 to recent in the Nile Delta aquifer from different sources

There were many studies over years that contributed to many important results and conclusions. Throughout those results, researchers have made some recommendations that contribute to solving the problem of groundwater and how to preserve it. According to Samak (2007) any well will be dug in the future, should be at least, at 300 m distance from any adjacent well and by 1 km from any nearby canal and at ranges from 500 m to 1 km from the intake of any canal. Ghoraba et al. (2013) suggested drilling extraction wells in order to reduce the contamination and pump the contaminated groundwater out of the aquifer.

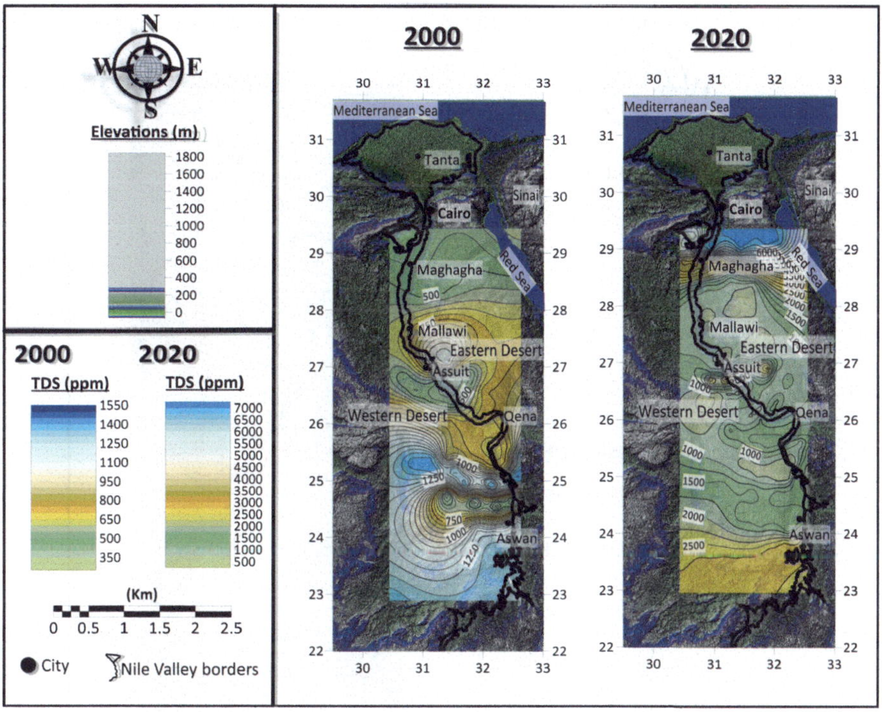

Fig. 16 TDS values from year 2000 to recent in the Nile Valley aquifer from different sources

11 Conclusions

Egypt is one of the most countries that facing many freshwater challenges, even that, as most of its land are deserts due to its location in the North African arid zone. Recently, the Ethiopian Dam constituted one of the major inevitable risks. Therefore, it was necessary for the government to head for using all other alternative freshwater resources rather than the total dependence on the Nile water, which will be certainly affected by these unavoidably potential risks.

Freshwater resources in Egypt vary between the Nile River, groundwater, rainfall water and the desalination water from the sea. Groundwater has a quantitative importance as the second largest resource of freshwater. So that, all the government's efforts are headed directly to optimal use and consumption smoothing of groundwater in order to use it in the sustainable development.

The current study is an attempt to summarize of the previous important studies on the Nile Valley and Nile Delta aquifer through presenting the latest results and offers suggestions to improve water quality and rationalization. All the previous studies have shown that the consumption of groundwater it should be utilized wisely and according to well-studied management plans.

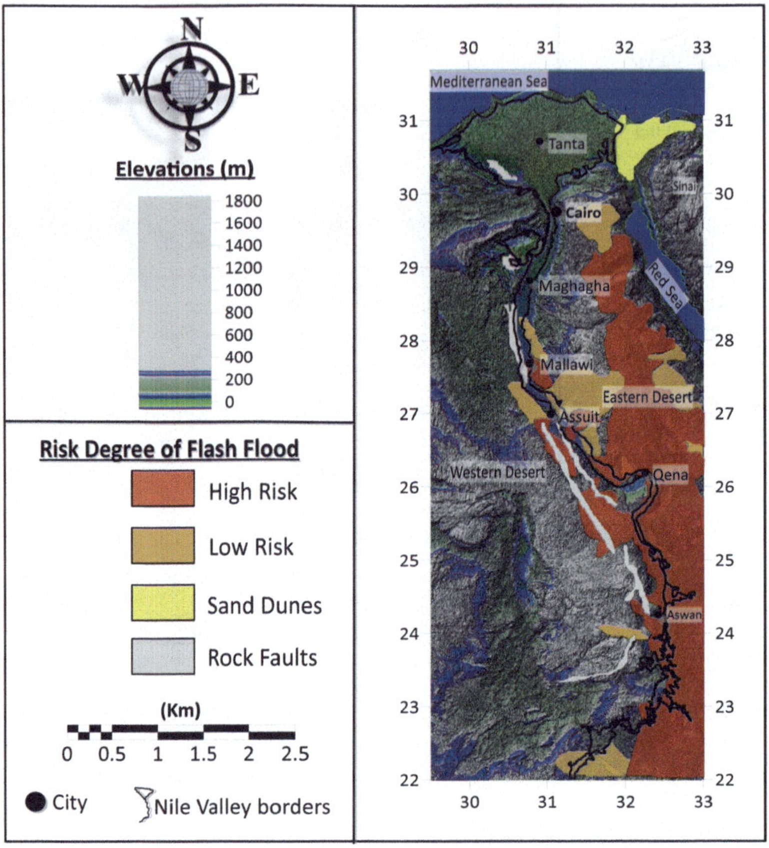

Fig. 17 A map shows the risk degrees of flash flood along the Nile Valley (after Salim 2012)

The results over the recent decades have shown that:

- The level of groundwater increased significantly through the last 20 years, especially in the middle Egypt towards Fayoum. However, the rising is slightly higher in the Delta region.
- There is an alternative rising of TDS and salinity rates because of seawater, which is getting higher in the north coast.

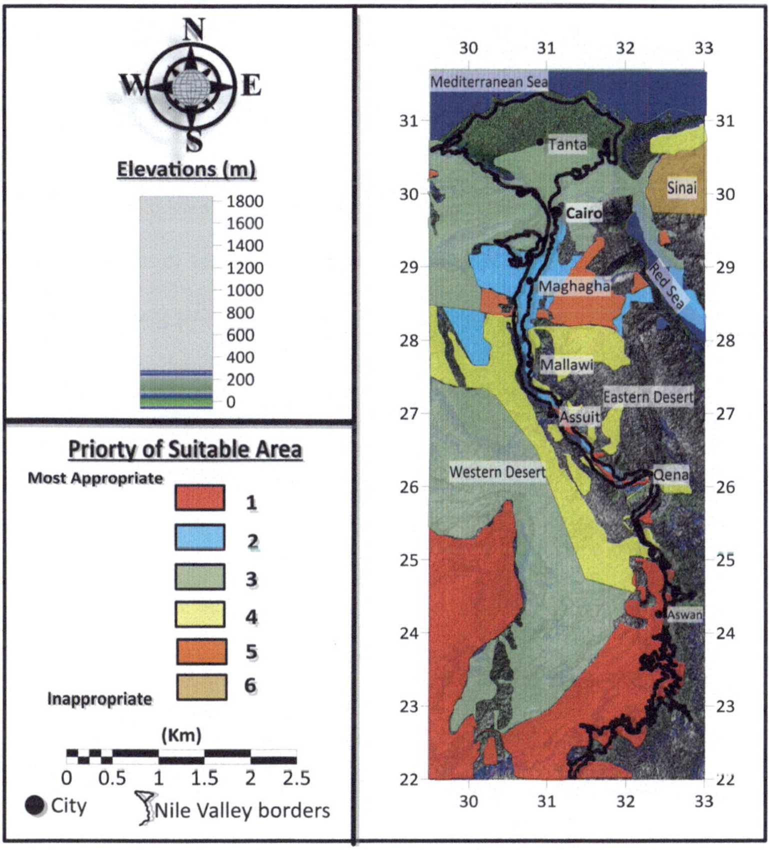

Fig. 18 A map shows the most appropriate areas for extraction suitable water along the Nile Valley (after Salim 2012)

12 Recommendations

The recent study recommends with a susbentantial water conservation policy that should be considered by the Egyptian government in order to face the current challenges. This can be carried out through establishing social awareness plans about water conservation.

References

Abdalla F, Moubark K (2018) Assessment of well performance criteria and aquifer characteristics using step-drawdown tests and hydrogeochemical data, West of Qena area, Egypt. J Afr Earth Sci 138:336–347. https://doi.org/10.1016/j.jafrearsci.2017.11.023

Abdel Latif A, El Kashouty M (2010) Groundwater investigation in Awlad Salameh, Southern Sohag, Upper Egypt. Earth Sci Res J 14(1):63–75

http://www.scielo.org.co/scielo.php?pid=S1794619020100001000005s&script=sci_arttext&tlng=en

Abdel-Dayem S (2011) Water quality management in Egypt. Int J Water Resour Dev 27(1):181–202. https://doi.org/10.1080/07900627.2010.531522

Abdelhalim A, Sefelnasr A, Ismail E (2020) Response of the interaction between surface water and groundwater to climate change and proposed megastructure. J Afr Earth Sc 162:103723. https://doi.org/10.1016/j.jafrearsci.2019.103723

Abdelkareem M, El-Baz F (2016) Mode of formation of the Nile Gorge in northern Egypt: a study by DEM-SRTM data and GIS analysis. Geol J 51(5):760–778. https://doi.org/10.1002/gj.2687

Abd Ellah RG (2020) Water resources in Egypt and their challenges, Lake Nasser case study. Egypt J Aquat Res 46(1):1–12. https://doi.org/10.1016/j.ejar.2020.03.001

Abdel-Lah A, Shamrukh M (2001) Impact of septic system on ground water quality in a Nile valley village, Egypt. Paper presented at the Sixth International Water Technology Conference, IWTC. https://scholar.google.com.eg/scholar?hl=en&as_sdt=0%2C5&q=Abdel-Lah%2C+A.%2C+%26+Shamrukh%2C+M.+%282001%29.+Impact+of+septic+system+on+ground+water+quality+in+a+Nile+valley+village%2C+Egypt.+Paper+presented+at+the+Sixth+International+Water+Technology+Conference%2C+IWTC&btnG

Abdel-Shafy HI, Mansour MS (2013) Overview on water reuse in Egypt: present and future. Sustain Sanitation Pract 14:17–25

Abdel-Shafy HI, Kamel AH (2016) Groundwater in Egypt issue: resources, location, amount, contamination, protection, renewal, future overview. Egypt J Chem 59(3):321–362. https://doi.org/10.21608/ejchem.2016.1085

Abdelwahab R, Omar M (2011) Wastewater reuse in Egypt: opportunities and challenges. Arab World. Arab Water Council Report https://scholar.google.com.eg/scholar?hl=en&as_sdt=0%2C5&q=Abdelwahab+R%2C+Omar+M+%282011%29+Wastewater+Reuse+in+Egypt%3A+Opportunities+and+Challenges.+Arab+World.+Arab+Water+Council+Report&btnG=

Abu Hatab A, Cavinato MER, Lindemer A, Lagerkvist C-J (2019) Urban sprawl, food security and agricultural systems in developing countries: a systematic review of the literature. Cities 94:129–142. https://doi.org/10.1016/j.cities.2019.06.001

Allam AR, Saaf E-J, Dawoud MA (2003) Desalination of brackish groundwater in Egypt. Desalination 152(1–3):19–26. https://doi.org/10.1016/s0011-9164(02)01044-5

Alley WM, Reilly TE, Franke OL (2020) Sustainability of ground-water resources. The United States Geological Survey (USGS), Report available at https://pubs.usgs.gov/circ/circ1186/html/intro.html

Arab Water Council (2012) The second Arab state of the water report. Available at www.arabwatercouncil.org/images/Publications/Arab_state/2nd_Arab_State_of_the_Water_Report.pdf

Ashour M, El Attar S, Rafaat Y, Mohamed M (2009) Water resources management in Egypt. J Eng Sci 37(2):269–279

Badawy HA (2009) Effect of expected climate changes on evaporation losses from Aswan High Dam Reservoir (AHDR). In: Thirteenth international water technology conference, IWTC 13 2009. Hurghada, Egypt

Barber W, Carr DP (1980) Water management capabilities of the alluvial aquifer system of the Nile Valley, Upper Egypt. Master Water Plan Technical 1, Report, Cairo; Legrain G (1929) Le temples de Karnak, Bruxelles, lip. https://scholar.google.com.eg/scholar?hl=en&as_sdt=0%2C5&q=Barber+W%2C+Carr+DP+%281980%29+Water+Management+capabilities+of+the+alluvial+aquifer+system+of+the+Nile+Valley%2C+Upper+Egypt.+Master+Water+Plan+Technical+1%2C+Report%2C+Cairo.+Legrain%2C+G.%2C+1929%2C+Le+temples+de+Karnak%2C+Bruxelles%2C+lip&btnG=

Brown P, El Gohary F, Tawfic MA, Hamdy EI, Abdel-Gawad S (2003) Nile river water quality management study. Egypt Water Policy Reform, United States Agency for International Development, Egypt

CEDARE (2011) Water use efficiency and economic approach in Egypt. Plan Bleu, UNEP/MAP Regional Activity Centre. http://cimalight.video/view.php?vid=2b1fb9b36
CEDARE (2014) Egypt Water Sector M&E Rapid Assessment Report. Monitoring & Evaluation for Water in North Africa (MEWINA) Project, Water Resources Management Program. http://web.cedare.org/wp-content/uploads/2005/05/Egypt-Water-Sector-Monitoring-and-Evaluation-Rapid-Assessment-Report-Final.pdf
Central Agency for Public Mobilization and Statistics (CAPMAS) (2017) available at: http://www.capmas.gov.eg/
Central Agency for Public Mobilization and Statistics (CAPMAS) (2020) Statistics yearbook. https://www.capmas.gov.eg/
Dawoud MA, Ismail SS (2013) Saturated and unsaturated River Nile/groundwater aquifer interaction systems in the Nile Valley, Egypt. Arab J Geosci 6(6):2119–2130. https://doi.org/10.1007/s12517-011-0483-4
Djuma H, Bruggeman A, Eliades M, Lange M (2016) Non-conventional water resources research in semi-arid countries of the Middle East. Desalin Water Treat 57(5):2290–2303. https://doi.org/10.1080/19443994.2014.984930
Drainage Research Institute (2010) Monitoring and analysis of drainage water quality project, drainage water status in the Nile Delta yearbook 97/98. Technical No. 52. https://asejaiqjsae.journals.ekb.eg/article_50350.html
Economic Conversance Service (EcoconServ) (2005) Study on status of the environment and relevant policies/measures in Egypt. https://www.env.go.jp/earth/coop/coop/c_report/egypt_h16/english/pdf/016.pdf
El Arabi N (2012) Environmental management of groundwater in Egypt via artificial recharge extending the practice to soil aquifer treatment (SAT). Int J Environ Sustain 1(3). https://doi.org/10.24102/ijes.v1i3.91
El-Aassar M, Hussien R, Ghoubachi S (2016) Groundwater quality and vulnerability assessment in the new reclamation areas, Assuit governorate, West Nile River, Egypt. J Am Sci 12(11). https://doi.org/10.7537/marsjas121116.01
Elbeih SF (2015) An overview of integrated remote sensing and GIS for groundwater mapping in Egypt. Ain Shams Eng J 6(1):1–15. https://doi.org/10.1016/j.asej.2014.08.008
Elbeih SF (2016) Production of an atlas for hydrogeological maps of the Southwestern Egypt. Engineering Applications and Water, Project. http://www.narss.sci.eg/webroot/attachments/Final_Atlas_NARSS_Link.pdf
Elnashar WY (2014) Groundwater management in Egypt. IOSR J Mech Civ Eng 11:69–78. https://doi.org/10.9790/1684-11446978
El-Rawy M, Ismail E, Abdalla O (2019) Assessment of groundwater quality using GIS, hydrogeochemistry, and factor statistical analysis in Qena Governorate, Egypt. Desalin Water Treat. https://doi.org/10.5004/dwt.2019.24423
El-Rawy M, Abdalla F, El Alfy M (2020) Water resources in Egypt. In: Hamimi Z, El-Barkooky A, Martínez Frías J, Fritz H, Abd El-Rahman Y (eds) The geology of Egypt. Regional Geology Reviews. Springer, Cham. https://doi.org/10.1007/978-3-030-15265-9_18
El-Rawy M, Moghazy HE, Eltarabily MG (2021) Impacts of decreasing Nile flow on the Nile Valley aquifer in El-Minia Governorate, Egypt. Alexandria Eng J 60(2):2179–2192. https://doi.org/10.1016/j.aej.2020.12.037
El Tahlawi MR, Farrag AA, Ahmed SS (2008) Groundwater of Egypt: "an environmental overview." Environ Geol 55:639–652. https://doi.org/10.1007/s00254-007-1014-1
El-Tantawi AM, Anming B, Liu Y, Gamal G (2021) An assessment of rainfall variability in northern Egypt. Arab J Geosci 14(13). https://doi.org/10.1007/s12517-021-07272-3
Farid MS (1980) Nile Delta groundwater study. MSc Thesis, Faculty of Engineering, Cairo University, Egypt
Farid MSM (1985) Management of groundwater system in the Nile Delta. PhD Thesis, Faculty of Engineering, Cairo University, Cairo, Egypt

Food and Agriculture Organization of the United Nation (FAOSTAT) (2013) Monitoring of climate change risk impacts of sea level rise on groundwater and agriculture in the Nile Delta available at Country Profile Egypt. http://www.fao.org/fileadmin/user_upload/agwa/docs/NIP_Egypt_Final1.pdf

Gedamy Y, Abdulhady Y, Zaghlool E (2019) Hydrochemistry of the Eocene aquifer at the desert fringes of west El-Minya Governorate, Egypt. Curr Sci Int 8(4):734–763. https://doi.org/10.36632/csi/2019.8.4.15

Ghoraba SM, Zyedan BA, Rashwan IMH (2013) Solute transport modeling of the groundwater for quaternary aquifer quality management in Middle Delta, Egypt. Alexandria Eng J 52:197–207. https://doi.org/10.1016/j.aej.2012.12.007

Goode DJ (1982) Modeling phreatic aquifers and lakes as boundary conditions to horizontal flow aquifers. Massachusetts Institute of Technology, Department of Civil Engineering

Hamza M, Aly A, Awad M, Nada AA, El-Samie SGA, Sadek MA, Salem WM (1999) Estimation of recharge from Nile aquifer to desert fringes at Wadi Qena, Egypt. In: International symposium on isotope techniques in water resources development and management, Vienna, Austria, International Atomic Energy Agency Report 10–14. http://www.iaea.org/inis/collection/NCLCollectionStore/_Public/30/025/30025739.pdf

Hefny K, Shata A (2004) Underground water in Egypt. Ministry of water supplies and irrigation, Cairo, Egypt, p 295

Ibrahim M, Al-Zyoud S, Elhaddad E (2018) Surface water quality monitoring for river Nile, Egypt using GIS-techniques. Open J Geol 8(2):161–173. https://doi.org/10.4236/ojg.2018.82010

Information and Decision Support Center (IDSC) (2013) available at: https://www.devex.com/organizations/information-and-decision-support-center-idsc-24475

Ismaila E, El-Rawy M (2018) Assessment of groundwater quality in West Sohag, Egypt. Desalin Water Treat 123:101–108. https://doi.org/10.5004/dwt.2018.22687

Mahmod WE, Watanabe K (2014) Modified grey model and its application to groundwater flow analysis with limited hydrogeological data: a case study of the Nubian Sandstone, Kharga Oasis, Egypt. Environ Monit Assess J 186:1063–1081. https://doi.org/10.1007/s10661-013-3439-1

Mahmoud HH, Tawfik MZ (2015) Impact of the geologic setting on the groundwater using geoelectrical sounding in the area southwest of Sohag-Upper Egypt. J Afr Earth Sci 104:6–18. https://doi.org/10.1016/j.jafrearsci.2015.01.004

Ministry of Environment of Egypt (2016) available at: http://www.eeaa.gov.eg/en-us/home.aspx

Ministry of Water Resources and Irrigation (MWRI) (2005) Water for the future. National Water Resources Plan for Egypt. Book. Available at: http://extwprlegs1.fao.org/docs/pdf/egy147082.pdf

Ministry of Water Resources and Irrigation (MWRI) (2010) A summary of the Egypt national strategy for the development and management of water resources until 2050. https://web.archive.org/web/20190327043515/http://www.mwri.gov.eg/

Ministry of Water Resources and Irrigation (MWRI) (2013) Proposed climate change adaptation strategy for the ministry of water resources & irrigation Egypt. Available at: www.eeaa.gov.eg/portals/0/eeaaReports/CCRMP/7.%20CC%20Water%20Strategy/CC%20Final%20Submitted%208-March%202013%20AdptStrtgy.pdf

Ministry of Water Resources and Irrigation (MWRI) (2017) Egypt and water: facts and numbers. Report, p 15 extwprlegs1.fao.org › docs › pdf › egy147082

Ministry of Water Resources and Irrigation (MWRI) (2020) available at: https://www.mwri.gov.eg/

Mirzavand M, Ghasemieh H, Sadatinejad SJ, Bagheri R (2020) An overview on source, mechanism and investigation approaches in groundwater salinization studies. Int J Environ Sci Technol. https://doi.org/10.1007/s13762-020-02647-7

Moawad A (2012) Water and sanitation status in Egypt. National Launching Workshop on Water Sector M&E in Egypt, Presentation https://scholar.google.com.eg/scholar?hl=en&as_sdt=0%2C5&q=Moawad+A+%282012%29+Water+and+Sanitation+Status+in+Egypt.+National+Launching+Workshop+on+Water+Sector+M%26E+in+Egypt%2C+Presentation&btnG=

Mohamed HI (2007) Egypt's water resources management under pressure of development. Paper presented at the First Eng. Conference for young researchers, Faculty of Engineering, Assiut University, Assiut, Egypt

Molle F, Gaafar I, El-Agha DE, Rap E (2016) Irrigation efficiency and the Nile Delta water balance, water and salt management in the Nile Delta: report International Water Management Institute and Australian Center for International Agriculture Research No. 9. http://www.semide.net/media_server/files/semide/documents/database/water-andsaltmanagementniledelta/Rep.9%20Irrigation%20Efficiency%20and%20the%20Nile%20Delta%20Water%20Balance.pdf

Morsy WS (2009) Environmental management to groundwater resources for Nile Delta region. PhD Thesis, Faculty of Engineering, Cairo University, Egypt

Münnich K, Vogel J (1962) Radioactive dating Proc. Athens IAEA, p 189

National Water Resources Plan for Egypt (NWRC) (2003) WL/DELFT hydraulics, facing the challenge, discussion, p 3

Nofal E, Amer M, El-Didy S, Fekry A (2015) Seawater intrusion in Nile Delta in perspective of new configuration of the aquifer heterogeneity using the recent stratigraphy data. J Am Sci 11(6):281–292

Nofal ER, Fekry AM, Ahmed MH, El-Kharakany MM (2018) Groundwater: extraction versus recharge; vulnerability assessment. Water Sci 32(2):287–300. https://doi.org/10.1016/j.wsj.2018.07.002

Omran E-SE (2017) Land and groundwater resources in the Egypt's Nile Valley, delta, and its fringes. In: The Handbook of environmental chemistry. https://doi.org/10.1007/698_2017_64

Radwan L (1997) Farmer responses to inefficiencies in the supply and distribution of irrigation requirements in delta Egypt. Geogr J 163(1):78–92. https://doi.org/10.2307/3059688

Research Institute for Groundwater (RIGW) (2006) National water quality monitoring component, NAWQAM, use of isotopes in studying the salinity distribution and salt water interface in the delta aquifer

Salim MG (2012) Selection of groundwater sites in Egypt, using geographic information systems, for desalination by solar energy in order to reduce greenhouse gases. J Adv Res 3(1):11–19. https://doi.org/10.1016/j.jare.2011.02.008

Sallam OM (2018) Vision for future management of groundwater in the Nile Delta of Egypt after construction of the Ethiopian Dams. Hydrol Curr Res 9(3)

Samak SAMA (2007) Groundwater management in Delta, MSc. Thesis, Faculty of Engineering, Zagazig University, Egypt

Selim S (1995) Geological and hydrogeological studies of the quaternary aquifer in Aswan town area, Egypt. Egypt J Geol 39:631–645

Selim S, Khedr E, Falasteen A, Kamel E (2000) Reasons of local rise in groundwater level at Luxor City area, Upper Egypt. Paper presented at the 5th international conference on the geology of the Arab World, Cairo University. https://scholar.google.com.eg/scholar?hl=en&as_sdt=0%2C5&q=Selim+S%2C+Khedr+E%2C+Falasteen+A%2C+Kamel+E+%282000%29+Reasons+of+local+rise+in+groundwater+level+at+Luxor+City+area%2C+Upper+Egypt.+Paper+presented+at+the+5th+international+conference+on+the+geology+of+the+Arab+World%2C+Cairo+University&btnG=

Sherif M (1999) Nile Delta aquifer in Egypt. Chapter 17 in Seawater intrusion in coastal aquifers, concepts methods and practices, theory and application of transport in porous media, vol 14. Kluwer Academic Publishers, The Netherlands, pp 559–590. https://doi.org/10.1007/978-94-017-2969-7_17

The World Bank (2000) Available at https://www.worldbank.org/

Thorweihe U, Heinl M (2002) Groundwater resources of the Nubian aquifer system, NE Africa. Synthesis, Observatories du Sahara ET du Sahel, Paris

Todini E, O'Connell PE (1981) Water master plane. Hydrological simulation of Lake Nasser, analysis and results, Technical report 1(14), Ministry of Water Resources and Irrigation, Egypt

Winter TC, Harvey JW, Franke OL, Alley WM (1998) Ground water and surface water: a single resource. US Geological Survey Circular 1139, DIANE Publishing Inc. https://pubs.usgs.gov/circ/circ1139/pdf/circ1139.pdf
Youssef MM (2003) Structural setting of central and south Egypt: an overview. Micropaleontology 49:1–13. https://doi.org/10.2113/49.suppl_1.1
Zeidan BA (2017) Groundwater degradation and remediation in the Nile delta aquifer. In: The Nile Delta. Springer, pp 159–232. https://doi.org/10.1007/698_2016_128

Groundwater Modeling

Fundamentals of Groundwater Modeling Methods and a Focused Review on the Groundwater Models of the Nile Valley Aquifer

Mustafa El-Rawy, Wouter Zijl, Ali Salem, Ahmed Awad, Mohamed Galal Eltarabily, and Abdelazim M. Negm

Abstract Groundwater plays an essential role in the overall management of water resources. The demand for groundwater for municipal, agricultural, and industrial use has been grown steadily during the past decades. Groundwater modeling can be used to evaluate the quantity and quality of groundwater in aquifer-aquitard systems, also in relation to environment and climate change. Modeling allows for the construction of scenarios based on different groundwater abstraction rates, changes in human activities, and varying environmental conditions. Models can be based on simple analytical methods and comprehensive finite difference or finite element methods, dependent on the model's purpose and the available data. Conceptualization and data

M. El-Rawy (✉)
Civil Engineering Department, Faculty of Engineering, Minia University, Minia 61111, Egypt
e-mail: mustafa.elrawy@mu.edu.eg; mustafa.elrawy@su.edu.sa

Civil Engineering Department, College of Engineering, Shaqra University, Dawadmi, Ar Riyadh 11911, Saudi Arabia

W. Zijl
Deparment of Hydrology and Hydraulic Engineering, Vrije Universiteit Brussel, Pleinlaan 2, 1050 Brussels, Belgium
e-mail: vub@zijl.be

A. Salem
Earth Sciences, University of Pécs, Ifjúság útja 6, 7624 Pécs, Hungary
e-mail: alisalem@gamma.ttk.pte.hu

A. Awad
College of Water Resources and Hydropower Engineering, Yangzhou University, Yangzhou 225009, China

M. G. Eltarabily
Civil Engineering Department, Faculty of Engineering, Port Said University, Port Said 42523, Egypt
e-mail: eng_m_trabily@eng.psu.edu.eg

A. M. Negm
Water and Water Structures Department Engineering, Faculty of Engineering, Zagazig University, Zagazig 44519, Egypt
e-mail: amnegm85@yahoo.com; amnegm@zu.edu.eg

A. M. Negm and M. El-Rawy (eds.), *Sustainability of Groundwater in the Nile Valley, Egypt*, Earth and Environmental Sciences Library,
https://doi.org/10.1007/978-3-031-12676-5_3

collection strongly influences the choice of the numerical model. In the past finite difference and finite element methods were considered as completely different, but more recent developments closed the gap between these two methods. Generally, the time-independent data like spatially distributed hydraulic conductivities, etc. are not available at the start of the modeling. These parameters have to be determined from a time series of both measured and calculated quantities, for instance hydraulic heads, This process is generally called model calibration, or parameter estimation. Groundwater modeling is intending to represents reality in a simplified form, without making too simple assumptions for the purpose of the modeling study; for instance scenarios to investigate the response of an aquifer-aquitard system for a number of hypothetical phenomena like, for instance flash rains or excessive droughts. Several numerical groundwater flow models have been developed for different parts of the Nile Valley aquifer to assess the interaction between the surface water and the aquifer, to apply various management recharge and discharge scenarios. All studies have shown that groundwater modeling can be used successfully to help understand the Nile Valley aquifer's behavior.

Keywords Groundwater modeling · Modeling methods · MODFLOW · Calibration and sensitivity analysis · Boundary conditions · Nile valley

1 Background

Groundwater is of great importance in securing most water needs in several regions worldwide (Suring 2020). Many ambitious water resources plans in developing countries are based on the groundwater for achieving the targeted agricultural expansion for the increase of population. Switzerland, Romania, Hungary, Denmark, Belgium, and Austria are examples of such countries where groundwater resources exceed 70% of the total water consumption (Nriagu 2019). Groundwater resources are not independent while they represent a substantial element in the hydrological cycle. Therefore, groundwater is in continuous interaction with the surrounding water bodies and environment. It is also being affected by most of the changes in natural processes, land uses, and field conditions [e.g., agricultural practices (Awad et al. 2021), human activities, and urbanization (Zhao et al. 2013)]. The condition of a particular groundwater resource much depends on these interactions. Thus the best management of groundwater resources is that which considers, as much as possible, all parameters and interactions that can affect the groundwater (Zhou and Li 2011).

The most known and direct technique to assess a particular groundwater resource is through observational works. Particularly, groundwater tables and quality at the targeted domain or area can be monitored at the desired timeframe and time intervals, similar in the study Awad et al. (2020). While such observations cannot answer some questions about, for example, the long-term hydrologic responses of groundwater resources to some significant environmental phenomena, human activities, and even future climate changes. Therefore, groundwater models were developed to play an essential role in facilitating the consideration of different impacts of climate, environmental, and human conditions on groundwater resources, thus improving and enhancing groundwater management (Zhou and Li 2011; Kumar 2019). Groundwater modeling has been practiced around the world to fulfill the following:

- Simulation and analysis of various responses of the groundwater to the surrounding environmental conditions.
- Investigating and understanding different groundwater dynamics and flow patterns.
- Evaluating the various aquifers' processes of recharge, discharge, and storage.
- Predicting future impacts of human activities and climate changes on groundwater resources.
- Planning the collected field data and designing practical solutions for groundwater issues.

Therefore, groundwater modeling is considered a visualization tool through which key messages about groundwater resources can be communicated to the public and decision-makers. Many researchers around the worlds used groundwater modeling to fulfill, and not limited to, the following:

- Assessment of potential impacts of climate change on aquifers (Al-Maktoumi et al. 2018).
- Assessment of the efficiency of groundwater replenishment through natural reservoirs (Salem et al. 2020).
- Investigate the role of chemical fertilizers and pesticides in groundwater contamination (Srivastav 2020; Eltarabily et al. 2017).
- Assessment of how aquifers' conditions can be impacted under different groundwater extraction scenarios (Al-Maktoumi et al. 2016, 2020; El-Rawy et al. 2019a; Eltarabily et al. 2018; Negm 2018).
- Understanding the mechanism of groundwater and surface water interactions (Fleckenstein et al. 2010; Dawoud and Ismail 2013; El-Rawy et al. 2016, 2021a, 2021b).

Tian et al. (2015) coupled groundwater and surface water flow model (GSFLOW) with storm water management model to study the groundwater system along with the hydrologic cycle in the Zhangye Basin, northwest China. Polomčić and Bajić (2015) used the groundwater modeling to decide the best scenario of the dewatering process in the Buvač open cast mine, Bosnia and Herzegovina. Nagels et al. (2015) built a transient groundwater flow model to dewater the groundwater levels in the Schietveld region of Belgium and studied the relationship between hydrology and vegetation types' spatial distribution. Attard et al. (2016) developed a 3-D heat transport modeling approach to quantify underground structures' impact on urban groundwater temperature in the greater Lyon area, France. El-Rawy et al. developed a numerical groundwater flow model for the Zarqa River Basin (Jordan) to predict changes in the aquifer and Zarqa River under different recharge rate scenarios from the As Samra wastewater treatment plant (WWTP) and different groundwater pumping rates. The MODFLOW model was used by Sashikkumar et al. (2017) found suitable areas for artificial recharge structures to boost groundwater supplies in the

Kodaganar river basin, Dindigul district, Tamil Nadu. They evaluated its effectiveness using a combination of GIS and numerical groundwater modeling tools. Bishop et al. (2017) explored groundwater pathways, recharge heights, and nitrate sources on Maui, Hawaii, using numerical groundwater modeling. Al-Maktoumi et al. (2018) evaluated the impacts of climate change and the rise of the sea level on Oman's Samail and Jamma coastal aquifers numerically using MODFLOW and MT3DMS. Salem et al. (2018) developed a 3-D groundwater model using MODFLOW-2005 for the Drava floodplain, Hungry, for better understanding the Drava floodplain system's water budget under different lake replenishment scenarios.

El-Rawy et al. (2020a) developed analytical and numerical groundwater flow solutions for the femme-modeling environment in order to examine the effects of rising Aa River water levels on the groundwater aquifer in the Nete River watershed, Belgium. Katpatal et al. (2021) used the GIS and MODFLOW model to simulate future groundwater scenarios in Maharashtra, India, to overcome the drastic decrease in groundwater levels in the study area. The MODFLOW and RT3D models were used by Esfahani et al. (2021) to simulate diffusion and reaction in aquifers without discretizing low permeability zones. Zijl and El-Rawy (2019) studied the evolution from a steady to a steady mixing zone between two groundwater flow systems with varying concentrations. Kenda et al. (2018) implemented machine learning methodologies to model the changes in groundwater levels in the Ljubljana polje aquifer, Slovenia. Meredith and Blais (2019) constructed a groundwater model in the alluvial gravel aquifer of the Bighorn River Valley north of Hardin, Montana, to quantify groundwater recharge's relative importance infiltration of flood-applied irrigation water versus leakage from irrigation canals. Sathe and Mahanta (2019) used the numerical groundwater modeling software to develop a 3-D transient state predictive (groundwater flow and contaminant transport) conceptual model for two arsenic-contaminated regions, to evaluate the groundwater flow and arsenic contaminant transport. An adaptive neuro-fuzzy inference system was trained and optimized by Kisi et al. (2019) to model groundwater quality variables in Iran. El-Rawy et al. (2019a) and Al-Maktoumi et al. (2020) used MODFLOW to develop a groundwater flow and transport model to explore the feasibility of managed aquifer recharge (MAR) utilizing treated wastewater (TWW) in the Jamma coastal aquifer and Al-Khod aquifer in Oman. Zijl and El-Rawy (2021) studied the relevance of the deep creep flow of the earth's viscous upper mantle and crust as a complement to the groundwater flow. Rahnama et al. (2020) applied the Groundwater Modeling System (GMS) software and MODFLOW model to study the variations in groundwater levels in Shahdad Aquifer, Iran. Another groundwater model was developed by Khadim et al. (2020) using the MODFLOW-NWT in the Gilgel-Abay Catchment, Upper Blue Nile, Ethiopia, where there is data scarcity. They proposed an approach that can combine different types of datasets (e.g., reanalysis products, satellite data, citizen science data, etc.) to overcome the data-scarce issue. To alleviate water shortage difficulties in the Drava River floodplain, Salem et al. (2020) created a combined surface water (Wetspass-M) and groundwater (MODFLOW-NWT) model.

This chapter presents a review of groundwater flow models' fundamentals, including the boundary conditions, calibration methods, validation, sensitivity analysis, model uncertainties, and limitations of numerical modeling techniques; in this context attention is paid to more recent developments closed the gap between the finite difference and the finite element method. In addition, this chapter presents a review of the groundwater modeling applications on the Nile Valley aquifer in Egypt.

2 Groundwater Modeling

2.1 Introduction

Groundwater resources are essential for sustainable demographic and economic growth (Gleeson et al. 2012). Groundwater modeling is the way of representing reality, in a simplified form without making invalid assumptions or compromising the accuracy, to investigate the system response under certain phenomena or predict the system's behavior (Baalousha 2008). Groundwater models are a powerful tool for efficient management and planning of water resources, groundwater remediation, and protection. In general, models are conceptually describing the approximated physical systems using mathematical equations (Kumar 2004). A better understanding of the physical systems and the embedded assumptions in the mathematical equations are required to assess the usefulness or applicability of the model. These assumptions include the fluid dynamics, the anisotropy and heterogeneity of the bedrock or sediments within the aquifer, aquifer geometry, chemical reactions, and the contaminant transport mechanisms. Models should be considered as an approximation not as an exact representation of reality because of the uncertainties in the model's data and the assumptions embedded in the mathematical equations. Figure 1 shows the main steps of methodology in groundwater modeling. The first step in the modeling process is defining the model's purpose. Data collection represents a significant issue in the modeling process. The model's conceptualization is the fundamental step in modeling, followed by setting up the numerical model. Model calibration and validation, and sensitivity analysis can be performed after model completion. The last step is preparing and running simulations for prediction scenarios.

2.2 Defining the model's Objectives

The modeling approach, including the choice of model type, may vary depending on the modeling objectives. Groundwater models can be applied as predictive, interpretive, or generic tools (Anderson and Woessner 1992a, b). Predictive models are

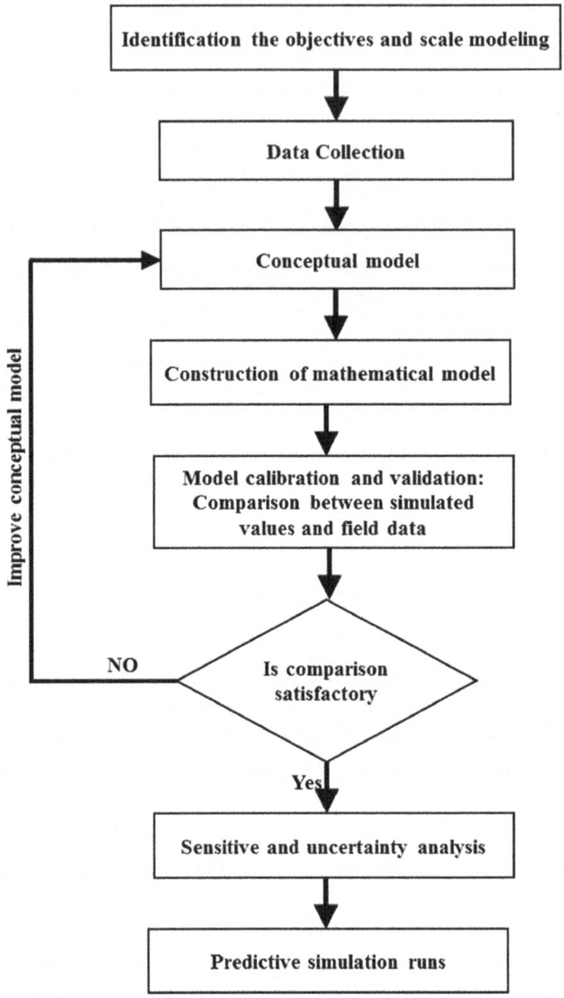

Fig. 1 A schematic stepwise diagram of groundwater modeling methodology

used to predict the effects of a proposed action on existing hydrogeologic conditions or to assess the future change in groundwater heads, flow rate, or solute concentration. Interpretive models are applied to investigate a particular case, study system dynamics, and evaluate groundwater flow or contaminant transport. Generic models are used to assess different scenarios of remediation schemes or management of water resource and identify regions' suitability for some proposed action. Note that calibration is a prerequisite for predictive models, whereas calibration is not required for interpretive and generic models. Baalousha (2008) summarized the objectives of groundwater modeling as:

- Predicting the temporal and spatial distribution of groundwater head and flow rate.

- Assessing the impcats of groundwater extraction on the flow regime and predicting the resulting drawdown.
- Evaluating the impact of human activities (e.g. agricultural activities, wastewater discharge, landfills, etc.) on groundwater quality.
- Investigating the effect of implementing various management scenarios quantitatively and qualitatively on groundwater systems.

2.3 Conceptual Model

A conceptual model represents the most important part of groundwater modeling; it is built on understanding how a groundwater system works. It consists of understanding the groundwater system's characteristics and spatio-temporal evaluation and provides a descriptive representation of the hydrogeologic system. Conceptualizing the groundwater system requires good information about hydrology, geology, hydraulic parameters, and boundary conditions. A good conceptual model should simplify the complexity of reality to achieve management requirements and modeling objectives (Bear and Verruijt 1987). Simplification relies on the objectives, the amount of available data, the scale of the model, and the current understanding level. The conceptual model describes factors that include:

- Model domain and aquifer geometry
- Time-independent aquifer/aquitard parameters such as hydraulic conductivity, hydraulic resistance, transmissibility, porosity, specific yield (effective porosity), specific storage, etc.
- Boundary geometry and boundary conditions
- Evapotranspiration and groundwater recharge
- Identification of sources and sinks
- Water balance.

The model's conceptualization is an iterative process that can identify the data gaps which have to be filled by further data gathering for model improvement. Remote sensing data can be used in building the conceptual model, particularly in situations of data scarcity where data-bases and maps may not be adequate. Hydrogeological data should be reliable and sufficient to some extent based on the targeted accuracy level to construct a proper conceptual model.

2.3.1 Boundary Conditions

Defining model boundaries represents the most critical step in building a reliable and accurate numerical groundwater model (Franke et al. 1987). The boundaries and the conditions that have to be applied as boundary conditions are strongly dependent on the hydrogeological conditions in the model domain. To achieve adequate conditions, the following inputs are used:

(i) Surface data and information from different available maps (topographic, hydrogeological, geological, soil, etc.) and remote sensing images for defining spatial distribution of impermeable lithological contacts, groundwater discharge zones, surface water bodies (river, an artificial reservoir, pond, canal, etc.), springs, geological faults, etc. These surface data can help in defining the recharge or no-flow boundaries (Singhal and Gupta 2010).
(ii) Subsurface data and information based on boreholes, geophysical surveys, geological sections, etc. These are used for defining subsurface hydrogeological characteristics such as the thickness of different aquifers and aquitards, weathering depths, subsurface salinity variations, buried channels, faults in basement rocks, etc.. Specifying boundary conditions is required to solve groundwater flow equations and represents the first step in model conceptualization. Improper representation of boundary conditions influence the solution and leads to incorrect results. Boundary conditions can be classified into three main types:
    (a) Type I boundary, called Dirichlet condition, represents specified head.
    (b) Type II boundary, called Neumann condition, represents specified flow rate.
    (c) Type III boundary, called Cauchy condition, represents head-dependent flow rate.

In particular, the following types of boundaries are described:

- **Constant head boundary**: This is a subset of the specified head boundary in which the head or concentration remains constant over time.
- **Specified head boundary**: This is a generalization of the constant head boundary, which has to be imposed when head or concentration can be identified as a function of location and time. Rivers and streams represent examples of the specified head boundary.
- **No flow boundary**: This is a special form of a specified flow or flux boundary. It has to be imposed at a surface where the normal component of the flux equals zero. Examples are: the impervious base of the basin (aquifer/aquitard system), groundwater divides, impermeable subsurface layers, and impermeable faults (see Fig. 2).
- **Specified flux boundary**: This is a generalized form of the no-flow boundary. This condition has to be imposed on the boundaries where the normal component of the flow rate (flux) is known as a function of location and time. For instance, it describes spring flow, measured flow from surface water bodies, and seepage to and from bedrock underlying the aquifer/aquitard system.
- **Head-dependent flux boundary**: This condition has to be imposed on boundaries through which the flux depends the head neighboringto that boundary. A semi-confined aquifer is one in which the water head is determined by the flux through the semi-constricting layer and semi-permeable or leaky faults (Fig. 2). It shows examples of these types of boundaries.

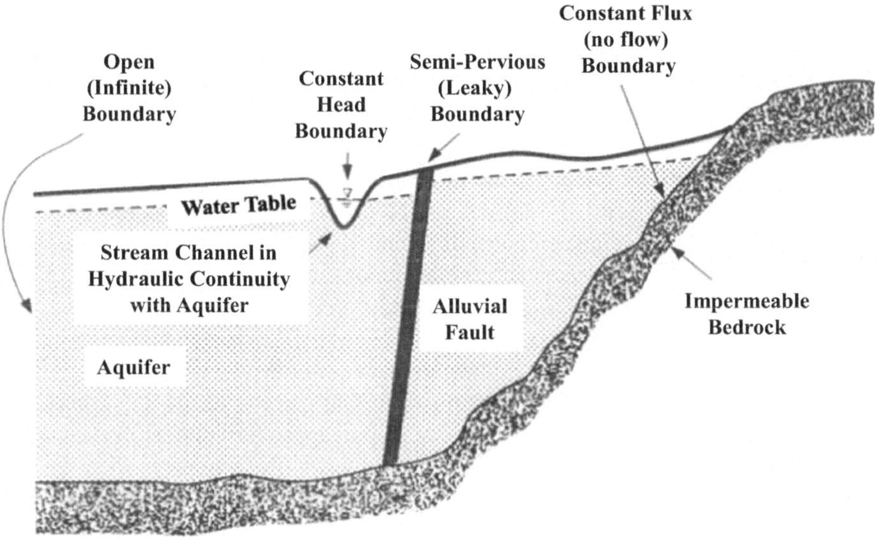

Fig. 2 Typical model boundary conditions *Source* Roscoe Moss Company (1990). http://ecoursesonline.iasri.res.in/mod/page/view.php?id=1852

Strictly speaking, there is a fourth type of boundary condition that does not correspond with the Dirichlet, Neuman or Cauchy types. The water table (Fig. 2) or phreatic surface, which is the top boundary of the groundwater model domain, is a moving boundary, going up and down during the seasons, rising during rainfall and falling in drought periods. On that case, two additional types of boundary conditions have to be imposed: a kinematic condition and a dynamic condition. Although the character of this top boundary is often overlooked, it is important to consider its conceptual consequences in more details. The kinematic boundary condition describes the evolution in time of the water table height, function $H(x, y, t)$, with respect to vertical height z, (Strictly speaking, there is a fourth boundary, namely the water table or phreatic surface, which is the top boundary of the groundwater model domain. It is a moving boundary, going up and down during the seasons, rising during rainfall and falling in drought periods. On that boundary two boundary conditions have to be imposed, a kinematic condition and a dynamic condition. Although this top boundary is often overlooked, it is important to consider its conceptual consequences. Strictly speaking, there is a fourth boundary, namely the water table or phreatic surface, which is the top boundary of the groundwater model domain. It is a moving boundary, going up and down during the seasons, rising during rainfall and falling in drought periods. On that boundary two boundary conditions have to be imposed, a kinematic condition and a dynamic condition. Although this top boundary is often overlooked, it is important to consider its conceptual consequences.

The kinematic condition describes the evolution in time, t, of the water table height, H, with respect to the vertical coordinate, z. That is, $z = H(x, y, t)$ in which x

and y are the horizontal coordinates. In this condition the specific yield (a parameter) and the recharge rate (a flux condition) have to be specified. It is important to note that water table height H is only a function of the horizontal coordinates and of time; H is independent from the vertical coordinate.

The dynamic condition is essentially the condition that on the water table the groundwater pressure is equal to the atmospheric pressure. The hydraulic head (briefly, the head) is essentially defined as the difference between actual pressure and hydrostatic pressure; this pressure difference is scaled in such a way that it is expressed as a height, head $h(x, y, z, t)$, a function of the horizontal and vertical coordinates as well as time,. The dynamic condition can then be written as $h(x, y, z = H, t) = H(x, y, t)$.

Only under the assumption that in the phreatic aquifer head h is independent from vertical coordinaten z it is possible—but not desirable—to equate the functions h and H. This assumption is known under the name Dupuit approximation. However, when dealing with general three-dimensional flow problems—for instance when dealing with Thótian flow systems—it is necessary to make a clear distiction between head h and height H in order to avoid serious mistakes. For more details see De Smedt and Zijl (2018) and Zijl and Nawalany (1993); also see Sect. 2.10.1.

2.4 Types of Models

There are several types of models available to model groundwater flow and contaminant transport. Models can be divided into three types: physical, analog, and mathematical.

2.4.1 Physical Models

Physical models such as sand tanks rely on developing the models in the laboratory to investigate specific contaminant transport or groundwater flow problems. Using these models, different hydrogeological phenomena such as artesian flow or the cone of depression can be investigated. In addition to flow, contaminant transport can be evaluated through physical models. Such models are easy to set-up and useful; however, they cannot be used to investigate complicated realistic problems.

2.4.2 Analog Models

The flow of electricity represents the most famous analog model. The electric analog depends on the similarity between Darcy's law relating head difference to groundwater flow rate and Ohm's law relating potential difference to electric current. Simple analog models can be set up easily to evaluate the groundwater flow. A detailed description of analog models can be found in (Anderson and Woessner 1992a, b; Fetter 2001).

2.4.3 Mathematical Models

Mathematical models represent a conceptual understanding of the groundwater system through a set of partial differential equations describing the flow and transport in the modeling domain where the equations are complemented with initial and boundary conditions. In addition, these equations require specification of spatially distributed time-independent parameters like hydraulic conductivity, storage coefficient, specific yield, porosity, etc., briefly the physical characteristics of the subsurface (aquifer/aquitard system). If both the time-independent pareameters (conductivities, resistances, etc.) and the time-dependent boundary conditions (recharge rates, well abstraction rates, etc.) are specified, the time-evolution of the heads and velocities can be determined from the mathematical model. This so-called forward model can be solved analytically in case of simple systems, or numerically in more complex systems.

2.5 *Solution of Mathematical Formulation*

As discussed in the preceding sections, the resulting equations of the mathematical model can be solved analytically in case of a simple system, or numerically in more complex systems. Some techniques use a mixture of analytical and numerical solutions.

2.5.1 Analytical Solutions

The primary benefits of using analytical solutions are that they are simple to implement and provide accurate and consistent results for simple problems. Because analytical solutions require many assumptions and simplifications, such as the homogeneity of a simple aquifer/aquitard sequence, they cannot handle a variety groundwater systems and are limited to simple ones. Some examples of analytical solutions include the Theis equation (Theis 1941) and Tóth's solution (Tóth 1962). Bear (1979) and Walton (1989) provide more information on analytical solutions to groundwater problems (1989). Darcy's law, in the form of its initial form, is the simplest equation describing groundwater flow. In this simplest form, Darcy's law is used to calculate one-dimensional groundwater flow through a section of an aquifer (Driscoll 1986):

$$Q = \frac{KA(h_1 - h_2)}{L} \tag{1}$$

where, Q is the flow rate (m^3/day), K is the hydraulic conductivity averaged over the length of the one-dimensional section of the aquifer (m/day), A is the area (m^2), $h_1 - h_2$ is difference in hydraulic head (m), L is the distance along the flowpath between the points, where h_1 and h_2 are measured (m). For the generalization of Darcy's law to two- and three-dimensional flow see Delleur 2006, Cushman and Tartakovsky (2016), and De Smedt and Zijl (2018).

2.5.2 Numerical Solutions

Numerical methods have been developed to overcome the limitations of analytical solution methods and to deal with the complexity of groundwater systems. Numerical models are based on numerical solutions of a system of coupled algebraic equations for discrete variables, e.g. heads in the nodal points of a discretized flow model, or solute concentrations in the nodes of a discretized transport model. Numerical groundwater models are generally based on the discretization of a model area into a great number of finite volume elements, in such a way that the fundamental conservation laws of mass, momentum and energy are honored for each volume element (Hölting and Coldewey 2019). Several types of numerical solution methods are available for groundwater flow and transport studies. Generally speaking, a distinction can be made between two types of approach: (i) the Finite Difference Method (FDM) and the related Finite Volume Method (FVM) and (ii) the Finite Element Method (FEM).

The Boundary Element Method (BEM) and Analytical Element Method (AEM) (Strack 1989) are based on analytical solutions combined with some form of disctetization—mainly discretization of the boundaries—and therefore result in a system of algebraic equations that has to be solved numerically. The advantage of BEM and AEM is that the flow around the wells can be solved very accurately because there is no need for grid refinement as required for FDM/FVM and FEM. On the other hand, FDM/FVM and FEM can easily handle all types of heterogeneity. For that reason FDM/FVM and FEM are generally considered as the most flexible types of numerical models (Anderson and Woessner 1992a, b; Igboekwe et al. 2008) and are therefore the most widely applied numerical methods. In summary: both in FDM/FVM and FEM, the subsurface is discretized (sub-divided) into a grid with a great many of small finite volumes. The heads and concentrations are calculated in nodal points, while the flow rates have to be considered at the grid faces between the grid volumes in order to analyze the flows through the subsurface (the aquifer/aquitard) system (Igboekwe and Achi 2011).

Finite Difference Methods

FDM has been widespread used in groundwater studies since the early 1960s. The finite difference method was alreadty studied by Gauss, Newton, Laplace and Bessel (Pinder and Gray 1977). The basic principle of FDM is to represent the derivative of the head function approximately at discrete points situated between adjacent head values near that point; in other words, derivatives are approximated by difference quotients. The distribution of nodes in FDM is regular (rectangular blocks) but the nodal spacing may either be uniform or non-uniform along the orthogonal coordinate system (Singhal and Gupta 2010). The accuracy of the FDM relies on grid size and uniformity. The approximation of the derivative improves as the grid spacing becomes smaller and smaller, and in the limit of zero grid spacing the numerical approximation becomes equal to the exact solution. For larger grid spacing the solution will deviate more from the exact solution. However, in the calculation of heads and velocities, even if they are relative inaccurate because of a coarse discretization, the mass balance will be honored accuurately provided that the system of algebraic equations is solved with sufficient accuracy. On the other hand, in transport models a coarse discretization gives rise to much larger numerical dispersion than the real physical dispersion. There are different approaches to finite difference approximations: the three most extreme forms are (i) central differences, (ii) forward differences, and (iii) backward diffrences.

In theory central differences provide the best results because in that case the truncation error has a second-order accuracy $(\Delta x)^2$ (Pinder and Gray 1970). However, in the most practical, and therefore most popular finite difference method—the block centered finite difference method—deviatins from the central differences are generally accepted because the rate of convergence to the exact solution does not apprecially deviate from the second-order convergence accuracy $(\Delta x)^2$. This result has been proved by Weiser and Wheeler (1988) (also see the discussion on the'face-centered finite element method' in Sect. 2.5.2.2). Irregular spacing is often needed to increase the accuracy at selected areas, especially near wells. As a rule of thumb for refining or expanding the finite-difference grid, the maximum multiplication factor should be 1.5 or smaller (Soderlind and Arevalo 2017). Local grid refinement, for instance around each individual well in a flow field, is not possible in finite difference grids. The best approach would be to apply around each well a refined tetrahedral grid (in 3D) or triangular grid (in 2D) in combination with the'face-centered finite element method'; see Sect. 2.5.2.2. However, for practical reasons application of a simple algebraic model relating grid block head to well head is generally used (see, for instance, Peaceman 1978).

The main advantage of the FDM, and in particular of the block-centered FDM, is the flux continuity, which means that the groundwater velocities are continuous at the faces between two adjacent grid blocks. Flux continuity is a prerequisite to obtain accurate solutions of transport problems and/or flow path tracking. In this sense finite difference methods are superior above conventional finite element methods (see Sect. 2.5.2.2). On the other hand, flux continuous methods are not head-continuous. The heads are continuous only in the centers of the grid faces, at the other locations of the faces the heads'jump' between two neighboring grid blocks. This is in contrast to conventional finite element methods, who are head-continuous. Another contrast is that the FDM under estimates the values of the hydraulic conductivities, while the conventional finite element method over estimates them.

The available finite difference models are easy to implement, well documented and proven to provide good results. The main disadvantage is that these methods do not fit properly to an irregular model boundary. Moreover, the grid's size (number of grid blocks) highly influences the computational effort and accuracy. MODFLOW is the most commonly used finite-difference groundwater model (Harbaugh and McDonald 1996).

Finite Element Methods

In the finite element method (FEM), the model domain is divided into many relatively small sub-regions, the volume elements (briefly the elements), which may have any shape. In each element head and velocity are represented by a relatively simple algebraic function of the spatial coordinates In addition, the FEM can handle any type of anisotropy including hydraulic conductivity tensors with off-diagonal components. In the FEM, the integral representation of a partial differential equation is obtained by one of the following approaches: variational principle, weighted residuals, or Galerkin's method. A detailed description of these method can be found in Pinder and Gray (1970). As the FEM uses irregular shapes of the elements, the nodes of the elements can be scattered arbitrarily in the domain, in concentrated or sparse patterns, to form various sizes of elements. This flexibility of the finite element grid enables a more realistic simulation of different boundaries (Singhal and Gupta 2010) and is useful for providing a close spatial approximation of irregular boundaries and for concentrating elements in areas where the considered variable is characterized by larger variations and where higher accuracy is required (Bear and Cheng 2010). Especially when dealing with grid refinement around wells, these advantages play a major role.

FEM requires more sophisticated mathematics and may provide more accurate results for a number of applications. The conventional finite element method is head-continuous over the faces (face elements) between two adjacent volumes (volume elements), which makes the method very suitable for soil-mechanical and geo-mechanical problems. However, because the conventional finite element method is not flux continuous, the conventional FEM is not suitable for transport problems, in contrast to volume centered finite difference methods.

A notable exception is the'face centered' finite element method (in mathematical terms: the'mixed-hybrid finite element method'). In this method the heads are calculated in the centers of the faces and are discontinous at the other points of the faces (in mathematical terms the face-centered heads are the'Lagrange multipliers'). The face centered finite element method is flux continuous and, therefore, suitable for transport problems. Grid refinement results in good convergenge to the exact solution The face-centered finite element method is algorithmically different from the block centered finite difference method (e.g. MODFLOW), but when refining the finite diference grid of the block centered finite difference method, this method converges in the same way to the solution obtained by the face-centered finite element method (Weiser and Wheeler 1988). Thanks to this result the modeling community has fully accepted the block centered finite difference method. It has good convergence properties; therefore there is no longer a reason to use the more complex point distributed finite difference method (Azis and Settari 1979, Sect. 3.5.1). In contrast to the conventional finite element method, the face centered method is under estimating the hydraulic conductivities. For more details about the face-centered finite element method or, in mathematicl languege, mixed-hybrid finite element method see the'pioneers' Chavent and Jaffré (1986) and Kaasschieter (1990), Kaasschieter and Huijben (1992); they applied and exempified the face-centerd finite element method for petroleum reservoir engineering and groundwater flow modeling, respectively.

Unfortunately, the highly mathematical presentation of this method has seriously hampered its understanding and acceptance by the hydrogeological community. Fortunately, Bossavit (1998, 2005) was able to present an alternative, much more transparant approach to replace the opaque mathematics and its terminology. Although Bossavit presented his theory in the context of electrical engineering, his approach could relatively easily be translated to groundwater flow; Trykozko (2001), Trykozko et al. (2001), Zijl (2005a; b).

In the finite element method the identification and construction of the input data set is more complicated than for a regular finite difference grid. Mesh design in the finite element method is crucial as it significantly influences the accuracy and convergence of the solution. It is highly recommended to keep the mesh configuration simple and to refine the mesh only at interesting areas where variables change rapidly, in particular near sources and sinks. It is better to keep the mesh configuration as simple as possible. To facilitate the construction of a finite element mesh it almost necessary to to use a pre- and post-processor. For further details on the use of FEM modeling, the reader is referred to Remson et al. (1971), Pinder and Gray (1977), and Huyakorn and Pinder (1983). The most common finite element based groundwater models are MODFE (Torak 1993), Femwater (Lin et al. 1997), and FEFLOW (Wasy 2005).

2.6 Model Calibration

The objective of calibration is to determine the time-independent parameters in the model; for instance the spatially distributed of hydraulic conductivities. This is achieved by feeding the model with input data, generally a time series of flow rates (of recharge rates, well rates, etc.) and comparing the computed output variables to the measured time-dependent heads, flow rates and concentrations with the values measured in some points (generally heads measured in observation wells. Popular approaches to calibration are: trial-and-error, indirect methods and direct methods.

To achieve a good fit between the observed and simulated values, adjustable time-independent parameters like hydraulic conductivity, storage coefficient, specific yield, river bed conductance) are gradually adapted to minimize the residual between observed and simulated values (Poeter and Hill 1997; Gupta et al. 1998). More specifically, in flow problems the hydraulic conductivities have to be adapted to match the time series of calculated heads with the time series of measured heads. In this approach, the groundwater model is not just the discretized mathematical model, but generally includes also the time series of measured recharge rates and well flow rates.

The trial and error procedure is time-consuming (Hill and Tideman 2007; Cao et al. 2006; El-Rawy 2013), especially when the number of unknown parameters is large. This method considers the matching history as a fundamental first step as it can provide the modeler much insight into the modeled area and how the time-evolution of the calibrated parameter values influences different parts of the model area and observation types (Anderson et al. 2015a, b).

Automatic methods like indirect calibration methods have been developed. Richard Cooley (Cooley 1977 and 1979; Cooley and Naff 1990) developed a pioneering inverse code using nonlinear regression, an approach later extended to the parameter estimation code MODINV (Doherty 1990), MODFLOWP (Hill 1992), UCODE (Poeter and Hill 1998), Poeter et al. 2005), and PEST parameter estimation (Doherty 2014a, b). These developments replaced MODINV in 1994 by the current PEST software suite which is now widely used in groundwater modeling. PEST calibration can be conducted in two ways: using zonation and using pilot points. The zonal approach is the most common one (XMS Wiki 2020). From a theoretical point of view indirect methods are superior regarding accuracy. However, there are practical limits to the applicability of indirect methods because they have a high computational complexity; i.e., the number of calculations (multiplications, etc.) increases quadratically with the number of parameters that have to be determined. Thus, according to some reviews, indirect automated methods take more time than the trial and error method (Hill and Tideman 2007).

Direct automated methods may be considered as an alternative to trial and error and automated indirect methods. Direct methods solve the model of equations, e.g. Darcy's law, mass balance and flux boundary conditions, to calculate data that can be measured, e.g. heads measured in observation wells. Then a 'filter' determines a weighted average between the measured and the calculated data. Initially, the measured data have the greatest weigh, but after a number of time steps for which the calculations and measurements have been performed the calculated data get greater weight and smaller uncertainty, until a limit determined by the model error, a measure of the hydrogeologist's trust in the model. The uncertainly is generally appreciably smaller than the spread (see epistemic uncertainty in Sect. 2.9). In this approach the filter also updates the parameters, e.g. the hydraulic conductivities, after each time step. In addition, the filter determines the spread (standard deviation) as well as the uncertainty in the thus-determined parameters. The Ensemble Kalman Filter (EnKF) is the most popular filter method; it is widely used in oceanography, meteorology, petroleum reservoir engineering and, to a lesser extent, in hydrogeology. Its computational complexity is almost linear; the number of calculations as proportional with the number of parameters that have to be determined, times a numerical factor. This factor is at least 100 and will become larger when a greater accuracy is required. As a consequence, EnKF is still a computationally expensive method for relatively simple problems.

For smaller problems, the Double Constraint Method (DCM) is an attractive alternative. In this method the groundwater flow is calculated by (i) a 'flux model' (Darcy and mass balance) with the measured fluxes as boundary condition and (ii) a 'head model' (again, Darcy and mass balance) with the measured heads-including the heads measured in the observation well-as boundary condition. It is important to note that in this approach the model is just the 'mathematical model'; it does not include the flow conditions like well flow rates and recharge data. This is in contrast with the 'conventional model' to be calibrated, which a flux model is including the time series of well flow rates and recharge data. According to Darcy's law (Eq. 1), the absolute value of the fluxes is determined by the flux model divided by the absolute value of head differences which are determined by the head values and then the calibrated hydraulic conductivities (e.g. in the grid bocks or in a cluster of grid blocks). These conductivities can then be used for a second flux run and head run of the models, which results in improved hydraulic conductivities, and so on until convergence. When a time series of head and flux measurements is available, the DCM can be complemented with a simple linear Kalman filter to estimate the standard deviation and the uncertainty in the resulting conductivities.

Considering a flux model, the conductivities calibrated by an indirect or direct method based on a finite difference model (or by a face centered finite element model), will be larger than the real field conductivities. On the other hand, applications of a conventional finite element model result in smaller conductivities than the real field conductivities. However, because the double constraint method is based on both a flux models and a head model, it does not calibrate one of the models. Instead, this method yield an estimation of the real field conductivities (El-Rawy 2013; El-Rawy et al. 2010, 2011, 2015a, b, 2018; Zijl et al. 2017, 2018a, b; De Smedt et al. 2018).

Several statistical indices have been recommended for assessing the performance of a model. The Mean Error (ME), Mean Absolute Error (MAE), Root Mean Square Error (RMSE), and Nash–Sutcliffe efficiency (NSE) are often used. They are presented mathematically as follows in an example of observed head data:

$$\mathrm{ME} = \frac{1}{n}\sum_{i=1}^{n}(h_m - h_s)_i \tag{2}$$

$$\mathrm{MAE} = \frac{1}{n}\sum_{i=1}^{n}|h_m - h_s|_i \tag{3}$$

$$\mathrm{RMSE} = \left[\frac{1}{n}\sum_{i=1}^{n}(h_m - h_s)_i^2\right]^{0.5} \tag{4}$$

$$\mathrm{NSE} = 1 - \frac{\sum_{i=1}^{n}(h_m - h_s)_i^2}{\sum_{i=1}^{n}\left(h_m - \overline{h}_m\right)_i^2} \tag{5}$$

where n is the number of observations; h_m is the observed groundwater level, h_s is the simulated groundwater level and $\overline{h}_m$ is the mean observed groundwater level. ME provides a general description of the model bias as both positive and negative differences are involved in this mean; the errors may eliminate each other and thus decrease the overall error (Anderson et al. 2015a, b). MAE measures the average error in the model. RMSE is the average of the squared differences in observed and simulated heads. NSE is applied to compare individually observed and simulated hydrographs in transient models; NSE ranges in values from 1 to ∞, where values close to 1 show a good fit. Equations (2–5) make sense for a conventional model', i.e. a model including the flux conditions. However, for a 'mathematical model' like the double constraint method, it would make more sense to apply Eqs. (2–5) to the conductivities instead of the heads. More specifically, for a time-independent parameter, e.g. the log-conductivity, calibration at a number of times (independently from earlier calibrations), the time-averaged value of the conductivity and its variance (spread) can be determined by well-known equations. The uncertainty is then equal to the spread divided by the square root of the number of calibrations). (Application of the log-conductivity instead of the conductivity because this uncertainty calculation is based on a Gaussian distribution). This approach has been developed and exemplified in great detail by El-Rawy (2013, 2015a); also see Sect. 2.9 on uncertainty analysis.

2.7 Model Validation: Acceptance or Rejection

Model validation is the following step after the model calibration. The model validation helps to check the performance of the calibrated model with any dataset. Since the calibration process includes changing different parameters (e.g. hydraulic conductivity, pumping rate, recharge, etc.) different sets of these parameter values may provide (almost) the same solution. Reilly and Harbaugh (2004) report that good calibration does not result in a good prediction. The validation process evaluates whether the calibrated model is applicable for any dataset. Based on the result of this evaluation it is decided whether the model is accepted or rejected and replaced with another model. In this regard, modern probabilistic Bayesian techniques, supplemented by Popper's falsification principle, are beneficial (Enemark et al. 2020) in increasing the confidence of conceptual models via an explorative systematic testing framework based on Popper-Bayes philosophy. Modelers typically divide available observed data into two datasets: one for calibration and one for validation (Abdelrady et al 2020). For more details about the groundwater model validation, see Anderson and Woessner (1992a; b), Davis et al. (1992); Henriksen et al (2003); Hassan (2003), (2004); Kori et al (2008); Du et al (2018); Abdelrady et al 2020).

2.8 Sensitivity Analysis

Sensitivity analysis is essential for calibration, risk assessment optimization, and data collection. It is a process of changing model input parameters through a reasonable range, evaluating the relative variation in the model's response (Kumar 2004) and measuring the effect of these variations on the model outputs. Sensitivity analysis quantifies the impacts of the uncertainty in the estimates of model parameters on model results and provides a basis for the choice which parameters have the greater impact on the output. Parameters that have a large influence on model results should receive the most attention during data collection and calibration. Furthermore, sampling location design and sensitivity analysis can be used to solve optimization problems. The most popular way of sensitivity analysis is based on finite differences, which are used to evaluate the changes in the model result after changing a few parameters (Baalousha 2008). This technique is used by the Parameter Estimation Package "PEST" (Doherty et al. 1994). Automatic differentiation was used in groundwater flow models for sensitivity analysis, and it produces precise results in comparison to finite difference approximations (Baalousha 2007). Sensitivity analysis can help in the selection of additional data that can be collected to elucidate how the modeled system works and to identify those parameters whose values must be specified most precisely during field investigations.

2.9 Uncertainty Analysis

Dependable groundwater modeling is required for effective groundwater resource management and planning. However, groundwater models are subject to several uncertainties in their predictions. The three main sources of uncertainty are epistemic uncertainty, aleatory uncertainty, and technological uncertainty (Pham and Tsai 2017). Aleatory uncertainty occurs as a result of randomness and is generally too small to be eliminated. Epistemic uncertainty is caused by a lack of data and knowledge (Hora 1996; Senge et al. 2014), and it can be effectively reduced by accumulating more informative field data. Technological uncertainty is caused by a lack of technology for converting known knowledge/data into valid groundwater models, and it can be significantly reduced by using better computing resources and numerical methods. Different sources of uncertainty in groundwater models include field data, the subsurface (aquifer/aquitard system) heterogeneity, and the assumptions underlying the mathematical modeling that increase the uncertainty of the model results (Baalousha and Köngeter 2006). There are several approaches for incorporating uncertainty into groundwater modeling. The most common method is stochastic modeling using the Monte Carlo or Quasi-Monte Carlo method (Liou and Der Yeh 1997; Kunstmanna and Kastensb 2006). Stochastic models, on the other hand, are time-consuming and require numerous computations. Some changes have been made to stochastic models to make them more deterministic, which reduces computational and time requirements. Zhang and Pinder (2003) modified Monte Carlo Simulation through Latin Hypercube Sampling, which reduces significantly the time requirements. Also see the work by El-Rawy (2013, 2015a), briefly presented at the end of Sect. 2.6.

2.10 Modeling Software/Codes

Numerous numerical groundwater modeling software have been developed and used widely based on various methods of solutions such as Visual Modular Three Dimensional Flow (Visual MODFLOW) (Anon 2000a, b), Finite Element subsurface FLOW system (FEFLOW) (Diersch 2005), Groundwater Modeling System (GMS) (Anon 2000a, b), a 2D and 3D uncertainty analysis geostatistics, and visualization software package (UNCERT) (Wingle et al. 1999). The GMS software is a graphical user interface for numerous groundwater flow models such as SEAM3D, FEMWATER, SEEP2D, MT3DMS, RT3D, MODFLOW (with many packages), MODAEM, MODPATH, and SEAWAT. A three dimensional particle-tracking program (MODPATH) Pollock (1990) and three dimensional mass transport modeling software MT3D (Zheng 1990) were developed to simulate mass transport in a groundwater system and is interfaced with MODFLOW to provide a plot of flow pathlines. Movement of Heat, Water, and Multiple Solutes in Variably Saturated Media (HYDRUS-1D and -2D) is conducted for the US Salinity Laboratory, USDA-ARS, Riverside (Simunek et al., 1999). Soil and Water Assessment Tool (SWAT) (Arnold et al. 1998; Arnold and Fohrer 2005) is supported by the USDA. Agricultural Research Service at the Grassland, Soil and Water Research Laboratory. MODFLOW is the most used software for simulating groundwater flow and contaminant transport (Fouad and Hussein 2018).

2.10.1 Modflow

The Modular Finite Difference Groundwater Flow Model (MODFLOW) was established by the US Geological Survey and is based on a block-centered FDM to simulate three-dimensional groundwater flow for both steady-state and transient state conditions. The model went through several versions, MODFLOW-88 (McDonald and Harbaugh 1988), MODFLOW-96 (Harbaugh and McDonald 1996), MODFLOW-2000 (Harbaugh et al. 2000; Hill et al. 2000), and the current version is MODFLOW-2005 (Harbaugh 2005). The MODFLOW model is based on the combination of two basic equations: the principle of momentum conservation (Darcy's law) and mass conservation. As a result, for groundwater with constant density (negligible density-driven flow), the groundwater flow through an anisotropic and a heterogeneous porous medium can be described by the partial-differential equation

$$\frac{\partial}{\partial x}\left(K_x\frac{\partial h}{\partial x}\right)+\frac{\partial}{\partial}\left(K_y\frac{\partial h}{\partial y}\right)+\frac{\partial}{\partial}\left(K_z\frac{\partial h}{\partial z}\right)\pm \mathrm{W}=S_s\frac{\partial h}{\partial t} \tag{6}$$

where h (L) is the hydraulic head in the porous medium; *Kx, Ky, and Kz* are anisotropic hydraulic conductivity for the porous medium in *x, y,* and *z* directions (LT^{-1}), *W* is the volumetric flux per unit volume at sources or sinks in the porous medium (T^{-1}), *Ss* is the specific storage of the porous medium (L^{-1}) and *t* is the time (T). MODFLOW cannot handle off-diagonal components of the anisotropy tensor.

At first glace, it seems that the parameter specific yield, *Sy*, is missing from Eq. (6). However, in MODFLOW the flow through the phreatic aquifer is based on the Dupuit approximation and, as a consequence, hydraulic head *h* may be replaced with water table height *H* (also see the last paragraphs of Sect. 2.3.1). This also allows us to replace *Ss* with *Sy*/*D* + *Ss*, where *D* is the thickness of the phreatic aquifer (or of the top layer in the MODFLOW model). The MODFLOW calculations in that layer are then based on the Dupuit-Forchheimer equation; for more details see De Smedt and Zijl (2018). Groundwater flow in which the specific storage terms are negligibly small, while the specific yield term plays an important role is called incompressible flow. Incompressible flow is often a good approximation for relatively shallow flow. Groundwater flow in which the specific yield term is neglected is called (quasi) steady flow.

Three packages are used in MODFLOW; Basic package, Hydrological packages, and Solver packages. The hydrological packages can further be divided into stress packages and internal flow packages. Detailed information for each package can be found in Hauber.

The MODFLOW model has been successfully applied in a large number of qualitative and quantitative groundwater studies because of its modular program structure, simple methods, and separate package to resolve special hydrogeological problems (Aghlmand and Abbasi 2019). For example, Visual MODFLOW and PMWIN are considered as popular software tools. They have been developed by GMS and are based on the MODFLOW program. The MODFLOW model with a graphical user interface (GUI) can be linked with a geographic information system (GIS) to provide a good visual environment for evaluating and managing groundwater resources (Wang et al. 2008). Using this facility, a model grid can be displayed on the monitor screen for graphically inputting the model parameters using menu options and cursor controls. This facility helps in creating input data files of the model to read and visualize the model output.

2.11 *Limitations of Modeling Techniques*

Despite all the sophistication in software and hardware, some simplifying assumptions are inevitable. Singhal and Gupta (2010) summarized the limitations of modeling techniques as follows: (a) estimating different aquifer parameters of groundwater systems, (b) techniques of input data acquisition, and (c) idealization and conceptualization of the groundwater behavior of the system, etc. Under these limitations, a model cannot be perfectly deterministic for all objectives. However, it can be applied to provide useful output for practical groundwater management and exploration.

3 Application of Numerical Groundwater Modeling in the Nile Valley

The aquifer systems in Egypt consist of five hydrological aquifers: the Nile Valley, Nile Delta, Eastern and Western desert, Coastal aquifer, and Sinia Aquifer (Ismail and El-Rawy 2018; Negm 2018; El-Rawy et al. 2020b, 2021c; El-Rawy and De Smedt 2020; Negm and Elkhouly 2021). The Nile Valley aquifer is replenished by seepage from river canals and return irrigation water (El Arabi 2012; El-Rawy et al. 2019a, 2021b). It is considered a renewable aquifer. The Nile Valley and the Nile Delta aquifers accounts about 7.5 billion cubic meters (BCM) yearly (El-Rawy et al. 2020b), which represents about 87% of the exploited groundwater in Egypt. The Nile Valley aquifer's estimated recharge rate is more than 3.5 million cubic meters (MCM) yearly, and the total groundwater storage is about 200 MCM per year (El-Rawy et al. 2020a, b). Some properties of the Nile Valley aquifer are given below (El Tahlawi et al. 2008):

- The top aquifer has a thickness of 0–20 m below the terrestrial surface.
- The aquifer has a saturated thickness of (10–200 m).
- Depth to the water table in the aquifer varies from 0 to 5 m.
- The aquifer's hydraulic conductivity varies from 50 to 70 m/day, and porosity ranges from 25 to 30%.

Based on the wide variety of groundwater modeling advantages mentioned above, many studies have been conducted to model groundwater resources in the Nile Valley to ensure better conditions for these resources. To analyze the performance of the bank filtration technique for water supply in Aswan City under different environmental conditions, Abdelrady et al. (2020) developed a hydrological model to assess the locations that are most appropriate for the installation of bank filtration wells and also to propose the best scenarios for managing the bank filtration fields. The model results showed that decreasing of Nile level (by 0.5–1.5 m) has a considerable effect on the bank filtration parameters (e.g., travel time, bank filtrate share) in the wells' onset operation.

Campos (2009) simulated the groundwater flow system at the West bank temples, Luxor city, by developing a groundwater conceptual flow model to assess the different water-related-damages on historical monuments in the study area. In the Qena governorate, Elsheikh et al. (2020) used the DRASTIC model to delineate the areas where the aquifer is vulnerable to waterlogging, head drop, and pollution in the new and old reclamation areas as well. In Esna City, Qena Governorate, El-Fakharany and Fekry (2014) developed a groundwater model based on monitored groundwater levels to assess the New Esna barrage's potential influences on groundwater resources in the study area.

Abdelshafy et al. (2019) used the PHREEQC model to characterize different hydrogeochemical characteristics of the groundwater aquifer in Sohag City and to understand the rock–water interactions. Ahmed (2009) used the 3D dimensional lithological modeling techniques to characterize and model the Quaternary aquifer system of the Sohag area. Shamrukh et al. (2001) used the groundwater modeling system to simulate the 3D dimensional flow of groundwater along with the contamination transport in the Tahta region, Sohag, to evaluate the potential effect of chemical fertilizers on groundwater resources in the study area.

For the Quaternary aquifer in Assiut Governorate, Sefelnasr et al. (2019) developed an integrated GIS-supported approach for constructing a 3D transient groundwater model. The model was created to investigate the most viable groundwater management option based on climatic, environmental, water demand, and developmental conditions. El-Rawy et al. (2021a) developed a groundwater model flow to study Assiut Quaternary aquifer's behavior under various recharge and discharge scenarios. The model was calibrated and a selectivity analysis was carried out. The findings demonstrate that increasing well pumping discharge by 1.25, 1.5, 2.0, and 2.5 times present pumping rates decreased groundwater heads by 17, 35, 72, and 110 cm, respectively. In comparison to the current situation, 22 and 46 cm reduced groundwater heads when a 0.5 and 1.0 m reduction in the River Nile's surface water levels were planned. Abdelhalim et al. (2019) used a numerical groundwater flow model to determine the different hydrogeological conditions of the Quaternary aquifer in Samalut city, El-Minia Governorate. The model was also used to determine the flow directions, calculate interaction between surface water and groundwater, and assess the aquifer's future response to some scenarios of increasing the groundwater extraction. El-Rawy et al. (2021b) investigated effects of the potential improper filling scenarios of the Grand Ethiopian Renaissance Dam on the Nile Valley aquifer in El-Minia Governorate. The study applied the MODFLOW numerical modeling of groundwater-surface water interaction for better understanding of the interaction mechanism between surface water and groundwater in the study area. Thus, it will be easy to assess the potential impacts on the groundwater levels if having a future decrease in Nile water levels.

4 Conclusions

We cannot live in an aquifer to see how its conditions and groundwater quantity change over long periods. Also, we cannot walk behind each particle of pollutants to trace whether and where it will settle in the aquifer. Through such groundwater modeling techniques, we can open the black-box that hinders us from understanding the mechanism of different aquifers' processes. By using groundwater modeling, researchers can evaluate the quantity and quality of groundwater in aquifers along with the assessment of how aquifers' conditions change under climate change, different groundwater extraction rates, changes in human activities, and a wide variety of environmental conditions.

Groundwater modeling is the way of representing reality, in a simplified form without making invalid assumptions or compromising the accuracy, to investigate the system response under certain phenomena or to forecast the performance of the aquifer system. Modeling the groundwater system is generally based on solving mathematical equations containing many parameters that characterize the system.

The modeling approach includes the choice of model type, the conceptualization of the model, boundary conditions, sensitivity analysis, calibration, validation, model uncertainty, and visualization. The choice of model type may vary depending on the modeling objectives. A conceptual model represents the most important part of groundwater modeling; it is built on the understanding of how a groundwater system works. The conceptualization of the model is an iterative process that can identify the data gaps that have to be filled by further data gathering to improve the model. Defining model boundaries represent the most critical step in building a numerical groundwater model. A sensitivity analysis is an important first step in the calibration process of a groundwater flow model. Model calibration is an important and essential step in groundwater modeling. It is usually carried out by comparing simulated hydraulic heads to observed hydraulic heads at a limited number of observation points. Model validation is the next step after calibration. The model validation helps to check the performance of the calibrated model with any dataset. Reliable groundwater modeling is required for successful groundwater resource management and planning. Groundwater models, on the other hand, are subject to a number of uncertainties in their predictions. To have a realistic simulation, the parameters' values should be adapted to their actual values. Parameters that have a large impact on model results should receive the most attention during data collection and calibration.

Several groundwater modeling applications have been developed for the Nile Valley aquifer, including various modeling objectives:

- Examine the effectiveness of the bank filtration technique for water supply in Aswan City under various environmental conditions. Investigate the interaction between the Nile River and the groundwater
- Evaluate the effects of the potential improper filling scenarios of the Grand Ethiopian Renaissance Dam on the Nile Valley aquifer
- Evaluate the different water-related-damages on historical monuments in the West bank temples, Luxor city
- Evaluate the potential effects of the New Esna barrage on groundwater resources
- Evaluate the potential effect of chemical fertilizers on groundwater resources in the Tahta region, Sohag.
- Investigate the most feasible option of groundwater management based on the climatic, environmental, water demand, and developmental conditions in the Quaternary aquifer in Assiut Governorate

- Study the behavior of the Nile Valley aquifer under various recharge and discharge scenarios.
- Determine the flow directions, calculate the recharge and discharge rates between surface water and groundwater, and assess the future response of the aquifer to some scenarios of increasing the groundwater extraction.

Based on the above mentioned, numerical groundwater modeling has been successfully used to achieve the various modeling objectives.

5 Recommendations

Despite all the sophistication in software, several simplifying assumptions are inevitable. Parameters with a high effect on model results should get the most attention in the data collection and the calibration process. The future research should be performed based on a continuous recording of groundwater level, water level changes in canals, an actual case of groundwater abstraction, actual representations of heterogeneity of the Nile Valley aquifer system. More field or laboratory works or designed well-pumping tests to measure/estimate the Nile Valley aquifer parameters is essential. Furthermore, groundwater models of the Nile Valley aquifer need to be developed considering the impacts of climate change on the streams-aquifer interactions. Additionally, the impacts of nitrate pollutants either in the irrigation canals' water or return agricultural flow on groundwater contamination could be studied. The interaction between the surface water and the groundwater aquifer can be assessed considering the lining of irrigation canals.

References

Abdelhalim A, Sefelnasr A, Ismail E (2019) Numerical modeling technique for groundwater management in Samalut city, Minia Governorate Egypt. Arab J Geosci 12:124

Abdelrady A, Sharma S, Sefelnasr A, El-Rawy M, Kennedy M (2020) Analysis of the performance of bank filtration for water supply in arid climates: case study in Egypt. Water 12:1816

Abdelshafy M, Saber M, Abdelhaleem A, Abdelrazek SM, Seleem EM (2019) Hydrogeochemical processes and evaluation of groundwater aquifer at Sohag City, Egypt. Sci Afr 6:e00196

Aghlmand R, Abbasi A (2019) Application of MODFLOW with boundary conditions analyses based on limited available observations: a case study of Birjand plain in East Iran. Water 11:1904

Ahmed AA (2009) Using lithologic modeling techniques for aquifer characterization and groundwater flow modeling of the Sohag area Egypt. Hydrogeol J 17:1189–1201

Al-Maktoumi A, El-Rawy M, Zekri S (2016) Management options for a multipurpose coastal aquifer in Oman. Arab J Geosci 9:1–14

Al-Maktoumi A, Zekri S, El-Rawy M, Abdalla O, Al-Wardy M, Al-Rawas G, Charabi Y (2018) Assessment of the impact of climate change on coastal aquifers in Oman. Arab J Geosci 11:1–14

Al-Maktoumi A, Zekri S, El-Rawy M, Abdalla O, Al-Abri R, Triki C, Bazargan-Lari MR (2020) Aquifer storage and recovery, and managed aquifer recharge of reclaimed water for management of coastal aquifers. Desalin Water Treat 176:67–77

Anderson MP, Woessner WW (1992a) The role of the postaudit in model validation. Adv Water Resour 15(3):167–173

Anderson MP, Woessner WW (1992b) Applied groundwater modeling: simulation of flow and advective transport. Academic Press, San Diego, Elsevier, p 381p

Anderson MP, Woessner WW, Hunt RJ (2015a) Applied groundwater modeling: simulation of flow and advective transport, 2nd ed. Academic Press, Elsevier, New York. ISBN 9780120581030

Anderson MP, Woessner WW, Hunt RJ (2015b) Model calibration: assessing performance, chap 9. In: Anderson MP, Woessner WW, Hunt RJ (eds) Applied groundwater modeling, 2nd edn. Academic Press, San Diego, pp 375–441

Anon (2000a) SSG Software. The Scientific Software Group, Washington http://www.scisoftware.com

Anon (2000b) Visual MODFLOW V.2.8.2 User's manual for professional applications in three-dimensional groundwater flow and contaminant transport modeling. Waterloo Hydrogeologic Inc, Ontario

El Arabi N (2012) Environmental management of groundwater in Egypt via artificial recharge extending the practice to soil aquifer treatment (SAT). Int J Environ Sustain 1

Arnold JG, Srinivasan R, Muttiah RS, Williams JR (1998) Large-area hydrologic modeling and assessment: Part I. Model development. J Am Water Resour Assoc 34(1):73–89

Arnold JG, Fohrer N (2005) SWAT2000: current capabilities and research opportunities in applied watershed modeling. Hydrol Process 19(3):563–572

Attard G, Rossier Y, Winiarski T, Eisenlohr L (2016) Deterministic modeling of the impact of underground structures on urban groundwater temperature. Sci Total Environ 572:986–994

Awad A, Eldeeb H, El-Rawy M (2020) Assessment of surface and groundwater interaction using field measurements: a case study of Dairut City, Assuit Egypt. J Eng Sci Technol 15:406–425

Awad A, Luo W, Zou J (2021) DRAINMOD simulation of paddy field drainage strategies and adaptation to future climate change in lower reaches of the Yangtze river basin. Irrig Drainage

Aziz K, Settari A (1979) Petroleum reservoir simulation. Applied Science Publishers, London, pp 135–139

Baalousha H, Köngeter J (2006) Stochastic modelling and risk analysis of groundwater pollution using FORM coupled with automatic differentiation. Adv Water Resour 29(12):1815–1832

Baalousha H (2007) Application of automatic differentiation in groundwater sensitivity analysis. In: Oxley L, Kulasiri D (eds) MODSIM 2007 international congress on modelling and simulation. Modelling and Simulation Society of Australia and New Zealand, pp 2728–2733, Dec 2007. ISBN: 978-0-9758400-4-7

Baalousha H (2008) Fundamentals of groundwater modelling. In: Konig LF, Weiss JL (eds) Groundwater: modelling, management and contamination. Nova Science Publishers, Inc.: New York, pp. 149–166; ISBN 9781604568325

Bear J (1979) Hydraulics of Groundwater. McGraw-Hill, New York, p 567p

Bear JJ, Cheng HDA (2010) Numerical models and computer codes. In: Modeling groundwater flow and contaminant transport. Theory and applications of transport in porous media, vol 23. Springer, Dordrecht. https://doi.org/10.1007/978-1-4020-6682-5_8

Bear J, Verruijt A (1987) Modeling groundwater flow and pollution. Springer, Berlin, p 432

Bishop JM, Glenn CR, Amato DW, Dulai H (2017) Effect of land use and groundwater flow path on submarine groundwater discharge nutrient flux. J Hydrol Reg Stud 11:194–218

Bossavit A (1998) Computational electromagnetism. Academic Press, Boston

Bossavit A (2005) Disctretization of electromagnetic problems: the "generalized finite difference approach". In: Ciarlet PG (ed) Handbook of numerical analysis, vol XIII. Elsevier, North Holland, Amsterdam

Campos E (2009) A groundwater flow model for water related damages on historic monuments—case study West Luxor Egypt. Vatten 65:247–254

Cao W, Bowden WB, Davie T, Fenemor A (2006) Multi-variable and multi-site calibration and validation of SWAT in a large mountainous catchment with high spatial variability. Hydrol Process 20:1057–1073

Chavent G, Jaffré J (1986) Mathematical models and finite elements for reservoir simulation. Elsevier, North-Holland, Amsterdam
Cooley RL, Naff RL (1990) Regression modeling of ground-water flow. U.S. Geological Survey Techniques of Water-Resources Investigations, 03eB4, p 232. http://pubs.usgs.gov/twri/twri3-b4/
Cooley RL (1977) A method of estimating parameters and assessing reliability for models of steady state groundwater flow, 1. Theory and numerical properties. Water Resour Res 13(2):318e324. http://dx.doi.org/10.1029/WR013i002p00318
Cooley RL (1979) A method of estimating parameters and assessing reliability for models of steady state groundwater flow, 2. Application of statistical analysis. Water Resour Res 15(3):603e617. https://doi.org/10.1029/WR015i003p00603
Cushman JH, Tartakovsky DM (eds) (2016) The handbook of groundwater engineering. CRC Press
Davis PA, Olague NE, Goodrich MT (1992) Application of a validation strategy to Darcy's experiment. Adv Water Resour 15(3):175–180
Dawoud MA, Ismail SS (2013) Saturated and unsaturated River Nile/groundwater aquifer interaction systems in the Nile Valley Egyt. Arab J Geosci 6:2119–2130
De Smedt F, Zijl W (2018) Two-and three-dimentional flow of groundwater. CRC Press, Boca Raton, p 61
De Smedt F, Zijl W, El-Rawy M (2018) Double constraint method for pumping test analysis. J Hydrol Eng 23(7):06018003
Delleur JW (ed) (2006) The handbook of groundwater engineering. CRC press
Diersch HJG (2005) WASY software FEFLOW, finite element subsurface flow and transport simulation system, user's manual. Berlin, Germany
Doherty J, Brebber L, Whyte P (1994) PEST—Model-independent parameter estimation. User's manual. Watermark Computing. Australia
Doherty J (1990) MODINV e suite of software for MODFLOW preprocessing, postprocessing, and parameter optimization. User's Manual: Australian Centre for Tropical Freshwater Research (various pagings)
Doherty J (2014a) PEST, Model-independent parameter estimationd user manual (fifth ed., with slight additions). Watermark Numerical Computing, Brisbane, Australia
Doherty J (2014b) Addendum to the PEST manual. Watermark Numerical Computing, Brisbane, Australia
Driscoll FG (1986) Groundwater and wells. 2nd Edition, St. Paul, Minnesota, Johnson Division
Du X, Lu X, Hou J, Ye X (2018) Improving the reliability of numerical groundwater modeling in a data-sparse region. Water 10(3):289
El-Rawy M (2013) Calibration of hydraulic conductivities in groundwater flow models using the double constraint method and the Kalman filter. Doctoral dissertation, Ph.D. thesis Vrije Universiteit Brussel, Brussels, Belgium
El-Fakharany Z, Fekry A (2014) Assessment of New Esna barrage impacts on groundwater and proposed measures. Water Sci 28:65–73
El-Rawy M, De Smedt F (2020) Estimation and mapping of the transmissivity of the Nubian sandstone aquifer in the kharga oasis, Egyt. Water 12(2):604
El-Rawy M, Zijl W, Batelaan O, Mohammed GA (2010) Application of the double constraint method combined with MODFLOW. In: Proceedings of the Valencia IAHR international groundwater symposium. Valencia, Spain, 22–24 Sept 2010
El-Rawy M, Mohammed GA, Zijl W, Batelaan O, De Smedt F (2011) Inverse modeling combined with Kalman filtering applied to a groundwater catchment, In: Proceedings of the MODFLOW and more conference, Golden, USA 6–8 June (2011)
El-Rawy MA, Batelaan O, Zijl W (2015a) Simple hydraulic conductivity estimation by the Kalman filtered double constraint method. Groundwater 53(3):401–413. https://doi.org/10.1111/gwat.12217
El-Rawy MA, Batelaan O, Zijl W (2015b) A simple method to apply measured flux and head data for the estimation of regional hydraulic conductivities. In: First EAGE/TNO workshop. European Association of Geoscientists & Engineers, p cp-450

El-Rawy M, De Smedt F, Batelaan O, Schneidewind U, Huysmans M, Zijl W (2016) Hydrodynamics of porous formations: simple indices for calibration and identification of spatio-temporal scales. Mar Petrol Geol 78:690–700. https://doi.org/10.1016/j.marpetgeo.2016.08.018

El-Rawy M, De Smedt F, Zijl W (2018) Zone-integrated double-constraint methodology for calibration of hydraulic conductivities in grid cell clusters of groundwater flow models. Trans Porous Media 122(3):633–645

El-Rawy M, Al-Maktoumi A, Zekri S, Abdalla O, Al-Abri R (2019a) Hydrological and economic feasibility of mitigating a stressed coastal aquifer using managed aquifer recharge: a case study of Jamma aquifer Oman. J Arid Land 11(1):148–159

El-Rawy M, Batelaan O, Buis K, Anibas C, Mohammed G, Zijl W, Salem A (2020a) Analytical and numerical groundwater flow solutions for the FEMME-modeling environment. Hydrology 7(2):27

El-Rawy M, Abdalla F, El Alfy M (2020b) Water resources in Egypt. In: The geology of Egypt. Springer, Cham, pp 687–711

El-Rawy M, Moghazy HE, Eltarabily MG (2021a) Impacts of decreasing Nile flow on the Nile Valley aquifer in El-Minia governorate Egypt. Alexandria Eng J 60:2179–2192

El-Rawy M, Makhloof AA, Hashem MD, Eltarabily MG (2021b) Groundwater management of quaternary aquifer of the nile valley under different recharge and discharge scenarios: a case study assiut governorate Egypt. Ain Shams Eng Univ 12(3):2563–2574

El-Rawy M, Abdalla F. Negm AM (2021c) Groundwater characterization and quality assessment in Nubian sandstone aquifer, Kharga Oasis, Egypt. In: Negm AM, Elkhouly A (eds) Groundwater in Egyptian deserts, Springer International Publishing (accepted for publication)

Elsheikh AE, Barseem MS, Sherbeni WA (2020) Susceptibility of shallow groundwater aquifers to water logging, case study Al Marashda area, West Nile Valley, Qena, Egypt, bulletin of the faculty of engineering. Mansoura Univ 40:1–28

Eltarabily MG, Negm AM, Yoshimura C, Saavedra OC (2017) Modeling the impact of nitrate fertilizers on groundwater quality in the southern part of the Nile Delta Egypt. Water Sci Technol Water Supply 17:561–570

Eltarabily MG, Negm AM, Yoshimura C, Takemura J (2018) Groundwater modeling in agricultural watershed under different recharge and discharge scenarios fort quaternary aquifer eastern Nile Delta Egypt. Environ Model Assess 23:289–308

Enemark T, Peeters L, Mallants D, Flinchum B, Batelaan OA (2020) Systematic approach to hydrogeological conceptual model testing, combining remote sensing and geophysical data. Water Resour Res 56(8):e2020WR027578

Esfahani SG, Valocchi AJ, Werth CJ (2021) Using MODFLOW and RT3D to simulate diffusion and reaction without discretizing low permeability zones. J Contam Hydrol 103777

Fetter CW (2001) Applied hydrogeology. Prentice Hall. 4th edn

Fleckenstein JH, Krause S, Hannah DM, Boano F (2010) Groundwater-surface water interactions: new methods and models to improve understanding of processes and dynamics. Adv Water Resour 33:1291–1295

Fouad M, Hussein EE (2018) Assessment of numerical groundwater models. Int J Sci Eng Res 9(6):951–974

Franke OL, Reilly TE, Bennett GD (1987) Definition of boundary and initial conditions in the analysis of saturated ground-water flow systems—An introduction: techniques of water-resources investigations of the United States Geological Survey, Book 3, Chapter B5, p 15

Gleeson T, Wada Y, Bierkens MFP, van Beek LPH (2012) Water balance of global aquifers revealed by groundwater footprint. Nature 488(197–200):2012. https://doi.org/10.1038/nature11295

Gupta HV, Sorooshian S, Yapo PO (1998) Toward improved calibration of hydrologic models: multiple and noncommensurable measures of information. Water Resour Res 34(4):751–763

Harbaugh A, McDonald M (1996) User's documentation for MODFLOW-96, an update to the U.S. Geological Survey modular finite-difference ground-water flow model: U.S. Geological Survey Open-File Report 96-485, p 56

Harbaugh AW (2005) MODFLOW-2005, the U.S. Geological survey modular ground-water model—the Ground-water flow process. U.S. Geological Survey Techniques and Methods 6-A16

Harbaugh AW, Banta ER, Hill MC, McDonald, MG (2000) MODFLOW—2000, the U.S. geological survey modular ground-water model—User guide to modularization concepts and the ground-water flow process, U.S. Geological Survey Open-File Report 00-92, p 121

Hassan A (2003) A validation process for the groundwater flow and transport model of the faultless nuclear test at central Nevada test area (No. DOE/NV/13609–27; DRI Pub. No. 45197). Desert Research Institute, University and Community College System of Nevada (US)

Hassan AE (2004) Validation of numerical ground water models used to guide decision making. Groundwater 42(2):277–290

Henriksen HJ, Troldborg L, Nyegaard P, Sonnenborg TO, Refsgaard JC, Madsen B (2003) Methodology for construction, calibration and validation of a national hydrological model for Denmark. J Hydrol 280(1–4):52–71

Hill MC (1992) A computer program (MODFLOWP) for estimating parameters of a transient, three dimensional, ground-water flow model using nonlinear regression. U.S. Geological Survey Open-File Report 91-484, p 358. http://pubs.er.usgs.gov/publication/ofr91484

Hill MC, Tideman CR (2007) Effective groundwater model calibration: with analysis of data, sensitivities, predictions, and uncertainty. Wiley-Interscience, p 455. ISBN 978-0-471-77636-9

Hill MC, Banta ER, Harbaugh AW, Anderman ER (2000) MODFLOW—2000, the U.S. geological survey modular ground-water model—User guide to the observation, sensitivity, and parameter-estimation processes and three post-processing programs, U.S. Geological Survey Open-File Report 00-184, p 210

Hölting B, Coldewey WG (2019) Numerical groundwater models. In: Hydrogeology. Springer Textbooks in earth sciences, geography and environment. Springer, Berlin. https://doi.org/10.1007/978-3-662-56375-5_9

Hora SC (1996) Aleatory and epistemic uncertainty in probability elicitation with an example from hazardous waste management". Reliab Eng Syst Saf 54:217–223. https://doi.org/10.1016/S0951-8320(96)00077-4

Huyakorn PS, Pinder GF (1983) Computational methods in subsurface flow. Academic Press, p 473

Igboekwe MU, Achi NJ (2011) Finite difference method of modelling groundwater flow. J Water Resour Prot 3:192–198. https://doi.org/10.4236/jwarp.2011.33025

Igboekwe MU, Gurunadha Rao VVS, Ok-wueze EE (2008) Groundwater flow modelling of Kwa Ibo river watershed Southeastern Nigeria". Hydrol Proc-Esses 22(10):1523–1531. https://doi.org/10.1002/hyp.6530

Ismail E, El-Rawy M (2018) Assessment of groundwater quality in West Sohag, Eypt. Desalin Water Treat 123:101–108

Kaasschieter EF (1990) Preconditioned conjugate gradients and mixed-hybrid finite elements for the solution of potential flow problems. Ph.D. thesis, Delft University of Technology, Delft, Netherlands

Kaasschieter EF, Huijben SJM (1992) Mixed-hybrid finite elements and streamline computation for the potential flow problem. Numer Methods Partial Differ Equ 8:221–266

Katpatal Y, Thakare P, Maunika S (2021) Groundwater modeling in a semi-confined aquifer using GIS and MODFLOW, in water management and water governance, Springer, Berlin, pp 179–187

Kenda K, Čerin M, Bogataj M, Senožetnik M, Klemen K, Pergar P, Laspidou C, Mladenić D (2018) Groundwater modeling with machine learning techniques: ljubljana polje aquifer, in Multidisciplinary Digital Publishing Institute Proceedings, p 697

Khadim FK, Dokou Z, Lazin R, Moges S, Bagtzoglou AC, Anagnostou E (2020) Groundwater modeling in data scarce aquifers: the case of Gilgel-Abay, upper blue nile Ethiopia. J Hydrol 590:125214

Kisi O, Azad A, Kashi H, Saeedian A, Hashemi SAA, Ghorbani S (2019) Modeling groundwater quality parameters using hybrid neuro-fuzzy methods. Water Resour Manage 33:847–861

Kori SM, Lashari BK, Memon NA (2008) Calibration and validation of groundwater flow model under scavenger wells operation. Mehran Univ Res J Eng Technol 27(2):213–228

Kumar CP (2004) Groundwater flow models. Technical Note, National Institute of Hydrology, Roorkee, (Uttaranchal), India

Kumar C (2019) An overview of commonly used groundwater modelling software. Int J Adv Res Sci Eng Technol 6:7854–7865

Kunstmanna H, Kastensb M (2006) Direct propagation of probability density functions in hydrological equations. J Hydrol 325(1–4):82–95

Lin HCJ, Richards DR, Yeh GT, Cheng JR, Cheng HP (1997) FEMWATER: a three-dimensional finite element computer model for simulating density-dependent flow and transport in variably saturated media; CHL-97-12; DTIC Document: Vicksburg, MS, USA

Liou T, Der Yeh H (1997) Conditional expectation for evaluation of risk groundwater flow and solute transport: one-dimensional analysis. J Hydrol 199(3–4):378–402

McDonald MG, Harbaugh AW (1988) A modular three dimensional finite difference groundwater flow model. Techniques of Water Resources Investigations, Chapter A1, US Geol. Surv, p 586

Meredith E, Blais N (2019) Quantifying irrigation recharge sources using groundwater modeling. Agric Water Manag 214:9–16

Nagels K, Schneidewind U, El-Rawy M, Batelaan O, De Becker P (2015) Hydrology and ecology: how Natura 2000 and Military use can match. Ecol Quest 21:79–85

Negm AM (Ed) (2018) Groundwater in the Nile Delta (73). Springer International Publishing. https://doi.org/10.1007/978-3-319-94283-4

Negm AM, Elkhouly A (2021) Groundwater in Egyptian Deserts, Springer International Publishing (accepted for publication)

Nriagu JO (2019) Encyclopedia of environmental health. Elsevier

Peaceman DW (1978) Interpretation of well-block pressures in numerical reservoir simulation. SPE J 18(3):183–194

Pham HV, Tsai FT (2017) Groundwater modeling. Chapter 48. In: Singh VP (ed) Handbook of applied hydrology. McGraw-Hill, New York, pp 48–1–48–8

Pinder G, Gray W (1970) Finite element simulation in surface and subsurface hydrology. Academic Press Inc. p 295

Pinder GF, Gray WG (1977) Finite element simulation in surface and subsurface hydrology. Academic Press, New York, p 295

Poeter EP, Hill MC (1997) Inverse models: a necessary next step in ground-water modeling. Ground Water 35(2):250–260

Poeter EP, Hill MC (1998) Documentation of UCODE, a computer code for universal inverse modeling, U.S. Geological Survey, Water-Resources Investigations Report 98-4080

Poeter EP, Hill MC, Banta ER, Mehl S, Christensen S (2005) UCODE_2005 and six other computer codes for universal sensitivity analysis, calibration, and uncertainty evaluation. U.S. Geological Survey Techniques and Methods, 6 A11, p 47. p 283. http://pubs.usgs.gov/tm/2006/tm6a11/

Pollock DW (1990) Documentation of computer programs to compute and display path lines using results from the US Geol. Surv. Modular three dimensional finite diffeence groundwater flow model. US Geol. Surv. Open-File Report 89, pp 381–388

Polomčić D, Bajić D (2015) Application of groundwater modeling for designing a dewatering system: case study of the Buvač Open Cast Mine, Bosnia and Herzegovina. Geol Croat 68:123–137

Rahnama M, Sohrabipour N, Barani G (2020) Investigation of groundwater level variations in Shahdad aquifer. Iran J Irrig Drainage 14:1650–1663

Reilly T, Harbaugh A (2004) Guidelines for evaluating ground-water flow. Scientific Investigations Report 2004–5038. U.S. Department of Interior, U.S. Geological Survey

Remson I, Hornberger GM, Molz FJ (1971) Numerical methods in subsurface hydrology. Wiley-Interscience, New York, p 389

Roscoe Moss Company (1990) Handbook of ground water development. Wiley, New York

Salem A, Dezső J, Lóczy D, El-Rawy M, Slowik M (2018) Modeling surface water-groundwater interaction in an oxbow of the Drava floodplain. EPiC Ser Eng 3:1832–1840

Salem A, Dezső J, Lóczy D, El-Rawy M, Loczy D (2020) Hydrological modeling to assess the efficiency of groundwater replenishment through natural reservoirs in the Hungarian Drava River floodplain. Water 12:250

Sashikkumar M, Selvam S, Kalyanasundaram VL, Johnny JC (2017) GIS based groundwater modeling study to assess the effect of artificial recharge: a case study from Kodaganar river basin, Dindigul district, Tamil Nadu. J Geol Soc India 89:57–64

Sathe SS, Mahanta C (2019) Groundwater flow and arsenic contamination transport modeling for a multi aquifer terrain: assessment and mitigation strategies. J Environ Manage 231:166–181

Sefelnasr AM, Omran AA-K, Abdel-Hak HA, El Tahawy WS (2019) GIS-based numerical modeling for the groundwater assessment: a case study in the Quaternary aquifer, Assiut Governorate, Egyt. Arab J Geosci 12:1–13

Senge RS, Bösner DK, Haasenritterb J, Hirsch O, Donner-Banzhoff N, Hüllermeier E (2014) Reliable classification: learning classifiers that distinguish aleatoric and epistemic uncertainty". Inf Sci 255:16–29. https://doi.org/10.1016/j.ins.2013.07.030

Shamrukh M, Corapcioglu MY, Hassona FA (2001) Modeling the effect of chemical fertilizers on ground water quality in the Nile Valley Aquifer Egypt. Groundwater 39:59–67

Simunek J, Wendroth O, van Genuchten MT (1999) Estimating unsaturated soil hydraulic properties from laboratory tension disc infiltrometer experiments. Water Resour Res 35(10):2965–2979

Singhal BBS, Gupta RP (2010) Groundwater modeling. In: Applied hydrogeology of fractured rocks. Springer, Dordrecht. https://doi.org/10.1007/978-90-481-8799-7_19

Soderlind G, Arevalo C (2017) Numerical methods for differential equations. Numerical analysis. Lund University (2008–2009) [retrieved on Aug. 3, 2017]. Retrieved from

Srivastav AL (2020) Chemical fertilizers and pesticides: role in groundwater contamination, in agrochemicals detection, treatment and remediation. Elsevier, pp 143–159

Strack ODL (1989) Groundwater mechanics. Prentice Hall Inc., Englewood Cliffs, New Jersey

Suring LH (2020) Freshwater: oasis of life—An overview

El Tahlawi M, Farrag A, Ahmed S (2008) Groundwater of Egypt: "an environmental overview." Environ Geol 55:639–652

Theis CV (1941) The effect of a well on the flow of a nearby stream. Am Geophys Union Trans 22(3):734–738

Tian Y, Zheng Y, Wu B, Wu X, Liu J, Zheng C (2015) Modeling surface water-groundwater interaction in arid and semi-arid regions with intensive agriculture. Environ Model Softw 63:170–184

Torak LJ (1993) A MODular Finite-Element model (MODFE) for areal and axisymmetric ground-water-flow problems, part 1--model description and user's manual: U.S. Geological Survey Techniques of Water-Resources Investigations, book 6, chap. A3

Toth J (1962) A theory of groundwater motion in small drainage basins in central Alberta. J Geophys Res 67(11):4375–4387

Trykozko A (2001) Numerical homogenization of the absolute permeability using the conformal-nodal and mixed-hybrid finite element method. Transp Porous Med 44:33–62

Trykozko A, Zijl W, Bossavit A (2001) Nodal and mixed finite elements for the numerical homogenization of 3D permeability. Comput Geosci 5:61–64

Walton W (1989) Analytical ground water modeling. Lewis Publishers, Chelsea, Michigan

Wang S, Shao J, Song X, Zhang Y, Hu Z, Zhou X (2008) Application of MODFLOW and geographic information system to groundwater flow simulation in North China Plain. China Environ Geol 55:1449–1462

Wasy GmbH (2005) FEFLOW: finite element subsurface flow and transport simulation system. Reference Manual. Wasy GmbH, Berlin

Weiser A, Wheeler MF (1988) On convergence of block-centered finite differences for elliptic problems. SIAM J Numer Anal 25:351–375

Wingle WL, Poeter EP, McKenna SA (1999) UNCERT: geostatistics, uncertainty analysis and contaminant transport modeling. Comput Geosci 25:365–376

XMS Wiki (2020) Online Help for GMS, SMS, and WMS. Available online: https://www.xmswiki.com/wiki/GMS:Parameters#Parameterizing_the_model (accessed on 07 Aug 2020)

Zhang Y, Pinder G (2003) Latin hypercube lattice sampling selection strategy for correlated random hydraulic conductivity fields. Water Resour Res 39(8):11-1/11–3

Zhao W, Lin J, Wang S-F, Liu J-L, Chen, Z-R, Kou W-J (2013) Influence of human activities on groundwater environment based on coefficient variation method, Huan jing ke xue= Huanjing kexue 34:1277–1283

Zheng C (1990) A modular three dimensional transport model for simulation of advection, dispersion and chemical reactions, of contaminants in groundwater systems. Environmental Protection Agency, Oklahoma, USA, R.S. Ken Environmental Research Laboratory

Zhou Y, Li W (2011) A review of regional groundwater flow modeling. Geosci Front 2:205–214

Zijl W (2005a) Face-centered and volume-centered discrete analogs of the exterior differential equations governing porous medium flow I: theory. Transp Porous Med 60:109–122

Zijl W (2005b) Face-centered and volume-centered discrete analogs of the exterior differential equations governing porous medium flow II: examples. Transp Porous Med 60:123–133

Zijl W, Nawalany M (1993) Natural groundwater flow. CRC Press, Boca Raton. p 321

Zijl W, El-Rawy M (2019) The evolution from an unsteady to a steady mixing zone between two groundwater flow systems with different concentrations. Alex Eng J 58(2):725–731

Zijl W, El-Rawy M (2021) Flow systems of the earth's viscous subsurface: a complement to groundwater flow systems. Ain Shams Eng J 12(1):775–788

Zijl W, De Smedt F, El-Rawy M, Batelaan O (2017) The double constraint inversion methodology. Eqn Appl Forward Inverse Model Groundwater Flow, Springer. https://doi.org/10.1007/978-3-319-71342-7

Zijl W, De Smedt F, El-Rawy, M, Batelaan O (2018a) The pointwise double constraint methodology. In: The double constraint inversion methodology. Springer, Cham, pp 35–55

Zijl W, De Smedt F, El-Rawy M, Batelaan O (2018b) Foundations of forward and inverse groundwater flow models. In: The double constraint inversion methodology. Springer, Cham, pp 15–33

Assessment of Groundwater Quality

Contamination Sources Along the Nile Valley, Egypt and Its Impact on Groundwater

Fathy Abdalla, Rafat El Attar, and Mohamed Shamrukh

Abstract In Egypt from Aswan to Cairo, the Nile receives discharge of wastewater from over 124 point sources along the banks of the Nile; more than 72 points are agricultural, while the rest are industrial drains. These disposed waters contain untreated or partially treated wastewater; accordingly, different contaminants might be identified in the water of the River. Organic pollutants can move from cultivated lands into the Nile water, as well as the discharged wastewater from villages and towns located at the surroundings. In instance, drinking water with high concentrations of dissolved organic carbon (DOC) and bromide needs specific and costly further treatment steps. All chemical disinfectants are to some degree poisons in the water treatment process, and the most common are oxidizers such as ozone and chlorine. The potential effects of anthropogenic activities including domestic activities especially sewage disposal practices, industrial and agricultural activities were reported in evaluated values within the Nile valley area by many authors. An evaluation of the hydrochemical and bacteriological characteristics of the surface and groundwater resources are necessary to assess the adverse impacts of wastewater disposal in the area. More Advanced methods for monitoring and detecting the source of contamination using remote sensing and GIS techniques are required.

Keywords Wastewater · Sewage disposal · Contamination · Anthropogenic activities Nile Valley · Agriculture · Egypt

1 Introduction

Sewage disposal practices, industrial and agricultural activities are representing the main potential threat for both water quantity and quality in Egypt. Where raw sewage water is a potential source of numerous pollutants, including pathogenic bacteria,

F. Abdalla (✉) · R. El Attar
Geology Department, South Valley University, Qena, Egypt
e-mail: fathy.abdallah@sci.svu.edu.eg

M. Shamrukh
Civil Engineering Department, Minia University, Minia, Egypt

A. M. Negm and M. El-Rawy (eds.), *Sustainability of Groundwater in the Nile Valley, Egypt*, Earth and Environmental Sciences Library,
https://doi.org/10.1007/978-3-031-12676-5_4

nutrients, toxic metals, and organic compounds (Gardner 1997; Geriesh et al. 2004; USEPA 2004; Abdalla and Shamrukh 2010, 2016). These dangerous pollutants can migrate directly or indirectly into groundwater and surface water bodies. Around 25% of human fecal content is bacteria, it contains approximately 3×10^7 coliform bacteria/100 mL. Those microbes differ in their ability to travel through the soil matrix (DeBorde et al. 1998) reaching the into the shallow groundwater aquifers. The presence of microorganisms like virsus and bacteria (fecal coliform bacteria e.g., Escherichia coli and total coliform bacteria) in water, as pathogenic indicator species, do not cause disease themselves but suggest that the water is polluted with pathogenic that causes waterborne disease (Macler and Merkel 2000; APHA 2012). In polluted water, total coliform levels are typically 10 times higher than the fecal coliform levels. Worldwide, pathogenic contamination of the water resources and related disease are a major water quality concern, the problem has been reported due to the transmission of pathogenic bacteria from sewage systems to water. Bacteria can cause many diseases, including diarrhea, dysentery, cholera, and typhoid fever (DeBorde et al. 1998; Powell et al. 2003; Lerner and Harris 2009). In case of groundwater production wells, hydraulic properties of the soil type plays a vital factor in the migration processes of pathogenic bacteria from sewage systems to groundwater, to avoid or mitigate the adverse effects of bacteriological pollution, wells must be positioned at a safe distance from the sources of contamination (Table 1).

Besides geogenic sources of contaminants, diverse mix of inorganic contaminants that potentially hazardous to water quality is coming from anthropogenic sources, including domestic waste, irrigation return-flow, and industrial effluents. These sources may include nitrate (NO^{3-}) and heavy metals potentially toxic, such as Pb, Cd, Zn, Fe, Mn, and Cr. Worldwide, the main and typical sources of nitrate in water are related primarily to discharges of agricultural and domestic wastewater (Andersen and Kristiansen 1983; Liu et al. 2005). Addition sources of nitrate in water bodies is the bacterial decomposition of the organic matter in sewage and animal waste. Nitrate content above the World Health Organization (WHO) guideline may cause health problems especially for infants, as nitrate interferes with the ability of the blood to carry oxygen and causes oxygen deficiencies that can lead to methemoglobinemia (ATSDR 2017).

Trace elements usually exist in small concentrations (1 mg/L) in water. According to the carcinogenic classifications of heavy metals in drinking water by the International Agency for Research on Cancer (IARC 2012) Cd, Cr, and Pb in water bodies are classified as carcinogenic substances, while Zn, Fe, and Mn are categorized as non-carcinogenic. Various organic compounds, including grease, detergent waste, cleaning solvents, gasoline, and prescription drugs are coming from the discharging

Table 1 Minimum recommended distance between a groundwater well and a contaminant source (Raghunath 1987)

Contamination source	Septic tank	Disposal field	Seepage pit	Cesspool
Recommended distance	15 m	30 m	30 m	40 m

of household wastewater, that may represent a threat to the water bodies. The main objectives of this chapter were to review the interaction between wastewater discharge from various sources and surface and shallow groundwater resources in the Nile Valley area as well as to identify the main potential sources of these contaminants in water. These were achieved by surveying the possible sources of contamination, which included discharges of raw sewage, agricultural runoff, and industrial effluents.

2 Study Area

The study area includes the section of Nile Valley from Aswan to Cairo, about 1000 km (Fig. 1). It is located between latitudes of 24° 5′ 20.1768″–30° 1′ 59.9988′ N and longitudes of 32° 53′ 59.3880″–31° 14′ 0.0024″ E. The area is classified as an arid zone, with arid and hot weather that ranges from 23 °C in the winter to 44 °C in the summer. The annual mean value of the wind speed is 4.06 m/s, in the winter, the relative humidity is 53%, while in the summer, it is 29%. Rainfall is scarce and infrequent throughout the year; however, sometimes random flash showers occur in wintertime. The mean annual rainfall value is approximately 5 mm/year and the rate of evapotranspiration is 185 cm/year (Abdalla et al. 2009). Human-related activities have a significant impact on the area, which is mostly a combination of residential, agricultural, and industrial usage. The region is characterized by cultivated land of the Nile flood plain with eight Sugar Factories at, Kom Ombo, Edfu, Armant, Qus, Deshna, Nag Hamadi, Girga and Abu Qurqas, Ferrosilicon factory at Edfu, Paper Factory in Qus City and the factory of "Egypt Edfu" company for pulp, writing and printing paper Edfu City (Figs. 1 and 6), mostly are located directly on the Nile bank. The majority of residential areas along the Nile Valley have no sanitation networks except the main cities. Thus, sewage is disposed of and stored into underground sewage rooms (Fig. 2) that have been designed to be in direct contact with groundwater (Abdel-Lah and Shamrukh 2001). These rooms have no isolated surfaces or lined beds to prevent pollutants from penetrating into the surrounding environment and mixing groundwater. In many parts of the study area, sewage is dumped into canals and drains or directly onto the ground (Fig. 3), pathogenic species such as bacteria and viruses can easily travel and to contaminate into the water bodies. Household raw sewage, agricultural waste, and industrial effluents are all examples for wastewater discharges in the study area.

3 Surface Water System in the Nile Valley

The Nile River, as well as the main irrigation canals and a number of smaller irrigation canals and drains, are the principal sources of water for the Nile Valley's surface water network. There are few numbers of barrages that constructed along the Nile to help water flow into irrigation main canals without major pump stations.

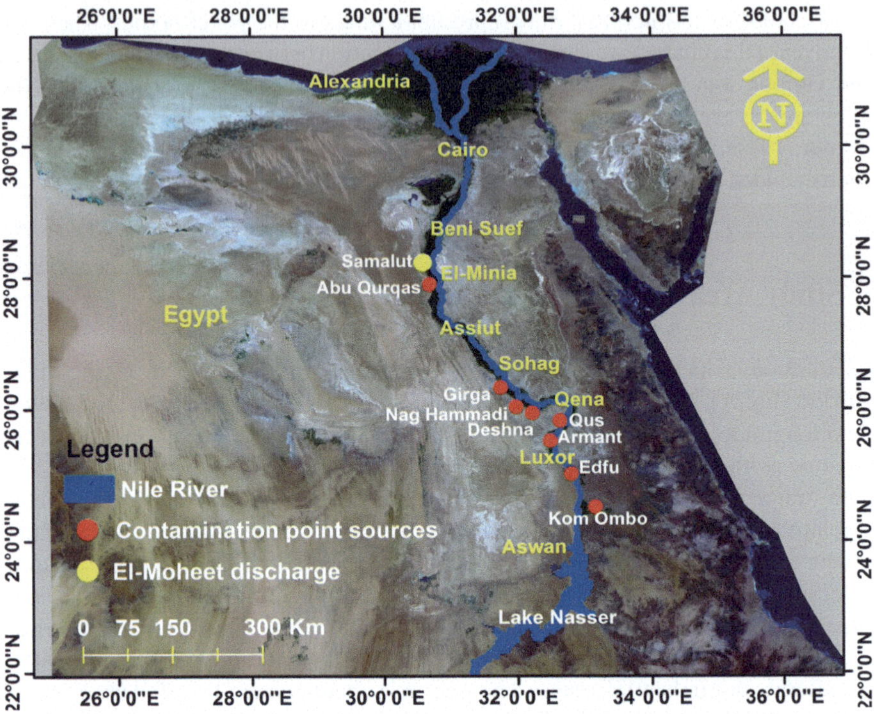

Fig. 1 Location map showing the contamination point sources in the Nile Valley

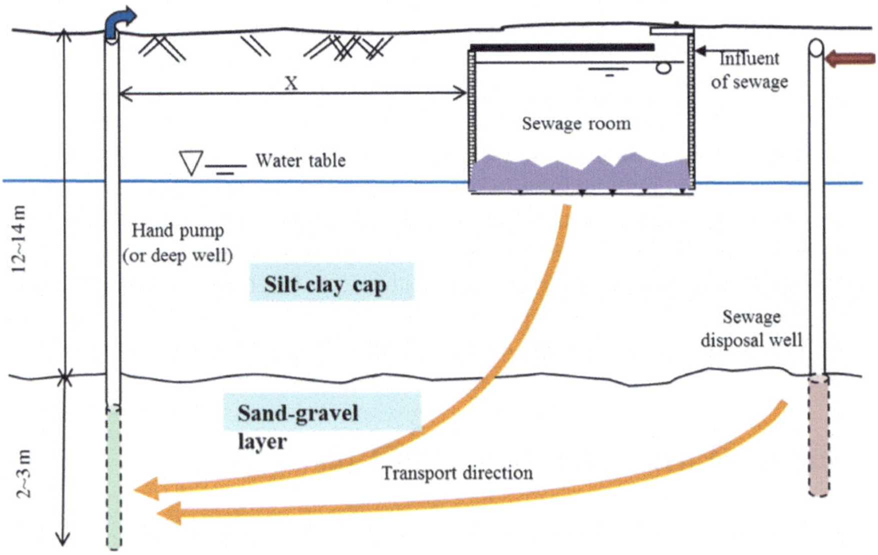

Fig. 2 Contamination pathway from sewage system used in Egyptian villages in Nile valley

Fig. 3 Field photographs showing sewage discharge directly over the ground at El-Salhyia wastewater treatment station, Qena area (25° 9′ 24″ N 32° 46′ 34″ E)

The Nile

At the High Dam in August, the Nile's mean water level is 169 m above mean sea level (asl), and the minimum water level in July is 163 m asl. In Qena area around 296 km from the High Dam, the Nile's mean water level in July is 72 m asl, and the minimum water level in January is 66 m asl. These water levels are lower than the groundwater table in the surrounding Quaternary aquifer; therefore, the Nile serves as the main drain for the aquifer (El-Rawy et al. 2020a, 2021a, b; Abdalla and Khalil 2018).

Irrigation canals

In the study area, there are several main canals that traverse the Nile Valley, the most important are El-Kalabia, Nag-Hammadai eastern and western canals and Ibrahimia canal, additionally, and the area is crossed by several small canals.

Agricultural drains

The irrigated land in the Nile Valley is protected by many drainages. The drainage water is reused for irrigation through discharges into the Nile and/or the main canals, either directly or indirectly from about 72 drains in the study area.

4 Contamination Sources

Sources of pollution of the Nile River were increasing, from discharge of wastewater, irrigation return-flows and the disposal of industrial wastewater. More than 72 agricultural drains throw directly into the Nile or into the major canals about 13.7 billion cubic meters annually of water loaded with pesticides and agrochemicals. This is along with the sanitation of villages that are deprived of service, adding: 872 million cubic meters annually of wastewater that is untreated or subjected to primary or partially treatment only at Upper Egypt and 4 drains in Damietta and Rashid branches, which are Al-Rahawi, Sabil, Tala and Omar Bey.

Fourteen power stations discharge about 4.2 billion cubic meters of cooling water in the river, and this water is chemically identical and violates only the rise in temperatures above the legally permissible limits. The discharge of industrial facilities as the most influencing the water quality, because it contains many organic and inorganic compounds, in addition to heavy metal. The river receives about 150 million cubic meters of industrial wastewater annually from 9 major industrial facilities for sugar and paper industries (Environmental Affairs Agency Report 2017; Abdalla and Shamrukh 2010, 2016). The point sources were relatively easy to recognize as the wastewater was discharged over the ground and was then allowed to flow into nearby surface waters.

4.1 Domestic Sources

In Egypt the majority of pollution sources are responsible for the absence of sanitation networks for 88% of the villages. It is projected that domestic wastewater is between 5.5 and 6.5 billion m^3 (BCM)/year, but only 2.97 BCM are subject to treatment processes. Of this treated portion, approximately 0.7 BCM / year is reused for agricultural purposes, of which 0.44 BCM is primarily treated, and the remainder (0.26 BCM) is subjected to secondary treatment (Abdel-Shafy and Abdel-Sabour 2006). The amount of treated wastewater that is discharged into the Nile River and its branches is about 1.06 BCM/year and is estimated at 5% of the total amount of drainage directly on the Nile, through treatment plants sanitation in Upper Egypt (Fig. 4), in addition to Al-Rahawi, Sabeel, Tala, and Omar Bey drains on Rashid and Damietta branches (Environmental Affairs Agency Report 2017).

The study area is considered as a high-density residential area, most of small towns and villages have no sewer access, people build in their homes pit latrines (single-chamber) made of red brickwork; thus, the main source of contamination was on-site sewage discharge processes, see Fig. 2. Such latrines are built in such a way that wastewater from the bottoms and walls can percolate, and higher permeability of soils means more percolation, leading to high loading levels of harmful materials and bacteria into groundwater. The amount of wastewater discharged depends on the total consumption of domestic water and the population density. Wastewater is

Fig. 4 Photographs of several types of domestic wastewater effluents: **a** Kima drain, discharge of treated wastewater from Aswan City **b** direct disposal of solid waste into surface water bodies (Qus canal) **c** direct sewage discharge (green color refers to algae cover), **d** garbage and solid waste discharge

also transported in large amounts by cars to local irrigation canals and drains or into neighboring cultivated areas (Abdalla et al. 2008).

4.2 Agricultural Sources

The Nile Valley is characterized by intense yearlong agricultural practices. Excessive use of chemical fertilizers and flash irrigation techniques (two to three times a month) lead to non-point sources of irrigation-return flow contamination (Abdalla and Khalil 2018). The irrigation-return flows are usually mixing with various pollutant and agro-chemicals, fertilizers. The main sources of nitrate contamination are due to incorrect use of agro-chemicals, fertilizers and manure in agricultural areas (Fig. 5). The fertilizers widely used include urea, calcium nitrate, ammonium nitrate, ammonium sulphate, and superphosphate. The number of major agricultural drains that flow to the Nile River is more than 72 drains, from Aswan to Cairo and the two branches of the Nile (Damietta and Rashid). The key types of contaminants found in agricultural drains are salts, nutrients (nitrogen and phosphorus), residues of pesticides, pests, toxic inorganic compounds, and organic compounds (Abdalla and Khalil 2018). The presence of nitrate above the permissible level set by the WHO and the Egyptian Higher Water Committee (EHCW) may be caused by the high use of fertilizers in neighboring agricultural fields and/or on-site wastewater discharge process. Egyptian farmers have also applied intensively nitrogenous fertilizers, manure, and other chemicals with flush irrigation since the Aswan High Dam construction in 1968. Consequently, groundwater pollution with nitrate poses a serious threat to groundwater supplies, especially in extensively irrigated areas such as the Nile Delta and the Nile Valley (Shamrukh et al. 2001; Abdalla 2005). Nitrate-rich surface runoff from agricultural fields in those areas can easily percolate into the groundwater.

Examples for agricultural drains in the study area

There are three main drains in the area between Nag Hammadi city to Abu Tesht city, Upper Egypt (El Attar 2010), from west to east Salam drain, the main Nag Hammadi drain and Bakhaness drain (Fig. 6). These main drains run from south to north (nearly parallel to the River Nile). In the area there are many relatively small drains which collect water from the cultivated land and throw their water into these main drains.

A—Salam drain

Fig. 5 Field photographs showing various types of point and non-point sources of irrigation contamination sources

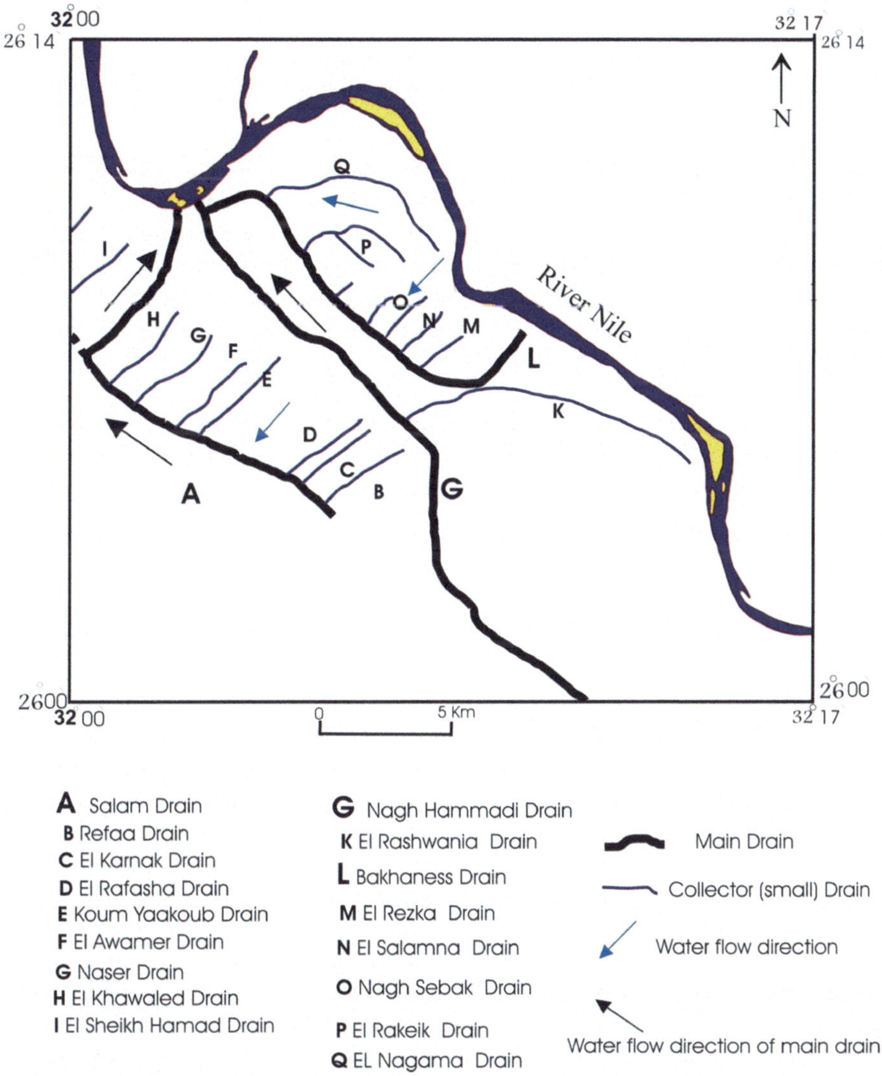

Fig. 6 Location map of agricultural drains in the area between nagh Hammadi city to Abu Tesht city, Upper Egypt

Salam drain extends from south to north (in the western part of the area). There are many small drains which are nearly perpendicular to the main drain. They can be arranged from south to north as follow:

B—Refaa drain	C—El Karnak drain
D—El Rafasha drain	E—Koum Yaakoub drain

F—El Awamer drain	G—Naser drain
H—El khawalled drain	I—El sheikh Hamad drain
J—Nag Hammadi drain	

Nag Hammadi drain extends from south to north (in the middle part of the area). There are many subsurface drains which debouch their water and sediments into Nag Hammadi drain, so that exist no surface drains on the location map. Except El Rashwania (K) drain which starts nearby Nag Hammadi City and extends toward the north and finally throws its water in the main channel of Naga Hammadi drain.

L—Bakhaness drain

Bakhaness drain starts beside Nag Hammadi barrage and extends to west and after that to the north and finally debouches its water to the river Nile. There are some drains which intersects this drain. From south to north they are:

M—El Rezka drain	N—El Salamna drain
O—Nagh Sebak drain	P—El Rakeik drain
Q—El Nagama drain	

These drains are usully affected by practices in the nearby villages such as sewage disposal of these villages and abattoir (slaughter house) leading to accumulation of different pollutants in sediment, surface and groundwater. Accordingly the chemistry of water bodies around these drains is affected by the chemistry of surface water in the drain and the impact of human activities in the nearby areas. To assess the water contamination with heavy metals from the heavy polluted agricultural drain, groundwater from wellsclose to the drain, drain water and sediment samples were collected and analyses for Fe, Mn, Zn, Cu and Pb (El Attar 2010; Abdalla et al. 2014). Some physiochemical parameters such as Electric Conductivity (EC), Total Dissolved Solids (TDS), pH and salinity have been measured in situ. Results of the hydrochemical analyses showed high concentrations of Fe and Mn in sediments and groundwater samples compared with that in surface water samples, reflecting the oxidation process due to the higher oxygen content of surface water. The results also showed high concentrations of Zn, Cu and Pb in sediment and surface water samples compared to groundwater samples of the study area.

Another example is El Moheet drain, El-Minia governrate (Fig. 7), it flows directly into the Nile River about 9000 m^3/day of polluted water. It is one of the most polluted drains due to agricultural, industrial, and sanitary wastewaters that dump in the drain, as the drain passes thousands of acres of an agricultural and residential area in the governorate (El Kashouty et al. 2011). The drain carries enormous amounts of contaminates, which exceeds the WHO guideline. Therefore, El Moheet drain is unsuitable for irrigation purposes (El Sayed and Kamel 2010).

Fig. 7 Disposal of liquid waste in El Moheet drain at El-Minia Governorate

El Moheet drain is located in El-Minia Governorate, extends for more than 150 km, starting from the Dayrut city of the Assuit Governorate in the south and ending at Samalot city in the north of El-Minia Governorate. The drain passes through hundreds of villages (approximately one million people), where its width ranges from 4 to 5 m and its depth of 7 m (Zaki et al. 2015). The environmental health administration in El-Minia Governorate reported that the pollution sources for the drain multiplied between (1) the wastewater treatment plant in Telahvillage, 90,000 m^3/day; (2) the wastewater treatment plant in Abu Qurqas city, whose disposal is spent on El Moheet drain, with 10,000 m^3/day; (3) Abu Qurqas sugar factory, whose waste is spent on El Moheet drain, with 5,000 m^3/day, as well as agricultural drainage and dead animals to be thrown into the drain, as shown in Fig. 7, which exposed the people of Dermawas, Mallawi Abu Qurqas, Samalut, and Bani Mazar cities, to get epidemic diseases (Elewa et al. 2013).

Several studies have examined the impact of the contaminated water in El Moheet drain on groundwater, agricultural areas, residential villages, and nearby irrigation canals (Bahr Youssefi, El Ibrahemia, and El Sabakha), such as El-Bakri et al. (1996). They emphasized the deterioration of groundwater quality in the areas surrounding the drain, and that is what El Sayed and Kamel (2010) confirmed as the groundwater around the drain contains higher toxic metal concentrations than River Nile, they derived mainly from agricultural wastewater (fertilizers, manures, and pesticides), and lithogenic dissolution. Also El Kashouty et al. (2011) investigated the environmental impact of human activities on the surface and groundwater systems in El-Minia Governorate, and they found that the COD and BOD_5 concentrations in El Moheet drain are higher than the drinking standards, then El Kashouty (2012) reported the toxic heavy metal concentrations were increased in El Moheet drain than those in River Nile and Ibrahimia canal. El-Azeim et al. (2016) concluded that soils irrigated by El Moheet drain water contaminated with B, Fe, Cu, Zn, Mn, Pb,

Cd, Cr and Se at Malawi city, Abo Qorkas city, El-Minia city, Samalut city; and Maghagha city. After investigating several areas in El-Minia Governorate. Sabet et al. (2017) observed concentrations of Cd and Pb above the permissible limits for drinking water, they mentioned the presence of these metals in water could cause many health problems for the residents of these areas.

The latest studies confirmed that the total hardness of the El Moheet drain and surrounding groundwater samples was very hard and varied from 110 to 571 ppm. Foremost of the water samples were unsuitable for domestic purposes, and that could have adverse impacts (Abou Heleika et al. 2018; Salman et al. 2019), in addition to recorded high concentration values of dissolved trace metals (Snousy et al. 2019; Morsi 2020). El-Rawy et al. (2020b) evaluated the potential risk of the Heavy Metals concentrations to soil pollution in a cultivation area in the Nile Valley (El-Minia area). The results indicated that the most affected soils are neighboring the El Moheet drain.

4.3 Industrial Sources

Industrial wastewater from industrial plants comes at the last rank in terms of the amount of liquid waste; it amounts to 1% of the total direct drainage on the Nile River, from eleven facilities. The amount of wastewater discharge was estimated at 547 million cubic meters/year; it discharged into the agricultural drains or the public sewerage network. In Upper Egypt almost of the industrial activities concentrated along the River Nile, and the main sources of industrial pollution are the eight Sugar Factories at, Kom Ombo, Edfu, Armant, Qus, Deshna, Nag Hamadi, Girga and Abu Qurqas, Ferrosilicon factory at Edfu, Paper Factory in Qus City and the factory of "Egypt Edfu" company for pulp, writing and printing paper Edfu City (Fig. 8), which are located directly on the Nile bank. After little to no treatment, these factories discharge their effluents into the Nile or to small lakes. Sugar Factories' main pollutants include high-carbohydrate organic matter; Paper Factory's include bagasse, oils, and grease. The wastewater discharged from these plants is high in demand for chemical oxygen (COD) and biochemical oxygen (BOD) as well as high in temperatures (Abdalla and Khalil 2018).

The Qus Paper Factory discharges industrial liquid waste into a poorly managed dam, situated near the Nile River. The chemical reactions between such contaminants produce a number of toxic substances. The plant discharges around 20,000 m^3/d of wastewater into the lake and the Nile (Abdalla and Khalil 2018). Different raw materials are used for paper production, including bagasse, and these materials are treated physically and chemically to eliminate lignins and produce white paper. Industrial wastewater containing various pollutants can infiltrate underground and reach groundwater, causing severe contamination (Abdalla and Khalil 2018).

Fig. 8 Photographs of several types of industrial wastewater effluents into the Nile: **a** Edfu sugar factory **b** Qus sugarcane factory **c** Ferrosilicon factory at Edfu, **d** Qus Paper Factory Lake

4.4 Dewatering Water

As an example, dewatering system around temples in Luxor city, where perforated drain pipes collect the near surface water and transported to manhole each 50 m via gravity. Twelve deep wells are drilled around the Karnak temple. Finally, the collected water drains into the River Nile through a pump station (Fig. 9). A vertical deep well drain directly into the River Nile. Similar dewatering network was applied around Luxor temple, but with a shorter length of 750 m. Six vertical deep wells were drilled inside Luxor temple and one well inside it to reduce the damage effects in the temples. The discharge rate is about 20,000 m^3/d and 10,000 m^3/d at Karnak and Luxor temple, respectively, from the groundwater are pumped to the Nile River. Because of dewatering process, the groundwater level decreases from 73 m to about 71.30 m (a.s.l) at Karnak temple and from 73 m to about 71 m (a.s.l) at Luxor temple (Abdalla et al. 2016).

4.5 Colling Water

The number of power stations that receive cooling water on the Nile directly is 14 stations along with 8 factories; the amount of violating water for 14 power stations

Fig. 9 Disposal of dewatering liquides into Nile at Karnak temple

is about 4.2 BCM/year, estimated at 22% of the total drainage over the Nile River (Environmental Affairs Agency Report 2017).

4.6 Impact of Contamination on Groundwater

Those sources of contamination on the study area made an impact on the groundwater quality directly or indirectly. Both domestic and agricultural waters have the direct qualitative and quantitative impacts. Industrial and colling sources have indirect impacts. Dewatering as has direct impact but localized, positive impact likely decreasing the groundwater table. Domestic sources represented by sewage rooms (see Fig. 2) in all rural locations in the study area made localized biological contamination on shallow groundwater as reported by Abdel-Lah and Shamrukh (2001). They reported elevated concentrations of E-Coli and fecal coliform on groundwater near those sewage rooms (i.e. septic tanks). Agricultural sources made groundwater table higher in the study area. Fertilizers used in agricultural activities degraded the quality of groundwater especially with nitrate contaminations (Shamrukh et al. 2001). Most of the industrial and colling sources are disposed or discharged into surface water streams (i.e. drains or Nile) in the study area. Their impact on groundwater water is made only when the elevation of water streams are higher than the groundwater table/heads But this is unlikely in the study area, elevation of drains and Nile are lower than the groundwater table/heads.

4.7 Protection and Legislations

Efforts made to protect the environment and natural resources in the study area. The Egyptian government established the Ministry of Environment (Egyptian Environmental Affairs Agency) to protect the whole environment and water resources in all Egyptian territories. The Ministry work implement and to force all activities to comply with environmental laws mainly Law 4/1994 for the Protection of the Environment that amended by Law 9/2009. In addition, the Ministry of Health and Population and the Ministry of Irrigation and Water Resources in Egypt support the efforts to protect the groundwater and surface water in the study area. Water samples are collected and analyzed by all those Ministries to evaluate the quality of surface water (i.e. Nile) and groundwater resources. Annual reports are published to present the state of the environment in Egypt.

5 Conclusions

The anthropogenic sources of contamination and its impact on the quality of Nile section from Aswan to Cairo were reported. There are 124 point source of contaminations that discharged into Nile in the study area. There are 72 agricultural point sources that discharge about 13.7 billion m^3/year. Industrial facilities including 7 sugar factories and 2 paper factories discharge about 150 million m^3/year, where 14 power stations discharge 4.2 billion m^3/year. Amounts of domestic sewage are discharged directly into the River Nile or indirectly into main drains that discharge into Nile and its branches is about 1.06 BCM/year. These in addition to the discharge from dewatering effluents and cooling water. All these sources have elevated chemical and biological contaminants that influence and degrade the quality of Nile directly. Elevated heavy metals, COD, BOD, hardness, fecal coliform and temperature are reported. Accordingly, urgent mitigation measures have to be implemented to maintain the sustainability of Nile water for different usage, along with more advanced techniques for monitoring and detecting the source of contamination using remote sensing and GIS techniques (El-Rawy et al. 2019a, b).

6 Recommendations

Some important recommendations should be take into consideration to reduce and minimize the adverse effects of discharging waste water into the Nile and its branches as follows:

(1) More intensive investigation for the whole area especially that dealing with the environmental pollution as a result of pollutants accumulation and detailed

distribution maps should be constructed for the most common pollutants along the Nile valley and delta.

(2) Emphasising the compliance with the environmental protection regulations especially the Environment's Laws regarding the wastewater disposal to minimize the risk of these pollutants.

(3) Effective coordination among geoscientists, environmentalists and planner/policy makers is important to resolve these problems and throw a light on the adverse effect of excessive use of organic (manures) and chemical fertilizers.

(4) Development and completion the sewage system network in the area and construct a drain water treatment stations at the downstream sites of these drains before mixing with the Nile water.

References

Abdalla F, Khalil R (2018) Potential effects of groundwater and surface water contamination in an urban area, Qus City, Upper Egypt. J Afr Earth Sc 141C:164–178

Abdalla F, Shamrukh M (2016) Quantification of river Nile/quaternary aquifer exchanges via riverbank filtration by hydrochemical and biological indicators Assuit, Egypt. J Earth Syst Sci 125–8:1697–1711

Abdalla F, Ismail I, Akawy A, Alakraby A, Kamal El-Din G (2008) Groundwater rising as environmental problem, causes and solutions: case study from Qus city, Qena, Egypt. Catrina 3(1):93–99

Abdalla F, Ahmed A, Omer A (2009) Degradation of groundwater quality of quaternary aquifer at Qena, Egypt. J Environ Stud 1:18–32

Abdalla F, Moubark M, Abdelkareem M (2016) Impacts of human activities on destruction of the archeological sites in southern Egypt using remote sensing and field data. J Environ Sci Manage (JESAM) 19(2):15–26

Abdalla F and Shamrukh M (2010): Riverbank filtration: developing countries choice for water supply treatment, Egypt case. In: 1st IWAYWP2010 conference

Abdalla F, Madkour H, Mohamed A, Ahmed A (2014) Heavy metal distributions in the quaternary aquifer as influenced by polluted agricultural drain, Nag-Hammadi area, Qena, Egypt. In: The 7th International conference for development and the environment in the Arab world pp 75–91

Abdalla F (2005): Contaminant transport in a fractured chalk aquifer at Sigerslev Denmark, as characterized by tracer techniques. Ph.D thesis, Geological institute, Faculty of mathematics and natural sciences, Bonn University, Germany

Abdel-Lah AK, Shamrukh M (2001) Impact of septic system on ground water quality in a Nile valley village, Egypt. Sixth international water technology conference, IWTC 2001. Alexandria, Egypt, pp 237–245

Abdel-Shafy H, Abdel-Sabour F (2006) Wastewater reuse for irrigation on the desert sandy soil of Egypt: Long-term effect. In: Hlavinek P et al (eds) Integrated urban water resources management. Springer, The Netherlands, pp 301–312

Abou Heleika M, Ismail E, Ahmed M (2018) Delineation of contamination zone using geophysical and hydrogeochemical methods around the El Moheet drain in the El Minia district Upper Egypt. Arab J Geosci 11:625

Agency for toxic substances and disease registry (ATSDR) (2017) Toxicological profile for nitrate and nitrite. https://www.atsdr.cdc.gov/toxprofiles/tp204.pdf

American Public Health Association (APHA) (2012) Standard methods for the examination of water and wastewater, 22nd edn. APHA, American water works association, Washington DC, p 1496

Andersen L, Kristiansen H (1983) Nitrate in groundwater and surface water related to land use in the Karup Basin. Denmark. Environ Geol 4:207–212

El Attar R (2010) Impacts of human activities and agriculture on drain sediments and water in Farshut and Abu Tesht, Qena, Egypt. Ph.D. thesis, South Valley University, Egypt, 426

DeBorde DC, Woessner WW, Lauerman B, Ball PN (1998) Virus occurrence and transport in a school septic system and unconfined aquifer. Groundwater 36(5):825–834

El Kashouty M (2012) The hydrogeochemical evolution and modeling of the quaternary aquifer west of the River Nile, El Minia district. Arab J Geosci 5:617–635

El Sayed E, Kamel A (2010) The hydrochemical characteristics and evolution of groundwater and surface water in the western part of the River Nile, El Minia district Upper Egypt. J Geosci, p 10

El Kashouty M, El Sayed E, Elewa A, Morsi M (2011) Environmental impact of anthropogenic activity on surface and groundwater systems in the western part of the River Nile, between EL-Edwa-Der Mawas area, El Minia Governorate, Upper Egypt. J Am Sci 7:956–961

El-Azeim A, Mohamed W, Hammam A (2016) Soil physiochemical properties in relation to heavy metals status of agricultural soils in El-Minia governorate, Egypt. J Soil Sci Agric Eng 7:423–431

El-Bakri A, Tantawi M, Hamza M, Awad M (1996) Estimation of groundwater inflow to irrigation drains in Minia, Upper Egypt, based on a deuterium and oxygen-18 evaporation pan technique. Hydrol Sci J 41:41–48

Elewa AM, El Sayed E, El Kashouty M, Morsi M (2013) Quantitative study of surface and groundwater systems in the western part of the river Nile, Minia governorate, Upper Egypt: water quality in relation to anthropogenic activities. Green J Phys Sci 3(6)

El-Rawy M, Fathi H, Abdalla F (2019a) Integration of remote sensing data and in situ measurements to monitor the water quality of the Ismailia Canal, Nile Delta, Egypt. Environ Geochem Health 1:20

El-Rawy M, Ismail E, Abdalla O (2019b) Assessment of groundwater quality using GIS, hydrogeochemistry, and factor statistical analysis in Qena Governorate, Egypt. Desalination and Water Treat 162:14–29

El-Rawy M, Abdelrahman MAE, Ismail E (2020b) Integrated use of pollution indices and geomatics to assess soil contamination and identify soil pollution source in Minia Governorate, Upper Egypt. J Eng Sci Technol 15(4):2223–2238

El-Rawy M, Moghazy HE, Eltarabily MG (2021a) Impacts of decreasing Nile flow on the Nile valley aquifer in El-Minia Governorate Egypt. Alexandria Eng J 60(2):2179–2192

El-Rawy M, Makhloof AA, Hashem MD, Eltarabily MG (2021b) Groundwater management of quaternary aquifer of the Nile valley under different recharge and discharge scenarios: a case study Assiut Governorate Egypt. Ain Shams Eng J 12(3):2563–2574

El-Rawy M, Abdalla F, El Alfy M (2020a) Water resources in Egypt. In: Hamimi Z et al. (eds.) The geology of Egypt, Regional geology reviews. https://doi.org/10.1007/978-3-030-15265-9_18 ©Springer Nature Switzerland AG

Environmental Affairs Agency (2017) Achievements report of the Egyptian ministry of environment for the year 2016 (in Arabic)

Gardner B (1997) Groundwater contamination via sewer infiltration and overflow. Groundwater pollution primer. CE 4594: Soil and groundwater pollution civil engineering department Virginia tech

Geriesh M, El-Rayes A, Ghodeif K (2004) Potential sources of groundwater contamination in rafah environs, North Sinai. In: Egypt proc 7th conference of geology of Sinai for development, Ismailia, pp 41-52

International Agency for Research on Cancer (IARC) (2012) A review of human carcinogens: metals, arsenic, fibres and dusts. In: International agency for research on cancer: monographs on the evaluation of carcinogenic risks to humans

Lerner DN, Harris H (2009) The relationship between land use and groundwater resources and quality. Land Use Pol 26:S265–S273

Liu AG, Ming JH, Ankumah RO (2005) Nitrate contamination in private wells in rural Alabama United States. Sci Total Environ 346:112–120

Macler AB, Merkel CJ (2000) Current knowledge on groundwater microbial pathogens and their control. Hydrogeol J 8:29–40

Morsi M (2020) Study of the hydrochemistry characters of surface water resources in minia district in comparing to its guideline for drinking water and domestic used-Minia Upper Egypt

Powell KL, Taylor RG, Cronin AA, Barrett MH, Pedley S, Sellwood J, Trowsdale SA, Lerner DN (2003) Microbial contamination of two urban sandstone aquifers in the UK. Water Res 37:339–352

Raghunath HM (1987) Groundwater pollution and legislation. In: Groundwater, 2nd edn. India

Sabet H, El-Gohary A, Salman S, Asmoay A (2017) Evaluation of surface water for different uses in the area between Abu Qurqas-dyer Mawas districts, El Minya Governorate, Egypt. Int J Innovative Sci Eng Technol 4(1)

Salman SA, Asmoay AA, El-Gohary A, Sabet H (2019) Evaluation of human risks of surface water and groundwater contaminated with Cd and Pb in the southern El-Minya Governorate Egypt. Drinking Water Eng Sci 12:23–30

Shamrukh M, Corapcioglu M, Hassona F (2001) Modeling the effect of chemical fertilizers on ground water quality in the Nile valley aquifer, Egypt. Groundwater 39(1):59–68

Snousy MG, Ismail E, Zaki R (2019) Trace element occurrence and distribution problems in the irrigation water at El-Minia district, North Upper Egypt. Arab J Geosci 12:582

US Environmental Protection Agency (USEPA) (2004) Potential sources of drinking water contamination index, Updated 16th Sept 2004. http://permanent.access.gpo.gov/lps21800/www.epa.gov/safewater/swp/sources1.html. Accessed 2 April 2007

Zaki R, Ismail EA, Mohamed WS, Ali AK (2015) Impact of surface water and groundwater pollutions on irrigated soil, El Minia province, Northern Upper Egypt. J Water Resour Prot 7:1467

Environmental Impacts of Treated Wastewater Contaminates on Groundwater Quality in the Nile River Valley, Egypt

Mustafa El-Rawy, Heba Fathi, Ahmed Abdelrady, and Abdelazim M. Negm

Abstract Increased demand for domestic water as a result of population growth improved living conditions increased water use in agriculture, and industrial development would increase the overall amount of wastewater. Untreated or partially treated wastewater discharge into open streams raises the risk of surface and groundwater contamination. Non-traditional solutions are urgently needed to prevent the degradation of groundwater quality caused by the long-term effect of the treated wastewater contaminants in Egypt. Thus, the aquifer vulnerability zones should be identified for the planning and implementation of the sustainability strategy pursued by all world countries to conserve groundwater resources. This chapter provided an overview about the effect of treated wastewater disposals on groundwater in the Nile River valley, an analysis the conclusions of the essential related studies, and the latest methodologies used for determining hazard zones of the Nile River quaternary aquifer. Satellite remote sensing and Geographical Information System techniques (GIS) have been worldwide developed to collect data needed to observe different groundwater in several regions across the world. Accordingly, these tools are lower cost and more useful instruments for measuring water quality parameters

M. El-Rawy (✉)
Civil Engineering Department, Faculty of Engineering, Minia University, Minia 61111, Egypt
e-mail: mustafa.elrawy@su.edu.sa; mustafa.elrawy@mu.edu.eg

Civil Engineering Department, College of Engineering, Shaqra University, Dawadmi, Ar Riyadh 11911, Saudi Arabia

H. Fathi
Jazan University, Jaza 45142, Saudi Arabia
e-mail: eng_hebafathi@yahoo.com

A. Abdelrady
Department of Water Management, Delft University of Technology, 2600 Delft, GA, The Netherlands
e-mail: A.R.A.Mahmoud@tudelft.nl

A. M. Negm
Water and Water Structures Department Engineering, Faculty of Engineering, Zagazig University, Zagazig 44519, Egypt
e-mail: amnegm@zu.edu.eg; amnegm85@yahoo.com

A. M. Negm and M. El-Rawy (eds.), *Sustainability of Groundwater in the Nile Valley, Egypt*, Earth and Environmental Sciences Library,
https://doi.org/10.1007/978-3-031-12676-5_5

in groundwater than in situ investigations where measurements are limited to select sampling points. The review and analysis of many studies concerning the regression of groundwater quality of quaternary aquifer in the Nile River valley (the most densely populated areas), showed that the main groundwater problems result from drains and the lack of monitoring control of the sewage system. Additionally, in this chapter, a case study was selected in North-East of Cairo due to the presence of effluent of the largest two traditional wastewater treatment plants in Egypt. The capability of the first plant (Gabal EL Asfar WWTP) is 2.5 million m^3/day, and the second plant (El Berka WWTP) is 600,000 m^3/day, in addition to the presence of Ismailia Canal, which is the most significant branches of the River Nile. The results indicated that the studied area was extremely polluted as a result of wastewater treatment plant effluent discharge. The results of the studies showed that BOD_5, COD, NH_4 and TSS were higher than the maximum acceptable limits, which negatively affected people's general health in the vicinity of these areas and the irrigation water. This chapter's outcomes may allow developers and decision-makers to decide the regions need critical and special treatment.

Keywords Groundwater Quality; Contaminations; Wastewater treatment plant (WWTP) · Water quality index (WQI) · Geographical information system (GIS) · Nile River Valley aquifer

1 Introduction

After the Nile River, groundwater is the second most important source of freshwater (Abdel-Shafy and Kamel 2016; El-Rawy et al. 2020a). The Nile River Valley's shallow aquifer is regarded as a renewable water source in Egypt, with water extracted primarily from shallow wells, as a result of flood irrigation; it is replenished from the Nile River by seepage from canals and deep percolation (El-Rawy et al. 2020a). It is a resource that is both renewable and productive (Attia et al. 1986). The Nile Valley aquifer system has a total storage capacity of about 200 BCM, while the Delta aquifer has a total storage capacity of about 300 BCM (Abdel-Shafy and Kamel 2016; El-Rawy et al. 2020a). Connectives use of surface and groundwater is common, especially during peak irrigation demand and on the periphery of the surface water irrigation network, where groundwater can be the only source as many areas in the world (MacAlister et al. 2013; El-Rawy et al. 2016).

One of the significant problems experiencing sustainable management is the potential threat to human health and the environment posed by industrial and domestic wastewater disposal sites into the surface water bodies (the Nile, irrigation canals, and drains) either directly or indirectly (Attia 1989). The primary pollutant products that resulted by these residential activities which caused by bacteria, organic and inorganic pollutants (Ahmed 2007).

This preliminary review on water resources has shown that groundwater contamination by seepage from wastewater effluents levels had already reached alarming levels. This may threaten the quality of the Nile valley and Delta aquifer in the long run, it will necessitate additional studies and awareness (Hefny et al. 1992). Domestic wastewater is approximated to be around 5.5 and 6.5 billion cubic meters per year, but only 2.97 billion cubic meters are subjected to treatment processes. About 0.7 BCM/yr of this treated share is reused for agricultural uses, with 0.44 BCM undergoing primary treatment. As sewage water infiltrates the shallow aquifer, the remainder (0.26 BCM) is subjected to secondary treatment (Abdalla and Khalil 2018). Pathogenic microbes, nutrients, toxic metals, and organic compounds can quickly move within the aquifer and contaminate the groundwater (Bauman and Schafer 1985; El Arabi 1999; El-Rawy et al. 2020a), and that may be caused many diseases, such as diarrhea, dysentery, cholera, and typhoid fever (Obeidat et al. 2013).

Previous studies conducted on water quality in Egypt confirmed that Nile River Valley groundwater receives contamination from several sources (Hefny et al. 1992; Ahmed 2008; Ashour et al. 2009; Abdalla et al. 2009; El Arabi 2012; Dawoud and Ismail 2013; Elnashar 2014; Gedamy2015; Abdel-Shafy and Kamel 2016; El-Aassar et al. 2016; Abdalla and Khalil 2018; Fathi and El-Rawy 2018; Ismail and El-Rawy 2018; El-Rawy et al. 2019a, b; El-Rawy et al. 2020c). Such as the excessive withdrawals, this increased salinity of groundwater with long-term consequences for water use and soil contamination, as well as bacterial contamination of groundwater from sewage water, which is also common in several area of the Delta (INECO 2009). There is a large amount of brackish groundwater in the Delta and contaminated from many leakages (mainly domestic and agricultural), salinity varies from 1500 ppm in the Nile Valley and the Southern Delta to 5000 ppm in the North Delta (Salim 2012; Masoud 2014).

The Delta region receives attention in environmental studies related to water quality due to the high population density that it reaches about 95% of the population in Egypt (El-Din 2013). The polluted aquifer is difficult, time-consuming, and expensive to reclaim; thus, not jeopardizing groundwater resources is a requirement of treated wastewater discharges. Rationales mentioned previously require researchers and specialists to find all solutions and scientific methods to develop the traditional groundwater assessment methodology, to protect it from pollutants, especially partially treated wastewater. There are several Physical parameters (pH and temperature), chemical parameters (Alkalinity, BOD_5, COD, TDS ... etc.), biological parameters (fecal coliform) must be performed using standard methods to assess the current status of groundwater quality. It is considered to be indicators of pollution due to organic wastes of municipal and industrial (APhA 1998; Patela and Vaghanib 2015). The traditional method for the assessment of groundwater quality is expensive and time-intensive, covering transportation, collection, and experimental analysis. Also, it does not provide a perfect overview of 'groundwater's predisposition (Simsek and Gunduz 2007; Balan et al. 2012). Furthermore, the traditional water sampling approach does not provide data about the accuracy of water for every position in the stream (Naithani and Pande 2012). Therefore, the use of mediators for spatial data principles can be constructive by rapidly providing systematical information. The most usual type of digital mediator technology currently is the Geographical Information System (GIS), this technology allows to merge layers of digital data from vary sources and drive the different layers related to each other (Zaporozec 2002).

The main objective of this chapter is to do reviews and summarize the most related studies and research that dealt with the impact of disposal of treated wastewater on groundwater and clarify this through a case study of the most extensive treatment plants in Egypt and the Middle East (Gabal EL Asfar WWTP with a capacity of 2.5 million m^3/day).

2 Contaminants Seepage Between Groundwater and Surface Water

Surface water in streams can frequently interchange with nearby groundwater. The groundwater-surface water interaction via water transfer between provides a significant pathway for chemical transfer between drains and aquifer systems (Winter 1998).

Due to groundwater interaction with surface waters, aquifers' contamination is strongly connected to polluted surface water. Pollution is more severe on the Nile River's edges and desert fringes, as well as in the shallow segments of the aquifers (Abdel-Shafy and Kamel 2016). The interactions between surface water bodies and groundwater are complex (Winter 1998; El-Rawy et al. 2016). The groundwater aquifer loses its water through seepage to the Nile River, abstraction, and evaporation from the surface (Hefny et al. 1992; El-Rawy et al. 2020a; El-Rawy et al. 2021a, b). Since the shallow aquifer is in hydraulic interaction with both the surface water irrigation system and the Nile River system, it can receive both contamination from surface water sources and is thus risky. The assessed methods of the surface water and groundwater interaction mechanisms can be classified into two groups: (1) analytical solutions and (2) numerical modeling (Nashed et al. 2014; El-Rawy et al. 2020b; El-Rawy et al. 2021a, b).

3 Environmental Impacts of Treated Wastewater Contamination

Treatment systems for wastewater have been built in response to adverse circumstances created by the disposal of untreated effluent into water bodies. This method aims at removing biodegradable organic substances, suspended and floating products, nutrients, and pathogens (Helmer et al. 1997). Nevertheless, the overloaded effect in peak periods causes the efficiency of removal of contaminations to decrease, and the possible pollution of groundwater aquifers in the area of disposal of wastewater treatment plants may be an environmental issue. Much research assessed the health risk of treated wastewater effluents regarding water quality intended for human use and crop productivity (Custodio 2013). So, incentive-based policies should be implemented to regulate the discharge of wastewater into surface streams.

Improper functionality of the wastewater treatment system may have a detrimental effect on the environment because it may accumulate pollutants. Even during the drying process and storage of the sludge, groundwater contamination is high potential due to the seepage from drying beds (Trpevska and Dodeva 2009). Uncontrolled disposals of treated wastewater can cause another significant adverse effect on groundwater. Zaporozec (2002) and Wahaab and Badawy (2004) described the impact of treated wastewater in groundwater by a further sub-classification as following:

(a) **Public Health**
A large number of water-related diseases are recorded by the groundwater research institute (RIGW) and the ministry of water resources and irrigation (MWRI). It is reported that several more communities suffer from problems related to other impacts of treated wastewater contamination, which influence several crops (Rashed et al. 1995; El-Arabi and Attia 1997; Antar et al. 2016; El Arabi 2012).

(b) **Biological Contaminations**
Significant quantities of coliform bacteria have been found in the Nile and its branches downstream of Cairo as a result of partial wastewater treatment (FAO 2016). The observed concentrations are higher than the level set out in Law 48 of 1982 concerning the safety of the Nile River and watercourses from contamination (Ezzat et al. 1999).

(c) **Inorganic substances**
The distribution of inorganic ions mostly is scarcely harmful to health. However, some could affect physical discomfort if consumed in high concentrations (e.g., sulfate); additionally, high concentrations of calcium and magnesium compounds cause water hardness (Fishman and Friedman 1989). Total dissolved solids (TDS) refers to the total ion concentration of minerals dissolved in water, and at levels of greater than 1500 mg/l, gastrointestinal irritation usually happen (Zaporozec 2002).

(d) **Heavy Metals**
Such contaminations are unfit for human consumption and harm the natural environment (Waly et al. 1987). High levels of different metals are noted in Lake Manzala (the final downstream of the main drains in the Nile Delta), and its detrimental effect on the lake's fish stocks has become a long-term threat to the ecosystem (Omran and El Razek 2012).

4 Wastewater Treatment Plants in Nile River Valley

The objective of wastewater treatment plants is to decrease the content of suspended particles (primary treatment), organic matter (secondary treatment), and nitrogen and phosphorus (tertiary treatment) before dumping treated wastewater into a water body or soil, thereby significantly reducing the negative environmental effects of wastewater discharges (Brion et al. 2015). Therefore, the traditional treatment process of any municipal wastewater executed shows an advanced process to meet the growing demand for water (Das and Das 2003). Another critical area is taking into account water and sanitation facilities in companies affiliated with Egyptian laws and regulations and administrative requirements (including human resources, occupational safety and health, operations, maintenance, and quality assurance). Law 48/1982 provides the "effluent-standards" for freshwater inland with the Ministry of Water Resources and Irrigation (MWRI). The major objective of the Law is to protect the Nile River and its waterways from contamination. It is necessary to increase the operating stations' rates to ensure that the goals of this sector are achieved in a sustainable manner (FAO 2016).

The 420 sewage treatment plants in Egypt are divided into 16% of which operates with the primary treatment system, 82% with secondary treatment, and 2% with tertiary treatment. It is divided according to the technology used: 189 plants with active sludge technology, 89 oxidation trenches, 75 with oxidation ponds, 26 with throbbing filters, and 41 with different technologies. The Egyptian Government is now expanding sewage treatment plants in Upper Egypt, and the plan includes the implementation of 52 treatment plants in Upper Egypt, with a total capacity of 1.147 million m^3/day, the amount of treated wastewater increased from 10.5 million m^3/day in 2014 to 12.8 million m^3/day in 2019. There are 119 existing sewage treatment plants in Upper Egypt, with a capacity of 1.8 million m^3/day (Holding Company for Water and Wastewater 2020).

The system of wastewaters in Greater Cairo is divided into two networks. (i) Regions located on the eastern side of the Nile collect most of their wastewater in the EI-Ameria pump station, after which it is directed to Gabal El Asfar WWTP (Fig. 1), 25 km northeast of Cairo. The wastewater stays 25 to 30 days for primary sedimentation in exposed basins. The sludge is picked, sun-dried, and then supplied to customers. The fluids are conveyed to irrigate 3000 feds, (ii) regions placed on the Nile's western bank carry their wastewater to Giza Pump station, moreover Zeinin and Abou- Rawash wastewater treatment plants (Osman et al. 2016).

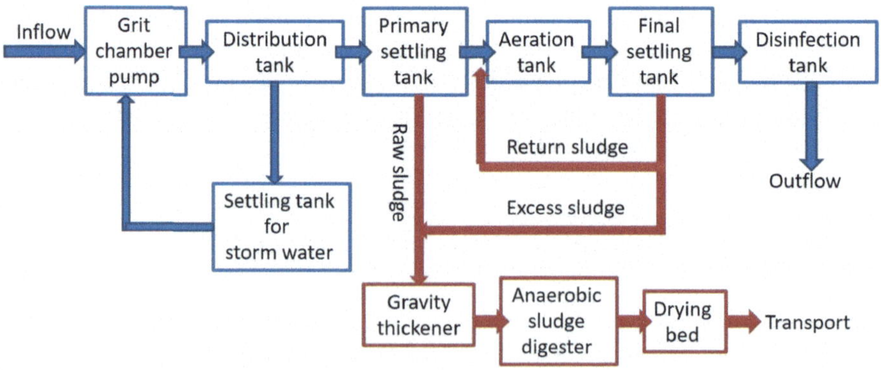

Fig. 1 Process arranging of Gabal El Asfar WWTP; source: authors' plot based on field visit to the plant

The treatment and disinfection of WWTPs in East of the Nile participated in the enhancement of the Eastern Delta region ecosystem and social environments (Abdel-Fattah and Helmy 2015).

5 Groundwater Assessment Methodologies

The essential objective of the groundwater quality assessment is to achieve an overall view of groundwater quality spatial distributions and time variations that occur, either progressively or under 'authorities' control. The advantage of complete and flexible design, as well as perfect programs, is that timely water quality management and/or contamination control mechanisms based on a systematic approach and useful water quality information can be taken (Chapman and Organization 1996).

There are four main techniques that can be used to assess the quality of water: (1) water quality index method (WQI), (2) trophic status index method, (3) statistical analysis method, and (4) biological analysis method (Elshemy and Meon 2011). The water quality index is the most common and used for several reasons that will be explained below.

The WQI is a premium method to understanding and summarizing vast amounts of water quality data. This is made by integrating complex information and expressions represent a composite impact of involved variables on the water quality in every water cycle. Therefore WQI becomes an important method for the monitoring and assessment of groundwater (Tiwari et al. 2014). Water quality indices are certified to demonstrate temporal and spatial differences in water quality, even at small concentrations, in an accurate and timely method (Sharma et al. 2014).

The flexibility and facility of the CCME WQI encourage many researchers to apply this method to characterize water quality (Hurley et al. 2012). This method has been developed by incorporating other variables which might corrupt the water quality or make water unqualified (Khan et al. 2004). Moreover, that is very obvious in several studies as an application of the CCME WQI on East Hammer Marsh (Al-Saboonchi et al. 2011), the Canadian Water Quality Index assessed the effluent quality of Shiraz wastewater treatment plant for agricultural irrigation, Iran (Baghapour et al. 2013), CCME water quality index for tracking water quality: Mackenzie river basin case, Canada (Lumb et al. 2006), and the Water Quality Assessment of Mahmoudia Canal in Northern West of Egypt (Abdullah and Hussona 2014).

Statistical factor analysis has been commonly applied to examine groundwater geochemistry (Lawrence and Upchurch 1982; Liu et al. 2003; Shuxia et al. Yu 2003; El-Rawy et al. 2019a, b). Factor analysis describes the correlations between the results of the dependent variables that are not explicitly observable. El-Rawy et al. (2019a, b) concluded that an integration of conventional hydrochemical analysis and GIS with factor analysis can help decision-makers understand the variables that govern groundwater chemistry and it may allow them to control quality of ground water.

Geographical Information System (GIS) can be an efficient and effective tool for monitoring, mapping, modeling and assessing water quality, identifying environmental changes, and assessing water existence (Balathandayutham et al. 2013; El-Rawy et al. 2019a, b). GIS can be applied in different groundwater assessments, such as assessing the groundwater resource hazard and recognizing the ecosystem and maintaining local or national water resources (Tiwari et al. 2014).

GIS technology enables to link with the satellite imagery system and serves as geodata management, by spectral, temporal and spatial information; in addition to the in situ measurements, it is achievable to combine the additional benefits of the two technologies (Asadi et al. 2011). Water quality assessment and study of many characteristics are becoming more convenient with the growing number of satellite imagery. Also, it gets to be cost-effective than conventional methods and efficient. The advantage of using remote sensing- dependent on discovered groundwater is to define all reasonably related features with quality, expansion, and recession of groundwater. It can be seen that remote sensing informed and supported economic development and groundwater management decisions (Ageeb et al. 2017; Usali and Ismail 2010).

During the previous 30 years, satellite remote sensing (RS) and Geographical Information System (GIS) techniques have been worldwide developed to collect data needed to observe different groundwater in several regions across the world (Staudenrausch et al. 2000). Accordingly, these tools are lower cost and more useful instruments for measuring water quality parameters in groundwater than in situ investigations where measurements are limited to select sampling points (Naithani and Pande 2012). Several investigators have used RS and GIS methods to assess water quality and have found many successful consequences (Mostafa and Soussa 2006; Alaguraja et al. 2010; Abuzied et al. 2016). They confirmed that combining RS and GIS with investigated water sampling and analysis might optimize and speed up the process for groundwater quality assessment with acceptable accuracy.

6 Investigations of Groundwater Contamination in Nile River Valley Aquifer

Researchers tested several groundwater quality variables in different governorates in the Nile Valley, such as Aswan, Luxor, Qena, Sohag, Assiut, El-Minia, and Cairo. Most of these governorates occupied an important historical position over the years, which led to an increase in population attractions, resulting in many environmental problems, especially in the field of sewage treatment and how to get rid of it. Locations of these governorates are illustrated in Fig. 2. Many rural residents in the Nile Valley use shallow hand pumps and deep extraction wells as a water source for drinking or irrigation, and several studies reported that used water is not appropriate for use due to the presence of fecal coliform and a seasonal and spatial variation in the other

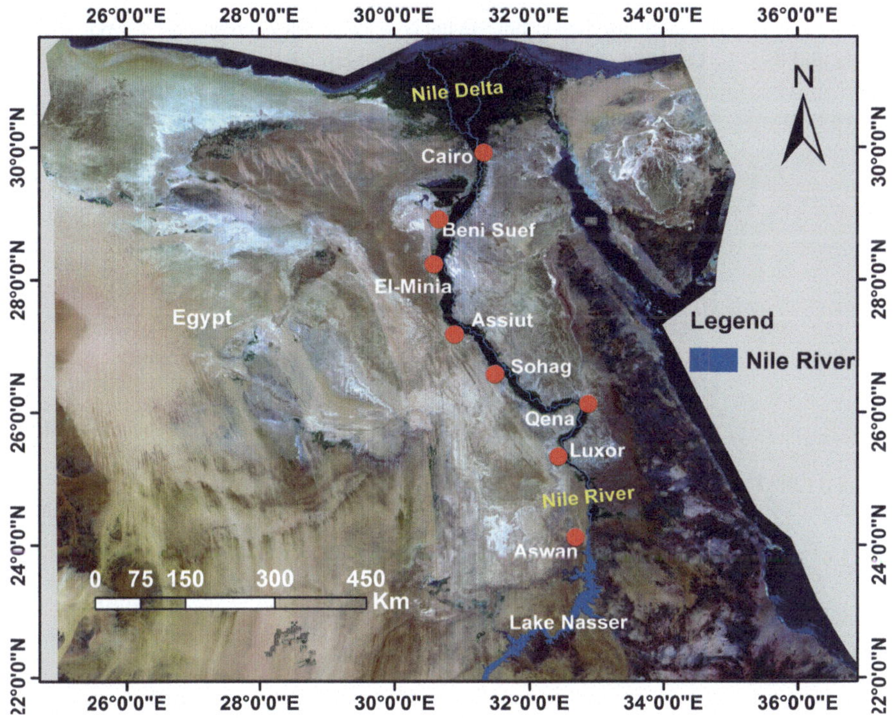

Fig. 2 Nile Valley governorates in Egypt

concentration of pollutants (Abdel-Lah and Shamrukh 2001; Abdel-Dayem 2011; Awad et al. 2020).

Each study analyzed and measured a variety of samples at various depths, resulting in a general descriptive view. Recent works (El-Rawy et al. 2019a, b, 2020b) have developed a framework for evaluating water quality using GIS technique and satellite data to develop a qualitative water quality maps. In turn, these data could be used to expand field surveys and monitor large locations on a repeated basis over a short period; these works can be summarized (from south—north of the Nile River valley). Some of these studies include the following (Table 1):

Many of Upper Egypt's drainage water flows into the Nile River, while much of it flows into the Delta in the northern lakes and the Mediterranean (El Arabi 2012). The MWRI identified major concern as health and safety and provided geographical maps showing high, medium, and low levels of pollution. Figure 3 describes the concentration of pollutants in the Delta.

According to several studies (El Arabi 1999; Sharaky et al. 2007; Al-Agha et al. 2015), the vulnerability of groundwater to contamination in the Nile Delta region indicates high groundwater frailty regions on the flood plain's fringes, where most agricultural land is distributed with some industrial and isolated rural areas; most groundwater in the flood plain with moderate to low vulnerability. Zaporozec (2002)

Table 1 Previous studies in different regions in the Nile Valley (in a chronological way)

Study and its date	Location of study	Objective	Methodology	Results
Ismail et al. (2005)	Luxor	Specify the reasons that cause groundwater and salinity levels to rise	Groundwater modeling	They investigated that Luxor Governorate incurs from the salt deposition, due to salt transport by capillary water from the relatively high salinity groundwater or inveterate water in the dense silty clay layer
Ahmed (2007)	Sohag	Assess the weakness of the Quaternary aquifer system to pollution from municipal, industrial, and agricultural sources	GIS-based models	The study revealed that the high and very high vulnerability characterizes 83% of the Quaternary groundwater aquifer to pollution with industrial, municipal and agricultural pollution
Ahmed (2008)	Sohag	Examine the groundwater quality in the Nile River valley along Sohag district and to demonstrate the relation between alluvium aquifer and surface water recharge source	In situ measurements and statically analysis	The groundwater samples were over-saturated concerning carbonate minerals as a result of excessive fertilizer use, and disposal of irrigation and wastewater
Abdalla et al. (2009)	Qena	Address the groundwater quality degradation due to the different pollution sources recorded in the study area	Water quality assessment and GIS analysis	The results revealed that groundwater sources are polluted by nitrates, phosphates, ammonia, and e.coli bacteria

(continued)

Table 1 (continued)

Study and its date	Location of study	Objective	Methodology	Results
Hemdan and Abdel Rady (2011)	Aswan	Studied the contamination of the Aswan quarternary aquifer owing to the leakage from the septic tanks of untreated wastewater into the aquifer	In situ measurements and statically analysis	An rise in the organic matter and nutrient concentrations was observed
Selim et al. (2014)	Aswan	Investigate the reasons of variations in groundwater levels in Aswan region, which is located in a subtropical arid region	In situ measurements and statically analysis	The abstracted water was found to be biologically polluted, which prompted the local water services company to shut down the Shallal region (Aswan) groundwater wells, and consequently raised the groundwater level by 1–3 m over the period from January 2009 to July 2013, which affects the city's infrastructure
El-Fakharany and Fekry (2014)	Esna city	Determine the cause of groundwater rise in Esna City and suggest mitigation measures to decrease and fix the problem	In situ measurements and numerical simulation techniques for the aquifer systems	They confirmed that groundwater problems result from drains and the lack of monitoring and control of the sewage system
Masoud (2014)	West of Nile Delta	Identify the most influential factors controlling groundwater evolution	Multivariate statistical and geo statistical techniques	The effect of evaporation in the eastern parts of their groundwater is more stated due to high salinity
Gad and Ismail (2015)	East of Nile Delta	Assess the groundwater quality	God, PRAST and DRASTIC merge and index methods	Indicated that groundwater is mainly contaminated with po_4 and no_3 at risk levels

(continued)

Table 1 (continued)

Study and its date	Location of study	Objective	Methodology	Results
Gedamy (2015)	Sohag	Identify the pollutants and determine of their sources	Major, minor, and trace constituents are analyzed chemically	The study results indicated that the high concentrations of iron, lead, chromium, aluminium, and zinc are because of seepage from municipal water and wastewater, reclamation projects, return flow after irrigation, and fertilizer overuse
Abdalla and Shamrukh (2016)	Assiut	Investigates the interaction between the Nile and the Quaternary aquifer	Hydrochemical and biological indicators	They concluded that disposal of domestic wastewater that contains many types of disease-causing organisms into the subsurface led to introduce contaminants, especially microbes, nitrate, and trace elements into the shallow aquifers
El-Aassar et al. (2016)	Assiut	The groundwater quality and vulnerability assessment	Drastic model	The resulting samples demonstrated the possibility of polluting groundwater with nitrate, iron, and manganese
Moneim et al. (2016)	El-Minia	Groundwater management in the Governorate	Numerical modeling	They concluded that aquifer is described by high dissolved solids caused by leaching sediments of the calcareous limestone

(continued)

Table 1 (continued)

Study and its date	Location of study	Objective	Methodology	Results
Ageeb et al. (2017)	West of Nile Delta	Estimate the groundwater level in the study area	Electric resistivity, merged remote sensing and GIS	The results of the delta groundwater chemical analysis show that salinity is well-compliant, in which salinity gradually increases with hydraulic gradient and vice versa
Bastawesy et al. (2017)	Aswan	Advising the best management plans to avoid the negative social and environmental affect of waste water regions	Remote sensing and GIS	They identified many of the growing environmental problems for water sources and soil due to most of the treated wastewater being discharged back into the Nile via the flash flood drain known as "Al Sail Drain
Morsi et al. (2018)	Assiut	Assessment of groundwater resources for irrigation purposes	Statically analysis	The study recommended avoiding the domestic wastewater and the industrial wastewater from direct disposal without treatment to the aquifer, irrigation canals, and River Nile
Abdalla & Khalil (2018)	Qena	Study the interaction of wastewater with surface and shallow groundwater resources in the study area, as well as potential anthropogenic contamination	Water quality assessment and GIS analysis	The high bacterial loads (59% of the groundwater samples) and some water pollution indicators confirmed that the contamination was caused by anthropogenic activities in the form of wastewater discharges. due to improper wastewater disposal practices

(continued)

Table 1 (continued)

Study and its date	Location of study	Objective	Methodology	Results
El-Rawy et al. (2019a, b)	Qena	Identify factors affecting groundwater quality in the research area	GIS and statistical analysis	They declared that most of the collected groundwater samples are not suitable for domestic uses due to a high level of hardness
Megahed & Farrag (2019)	Assiut	Create a GIS-based water potentiality spatial model to examine the various water quality classes in the study area	Integrated hydrochemical and bacteriological analyses in a geographic information system (GIS)	Poor water quality was observed along the Nile Valley boundary, due to the obvious influence of the adjacent Limestone plateau and Pliocene deposits
Awad et al. (2020)	Assiut	Assesse the surface and groundwater interaction	Field measurements and water quality assessment	They proved that the aquifer is recharged by the drains
Abdelmawgoud et al. (2020)	El-Minia	Assesse groundwater quality in El-Minia Governorate	Field measurements, GIS and hydrochemistry analysis	They indicated that 78% of collected groundwater samples are suitable for irrigation uses as per the Residual sodium carbonate, and 74% and 55.5% are suitable according to Na% and PI, respectively
El-Rawy et al. (2020b)	El-Minia	Evaluate the concentrations of heavy metals to determine their potential contribution to soil pollution in the Nile Valley	Field measurements, laboratory anaysis, and a geographic information system (GIS)	As a result, contaminated soils near the El Moheet drainage system have been found, involving drainage water from the Abu Qurqas Sugar Factory and partial wastewater treatment plans

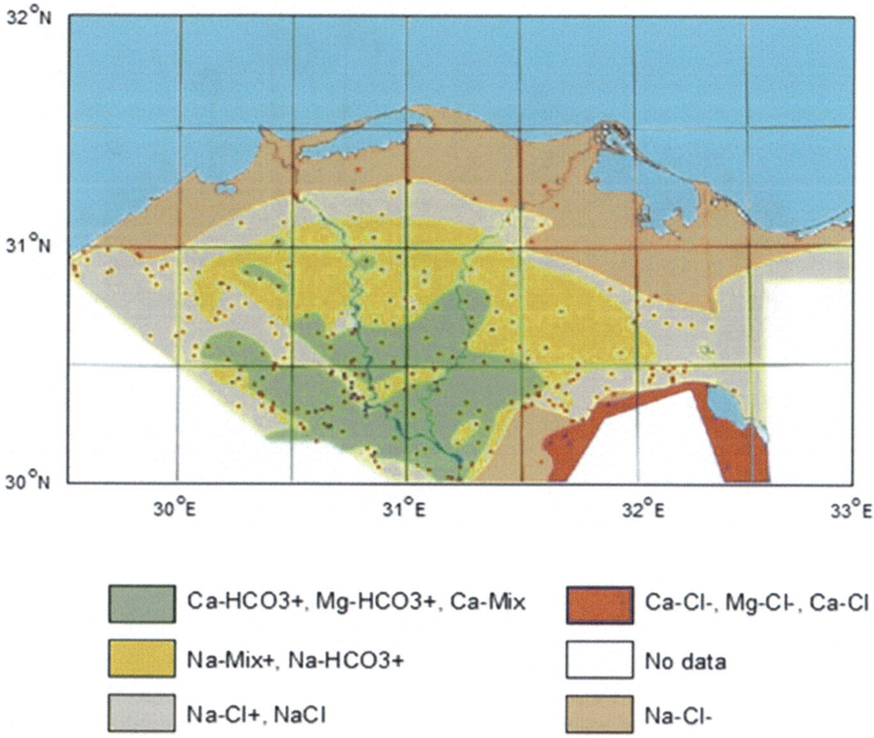

Fig. 3 Chemical groundwater types according to RIGW, Hydrogeological map of Egypt (Nile Delta), scale 1:500,000; RIGW (1992)

and El Tahlawi et al. (2008) identified the primary sources of groundwater contamination as domestic wastewater and agricultural disposal, water mismanagement, and mining and industrial activities. Moreover, due to the large part of the time involved in transporting groundwater pollutants, pollutants are likely to be clustered around the source (Harter 2003; Taha et al. 2004; Foster et al. 2013). Figure 4 shows the poor condition of the Gabal El Asfar wastewater treatment plant (WWTP), which is placed in the East Delta, and how this affects the environment in the surrounding areas, as mentioned above.

Recent research as Abd (2017) in Iraq and El-Zeiny and El-Kafrawy (2017) has created an approach of evaluating water quality from Landsat 8 data in Egypt. The studies indicated that a significant correlation between the characteristics of water quality and their radiance values in Multiple Linear Regression Equation Models. The latest study was conducted in the Delta region (El-Rawy et al. 2020c). It provided a novel method for obtaining accurate measurements from Landsat images for any narrow water stream (Canal or drain) to verify the effect of wastewater treatment plants in such areas on the primary water sources, either surface or groundwater. The study showed a sufficient percentage of error between satellite-estimated values and

Fig. 4 Mixing the treated wastewater of Gabal El Asfar WWTP with an agricultural drain, *Source* Authors' photo based on field visit (X: 31.345749; Y: 31.345749)

in situ observed compared to previous studies (Kloiber et al. 2002; Abd 2017). These results confirmed that the water sources in the area (surface and groundwater) were negatively affected by the disposal of treated wastewater and agricultural drainage. These results agreed with the previous studies results in many areas of the Nile delta that groundwater should be protected as an essential source of water, particularly in Egypt, along with basic science and viable planning for implementing an effective system for managing surface water quality.

6.1 Case Study: Evaluation of Groundwater Resources Nearby the Main Wastewater Treatment Plants

The evaluation was held in the surrounding area the significant two traditional wastewater treatment plants (El Gabal EL Asfar and El Berka) in Greater Cairo, as mentioned previously. The widespread use of groundwater for various purposes such as irrigation, drinking, and domestic activities in the nearby region has attracted attention and concerns about the environment issue in this region. Therefore, this study's main objective is to evaluate the groundwater quality in this region and to identify contamination sources associated with these wastewater treatment plants. Most previous studies focused on studying El Gabal El Asfar farm only (Rashed et al. 1995; El-Arabi et al. 1996; El-Arabi and Attia 1997; Gemail 2012). They concluded that the groundwater in the surrounding area might become at risk with time and warned from the surface leakage of wastewater, especially with many canals in the region.

Based on the above information, Fathi and El-Rawy (2018) developed a GIS-based index method as a universal assessment tool with all related physicochemical groundwater quality parameters. This will achieve to manage and succeed in the sustainable use of groundwater resources, also to help planning and decision-making for incorporated management regarding mitigating the expected negative environmental impacts.

Fourteen groundwater samples were collected from several locations (shallow or deep boreholes), as shown in Fig. 5, which are usually used for drinking and irrigation purposes in the study area, between August 2016 and January 2017. The samples were divided to unequal intervals along the study area in order to cover the entire region. These samples were distributed at uneven distances along the study area to cover the whole area. Such samples were subjected to laboratory analyzes at the El Berka wastewater treatment plant (the closest laboratory) to ensure more accurate results. The analyzes were carried out using standard methods (Rainwater and Thatcher 1960; APhA 1998).

According to WHO, Egyptian, FAO, and CCME standards parameters, the observations show that NO_3, Cl, SO_4, and TDS are suitable for irrigation and domestic uses, whereas NH_4, TSS, COD, and BOD_5 are higher than the maximum allowable limit due to contaminants in aquatic systems that are caused by seepage from wastewater drains in the study area, as shown in Fig. 6. All physicochemical data were handled and evaluated in order to calculate the Water Quality Index (WQI), determining its applicability for irrigation and domestic purposes in the study area. The results have confirmed the degradation of water quality and have already been confirmed (Rashed et al. 1995).

The Correlation matrix declares a high positive correlation for COD with BOD_5 and fecal coliform average and apparent positive correlation between TSS and TDS, alkalinity, and chloride. The result matched with the previous studies concerned with this study area, as (Gemail 2012) declared that the treated wastewater used for irrigation seeps through the upper sandy soil recharging the upper part of the Quaternary aquifer. This is causing a detrimental effect in the study area, leading to severe problems for plants due to increasing the soil microbial load, consequently increasing the spread of parasitic infections between the populations (Habbari et al. 2000; Ensink et al. 2005).

Previous results require spatial mapping to reveal the most concentrated locations and to identify potential causes and problems. GIS technique is used to obtain a spatial distribution map for all groundwater samples in the study area. The distribution of pH, TDS, TSS, and NO_3 in the study area is shown in Fig. 8a, b, c, and d, respectively. The pH varies from 7.22 to 8.0, with an average of 7.7, which is suitable for drinking (Fig. 7a). The TDS varies from 150.1 to 400 mg/l, which is also acceptable for drinking (Fig. 7b). The TSS ranges From 16 to 34 mg/l, with an average of 25.41 mg/l (Fig. 7c). The NO_3 varies from 7 to 43 mg/l, with an average of 16.88 mg/l, which is suitable for drinking (Fig. 7d). The correlation definite a robust positive correlation between TDS and TSS ($r = 0.83$), the high values showed in Gabal El Asfar farm

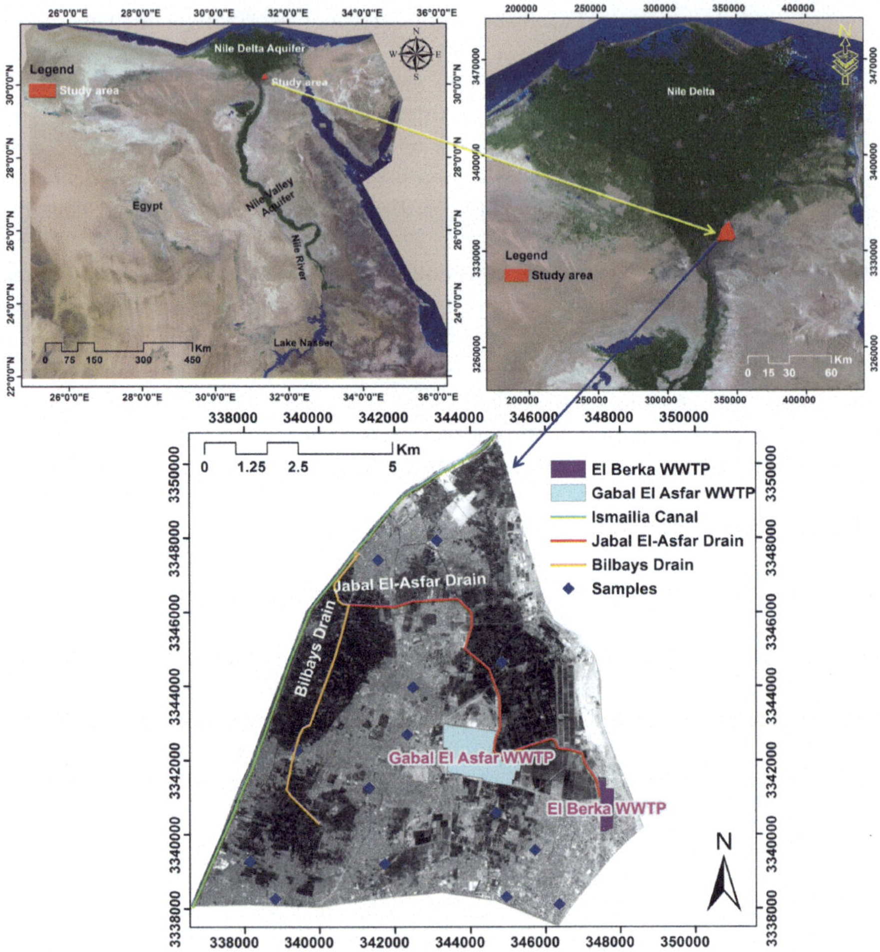

Fig. 5 Egypt map (upper left), Nile Delta map (upper right), and the lower map show the study area, and the location of El Berka and El Gabal EL Asfar WWTW, Ismailia canal, Jabal EL-Asfar and Bilbays drains, and groundwater samples

(Fig. 7b, c). This due to the evaporation of irrigation water (treated wastewater), this corresponds to the expectation of many researchers who have studied Gabal El Asfar farm as (Rashed et al. 1995; El-Arabi et al. 1996; El-Arabi and Attia 1997; Gemail 2012), where they hesitated from the use of treated sewage in irrigation and the long-term effects due to pollution problems in soils and plants.

The spatial distribution of BOD_5, COD, fecal coliform, and SO_4 in the research area is shown in Fig. 8.The BOD_5 ranges between 10 mg/l and 56 mg/l, while the COD ranges between 33 mg/l to 197 mg/l, which exceeded the max permissible range prescribed by Egyptian standards and WHO drinking water standards. Figure 8 indicates that the higher values of BOD_5, COD, and fecal coliform were in the junction

Fig. 6 The effect of Gabal El Asfar WWTP disposal on the drain; *Source* Authors' photo based on field visit (X: 30.236843; Y: 31.345749)

area of the major drains (Bilbays and Gabal El Asfar drains). The results indicated that BOD_5 has a robust positive relationship with COD and fecal coliform with a correlation of 0.96 and 0.90, respectively. SO_4 has a good relationship with BOD_5, COD, and fecal coliform with a correlation of 0.55, 0.54, and 0.62, respectively.

The distribution of NH_4 and CO_3 in the research area is shown in Fig. 9a, b, respectively. The NH_4 varies from 0.2 to 14 mg/l, which are unsuitable for drinking (Fig. 9a), and CO_3 values vary from 156.4 to 629 mg/l.

The linear regression model (Fig. 10) helped to predict more water quality parameters, saving a lot of time and costs to test a variety of water quality indicators that will aid decision-makers in developing and managing groundwater contamination problems in the study area.

7 Summary

The threats to citizen safety posed by the transfer of pathogenic bacterial substances from the sewage system to groundwater are recorded worldwide. Mixing treated wastewater with a drainage network requires stringent oversight to reduce adverse environmental impacts and negative effects on rural people's health and their significant impacts on groundwater quality. Egypt is primarily a desert region, and thus

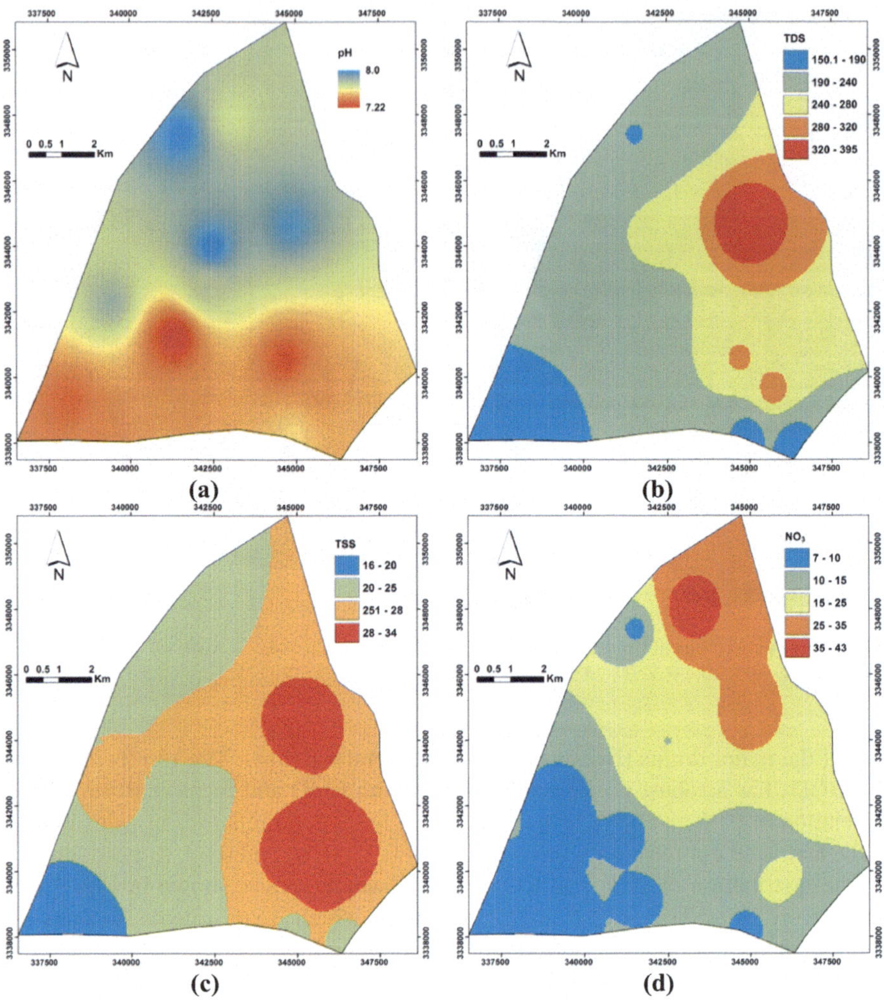

Fig. 7 Spatial distribution of **a** pH; **b** TDS; **c** TSS; and NO_3 in the study area

groundwater is a crucial factor in sustainable development and economic growth. This chapter is focused on evaluating groundwater pollution in Nile Valley due to the uncontrolled disposal of wastewater treatment plants by presenting a summary for the main studies and analysis concerned with this topic and, more specifically, the latest methodologies in the groundwater assessment such as MODFLOW, GIS, and Remote Sensing. These methods will result in more observation and a reduction in the formal investigation obstacles. This is what was reached and confirmed in the case study, whose results were reviewed.

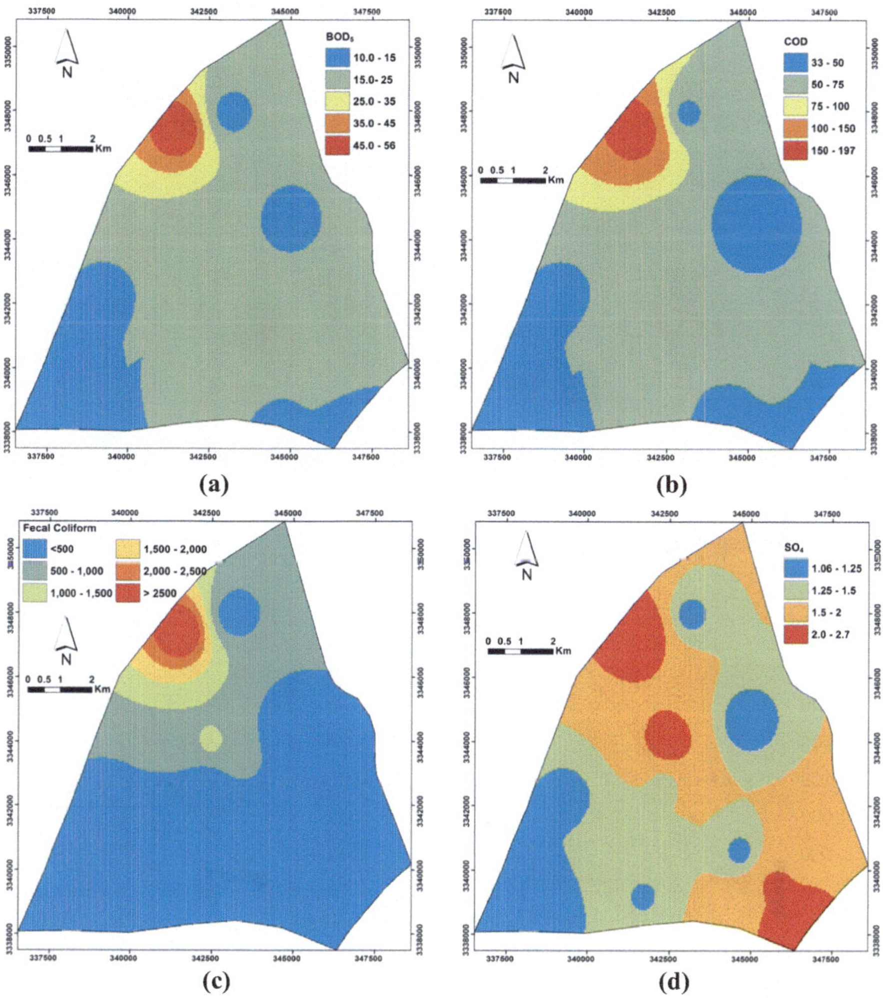

Fig. 8 Spatial distribution of **a** BOD_5; **b** COD; **c** fecal coliform, and **d** SO_4 in the study area

The main factors responsible for groundwater contamination are the improper development of wastewater treatment plants and the unregulated use of agrochemicals in the Nile Valley. Such richness did transitionally boost the microbial load and compositional changes of Nile water at the disposal points.

The previous studies indicated that many places in the Nile valley are heavily contaminated, and immediate action is needed to protect the Quaternary aquifer and reduce pollution. The present conditions indicated that groundwater in the study area would get progressively deteriorated unless immediate response is conducted.

The total bacteria were reported to be increased by 45–50%. The detection of pathogenic bacteria in Nile Valley villages including such fecal coliform in hand

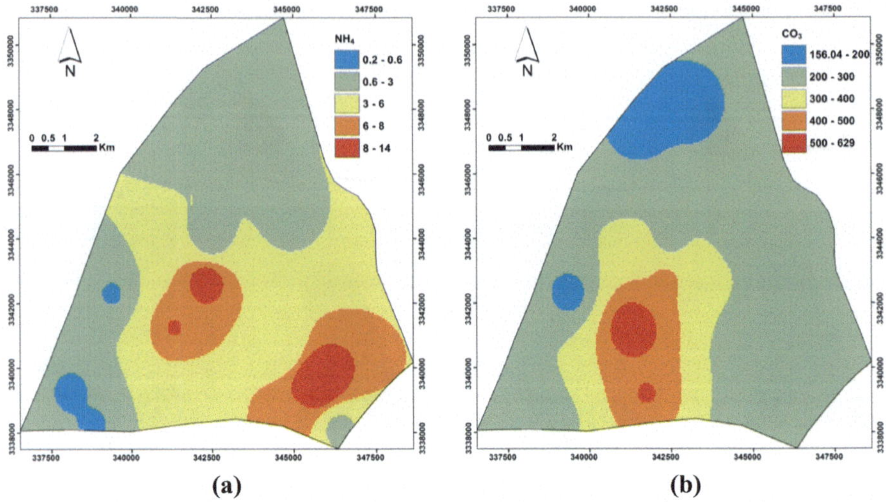

Fig. 9 Spatial distribution of **a** NH_4 and **b** CO_3 in the study area

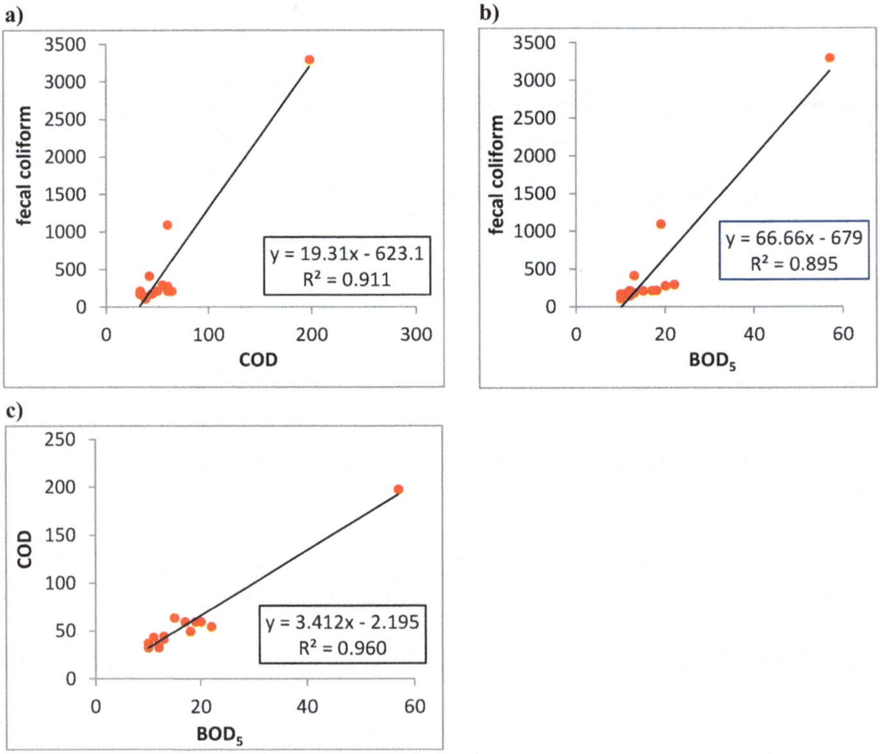

Fig. 10 **a–c** correlation linear regression between COD, BOD_5, and fecal coliform parameters in the study area

pump and well water indicates biological pollution from the sewerage system. It is not limited to only biological contamination, but also included heavy metals, where it reported that the groundwater samples in Northern Upper Egypt exhibit high values of some heavy metals. These heavy metals may be the result of anthropogenic leaching wastewater from several activities. In Sohag Governorate, it is indicated that the high quantities of lead, iron, aluminum, chromium and zinc are caused by seepage from municipal water and wastewater, reclamation operations, return flow after irrigation, and fertilizer overuse. The levels of a toxic substance (Cd and Pb) in groundwater in El-Minia Governorate surpassed the permissible limits of drinking water level established by the World Health Organization (WHO), and they warned that long persistence of these contaminants in the environment exacerbates its threat to human health.

There is another problem that has been stated by several researchers, namely the discharge of many agricultural drains directly to the Nile in the Upper Egypt region, and what is raising the problem is that these drains are filled with large amounts of wastewater that are directly deposited, such as Al Sail Drain in Aswan and El Moheet in El-Minia. Consequently, it is necessary to develop sewage treatment systems and applying the tertiary treated in the wastewater plants to avoid increasing pollution concentrations in groundwater.

Contaminants in the Delta region are relatively less than the Nile Valley, but the problem of groundwater affected by the disposal of treated wastewater and agricultural drainage. These results in many areas of the Delta agreed that groundwater should be protected as an essential source of water, particularly in Egypt, in addition to practical and adequate planning for implementing an effective system for managing surface water quality.

8 Conclusions

The obtained results concluded from the case study (the region surrounding the main two traditional wastewater treatment plants El Gabal EL Asfar and El Berka in Greater Cairo) findings lead to the following concluding remarks:

- Because the discharge of effluents from wastewater treatment plants, the studied drains were extremely contaminated.
- Despite the high efficiency of the wastewater treatment plants, the spatial distribution analysis of groundwater quality indicated that some of the collected samples were not satisfying the water quality standards prescribed by the WHO or the Egyptian standards such as the area around the plant's disposal and the convergence area of the main two drains (EL Gabal El Asfar and Beilbes).

9 Recommendations

To reduce the harmful impact of wastewater treatment on groundwater quality in the Nile Valley aquifer, it is strongly recommended that:

1. Continues groundwater quality monitoring to assess any deterioration in water quality for various uses, which decreases the possibility of contamination in aquifers.
2. It is necessary to continue and consider the assessment of water quality by GIS technique since it is instrumental in devising a cost-effective approach for routine groundwater monitoring
3. The Egyptian Government must lay down the Law that forbids the dumping of wastewater and wastewater treatment plants from reaching the study region without treatment.
4. Scientific planning is required to create an efficient groundwater quality management system in Egypt.
5. El Berka WWTP needs to upgrade performance, to add backup treatment units, especially during peak time, to reduce the possibility of pollutants discharging into the drains and subsequently into the groundwater.
6. Managed aquifer recharge (MAR) can be used to enhance the groundwater quality and growing the water resources by the injection of partially treated wastewater in a proper depth in the aquifer and recovery by planned extraction (Dillon 2009; Abiye et al. 2009; Al-Assa'd and Abdulla 2010; Page et al. 2018; El-Rawy et al. 2019a, b; Al-Maktoumi et al. 2020). Also, MAR can be used as a hydrological boundary to manage seawater intrusion in the Nile Delta (Missimer et al. 2012; Masciopinto 2013; Al-Maktoumi et al. 2015, 2016; El-Rawy et al. 2019a, b). MAR provides the cheapest water supply (Zekri et al., 2013). MAR will be the title of the next book on groundwater in MENA regions.

References

Abd MH (2017) Water quality evaluation for river reach by using the remote sensing, AL-Gharraf River Southern Iraq as a case study. Republic of Iraq. Middle Technical University, Master thesis

Abdalla F, Khalil R (2018) Potential effects of groundwater and surface water contamination in an urban area, Qus City, Upper Egypt. J Afr Earth Sc 141:164–178

Abdalla FA, Ahmed A, Omer AA (2009) Degradation of groundwater quality of quaternary aquifer at Qena. Egypt. J Environ Stud 1:18–30

Abdalla F, Shamrukh M (2016) Quantification of river Nile/Quaternary aquifer exchanges via riverbank filtration by hydrochemical and biological indicators, Assiut, Egypt. J Earth Syst Sci 125:1697–1711

Abdel-Dayem S (2011) Water quality management in Egypt. Int J Water Resour Dev 27:181–202

Abdel-Fattah M, Helmy A (2015) Assessment of water quality of wastewaters of Bahr El-Baqar, Bilbies and El-Qalyubia drains in East Delta, Egypt for irrigation purposes. Egypt J Soil Sci 55(3):287–302

Abdel-Lah A, Shamrukh M (2001) Impact of septic system on ground water quality in a Nile valley village, Egypt. In: Proceedings of the sixth international water technology conference. IWTC, Citeseer, pp 237–245

Abdelmawgoud AH, El-Rawy M, Moussa Abu Bakr A (2020) Evaluating the suitability of groundwater quality for drinking and irrigation purposes in El-Minia Governorate, Egypt. J Adv Eng Trends. https://doi.org/10.21608/JAET.2020.25561.1003

Abdel-Shafy HI, Kamel AH (2016) Groundwater in Egypt issue: resources, location, amount, contamination, protection, renewal, future overview. Egyptian J Chem 59:321–362

Abdullah AM, Hussona SE-D (2014) Water quality assessment of mahmoudia canal in Northern West Egypt. J Pollu Effects Control 1–10

Abiye TA, Sulieman H, Ayalew M (2009) Use of treated wastewater for managed aquifer recharge in highly populated urban centers: a case study in Addis Ababa Ethiopia. Environ Geol 58(1):55–59

Abuzied SM, Yuan M, Ibrahim SK, Kaiser MF, Seleem TA (2016) Delineation of groundwater potential zones in Nuweiba area (Egypt) using remote sensing and GIS techniques. Int J Signal Process Syst 4(2):109–117

Ageeb G, Ali R, El Semary M, Essa E (2017) Characterizing groundwater west of nile delta using electric resistivity

Ahmed M (2008) Geochemical and isotopic characteristics of Quaternary aquifer in Sohag District, Upper Egypt. Arab J Nucl Sci Appl 41:129–146

Ahmed AA (2007) Using contamination and pesticide DRASTIC GIS-based models for vulnerability assessment of quaternary aquifer at Sohag, Egypt. In: Proceedings of the The 5th Internernational Conference Geol Africa vol. 1, pp 29–47

Al-Agha DE, Closas A, Molle F (2015) Survey of groundwater use in the central part of the Nile Delta. Water and salt management in the Nile Delta: Report

Alaguraja P, Yuvaraj D, Sekar M, Muthuveerran P, Manivel M (2010) Remote sensing and GIS approach for the water pollution and management in Tiruchirappli Taluk, Tamil Nadu, India. IntJ of Environ Sci 1:66–70

Al-Assa'd TA, Abdulla FA (2010) Artificial groundwater recharge to a semi-arid basin: case study of Mujib aquifer, Jordan. Environ Earth Sci 60(4):845–859

Al-Maktoumi A, El-Rawy M, Zekri S (2016) Management options for a multipurpose coastal aquifer in Oman. Arab J Geosci 9(14):636

Al-Maktoumi A, Zekri S, El-Rawy M, Abdalla O, Al-Abri R, Triki C, Bazargan-Lari MR (2020) Aquifer storage and recovery, and managed aquifer recharge of reclaimed water for management of coastal aquifers. Desalin Water Treat 176:67–77

Al-Maktoumi A, El-Rawy M, Zekri S, Abdalla O (2015) Managed aquifer recharge using treated wastewater to mitigate seawater intrusion along the Jamma coastal aquifer, Oman. In: International congress of the international water association. Presented at the International Congress of the International Water Association, Amman, Jordan

Al-Saboonchi A, Mohamed A-RM, Alobaidy AHMJ, Abid HS, Maulood BK (2011) On the current and restoration conditions of the southern Iraqi marshes: application of the CCME WQI on East Hammar marsh. J Environ Protection 2:316

Antar A, George B, Shabana M et al (2016) Impact of subsurface drainage on salinity management in the Nile Delta, Egypt. In: Proceedings of the 4th African regional conference on irrigation and drainage (ARCID) "agricultural land and water management for sustainability under climate variability" organized by The Egyptian national committee on irrigation and drainage (ENCID) Aswan, Egypt, pp 26–28

APHA (1998) WEF (American Public Health Association, American Water Works Association, and Water Environment Federation). Standard methods for the examination of water and wastewater19

El Arabi NE (1999) Problems of groundwater quality related to the urban environment in Greater Cairo. IAHS Publication, pp 29–38

El Arabi N (2012) Environmental management of groundwater in Egypt via artificial recharge extending the practice to soil aquifer treatment (SAT). Int J Environ Sustain (3)

Asadi S, Raju M, Rani PN, Sekar S (2011) Pollution prevention study using remote sensing and GIS: A model study from visakhapatnam district, Andhra Pradesh. Int J Eng Sci Technol 3:27–35

Ashour.A, Attar.S E, Rafaat.Y, Mohamed.M (2009)Water Resources Management in Egypt. J Eng Sci, Assiut University 37: 269-79.

Attia FA, Allam MN, Amer AW (1986) A hydrologic budget analysis for the Nile Valley in Egypt. Groundwater 24:453–459

Attia F (1989) Use of groundwater for irrigation in the Nile Valley. Int J Water Res Dev 5:91--98

Awad A, Eldeeb H, El-Rawy M (2020) Assessment of surface water and groundwater interaction using field measurements: a case study of Dairut City, Assuit, Egypt. J Eng Sci Technol 15(1):406–425

Baghapour MA, Nasseri S, Djahed B (2013) Evaluation of Shiraz wastewater treatment plant effluent quality for agricultural irrigation by Canadian Water Quality Index (CWQI). Iran J Environ Health Sci Eng 10:27

Balan I, Shivakumar M, Kumar PM (2012) An assessment of groundwater quality using water quality index in Chennai, Tamil Nadu India. Chron Young Sci 3(2):146–150

Balathandayutham K, Valliammai A, Kavitha B, Mayilswami C (2013) Remote sensing and GIS application for groundwater quality mapping in palar sub basin, Tamilnadu, India. Int J Adv Life Sci (IJALS) 6:210–215

Bastawesy M, Shalaby A, Mohamed I (2017) Assessment of the hydrological and environmental problems of waste water disposal sites in aswan of egypt using remote sensing and GIS. Int J Hydro 1:17–23

Bauman B, Schafer W(1985) Estimating ground-water quality impacts from on-site sewage treatment systems

Brion N, Verbanck MA, Bauwens W, Elskens M, Chen M, Servais P (2015) Assessing the impacts of wastewater treatment implementation on the water quality of a small urban river over the past 40 years. Environ Sci Pollu Res 22:12720–12736

Chapman DV, Organization WH (1996) Water quality assessments: a guide to the use of biota, sediments and water in environmental monitoring. Published on behalf of UNESCO/WHO/UNEP, E&FN Spon

Custodio E (2013) Trends in groundwater pollution: loss of groundwater quality and related services. Groundwater Governance: a Global Framework for Action, GEF-World Bank-UNESCO. http://www.groundwatergovernance.org/fileadmin/user_upload/groundwatergovernance/docs/Themat

Das R, Das SN (2003) Impact of wastewater discharge on soil and ground water—a case study. J Sci Industr Res 62:207–211

Dawoud MA, Ismail SS (2013) Saturated and unsaturated River Nile/groundwater aquifer interaction systems in the Nile Valley Egypt. Arab J Geosci 6:2119–2130

Dillon P (2009) Water recycling via managed aquifer recharge in Australia. Bol Geol Min 120(2):121–130

El-Aassar M, Hussien R, Ghoubachi S (2016) Groundwater quality and vulnerability assessment in the new reclamation areas, Assuit governorate, West Nile River, Egypt. J Am Sci 12

El-Arabi N, Rashed M, Vermeulen A (1996) Environmental impacts of sewage water irrigation on groundwater. In: Proceedings of the 16th congress on irrigation and drainage, workshop on managing environmental changes due to irrigation and drainage. Cairo, Egypt, pp 102–114

El-Arabi, Attia (1997) Impacts of sewage—based irrigation on groundwater: an Egyptioncase. In: Towards sustainable irrigation in the mediterranean region, proceedings of international conference on water management, salinity and pollution control. Bari, Italy, Sept 1997, pp 215–228

El-Din MMN (2013) Proposed climate change adaptation strategy for the ministry of water resources and irrigation in Egypt. Joint programme for climate change risk management in Egypt

El-Fakharany Z, Fekry A (2014) Assessment of New Esna barrage impacts on groundwater and proposed measures. Water Sci 28:65–73

Elnashar WY(2014) Groundwater management in Egypt. IOSR J Mech Civil Eng (IOSR-JMCE) e-ISSN, 2278-1684

El-Rawy M, Zlotnik VA, Al-Raggad M, Al-Maktoumi A, Kacimov A, Abdalla O (2016) Conjunctive use of groundwater and surface water resources with aquifer recharge by treated wastewater: evaluation of management scenarios in the Zarqa River Basin, Jordan. Environ Earth Sci 75:1146

El-Rawy M, Al-Maktoumi A, Zekri S, Abdalla O, Al-Abri R (2019a) Hydrological and economic feasibility of mitigating a stressed coastal aquifer using managed aquifer recharge: a case study of Jamma aquifer, Oman. J Arid Land 11(1):148–159

El-Rawy M, Ismail E, Abdalla O (2019b) Assessment of groundwater quality using GIS, hydrogeochemistry, and factor statistical analysis in qena governorate, Egpt. Desalin Water Treat 162:14–29

El-Rawy M, Abdelrahman MAE, Ismail E (2020b) Integrated use of pollution indices and Geomatics to assess soil contamination and identify soil pollution source in El-Minia governorate, Upper Egypt. J Eng Sci Technol 15(4):2223–2238

El-Rawy M, Fathi H, Abdalla F (2020c) Integration of remote sensing data and in situ measurements to monitor the water quality of the Ismailia Canal, Nile Delta, Egypt. Environ Geochem Health 42:2101–2120. https://doi.org/10.1007/s10653-019-00466-5

El-Rawy M, Makhloof AA, Hashem MD, Eltarabily MG (2021a) Groundwater management of quaternary aquifer of the nile valley under different recharge and discharge scenarios: a case study assiut governorate, Egypt. Ain Shams Eng Univ 12(3):2563–2574

El-Rawy M, Moghazy HE, Eltarabily MG (2021b) Impacts of decreasing nile flow on the nile valley aquifer in El-Minia governorate, Egypt. Alexandria Eng J 60(2):2179–2192

El-Rawy M, Abdalla F, El Alfy M (2020a) Water Resources in Egypt. In: Hamimi Z, El-Barkooky A, MartínezFrías J, Fritz H, Abd El-Rahman Y (eds) The geology of egypt; regional geology reviews, Springer Nature, Cham, Switzerland, pp 687–711

Elshemy M, Meon G (2011) Climate change impacts on water quality indices of the southern part of Aswan High Dam Reservoir. Lake Nubia. In: Proceedings of the fifteenth international water technology conference, IWTC-15, 17

El-Zeiny A, El-Kafrawy S (2017) Assessment of water pollution induced by human activities in Burullus Lake using Landsat 8 operational land imager and GIS. Egypt J Remote Sens Space Sci 20:S49–S56

Ensink JH, Van Der Hoek W, Mukhtar M, Tahir Z, Amerasinghe FP (2005) High risk of hookworm infection among wastewater farmers in Pakistan. Trans R Soc Trop Med Hyg 99:809–818

Ezzat N, Elassiouty I, El-Gohary F, Kimm V, Sameh M (1999) Revision of Law 48 of 1982 for the protection of the Nile River and its waterways from pollution In: Report TIWBC (ed) United States agency for international development: international resources group, winrock international, nile consultants. (Contract No. PCE-I-OO-96–00002–00.)

FAO (2016) AQUASTAT country profile—Egypt. http://www.fao.org/3/i9729en/I9729EN.pdf

Fathi H, El-Rawy M (2018) GIS-based evaluation of water quality index for groundwater resources nearby wastewater treatment plants, Egypt. Pollut Res 37:105–116

Fishman MJ, Friedman LC (1989) Methods for determination of inorganic substances in water and fluvial sediments. US Department of the Interior, Geological Survey

Foster S, Chilton J, Nijsten G-J, Richts A (2013) Groundwater—a global focus on the "local ' resource." Curr Opinion Environ Sustain 5:685–695

Gad MIE-KMM, Ismail HM (2015) Groundwater vulnerability assessment using different overlay and index methods for quaternary aquifer of Wadi El-Tumilat, East Delta, Egypt. Asian Rev Environ Earth Sci 2:9–22

Gedamy YR (2015) Hydrochemical characteristics and pollution potential of groundwater in the reclaimed lands at the desert fringes, west of Sohag governorate-Egypt. Curr Sci Int 4:288–312

Gemail KMS (2012) Monitoring of wastewater percolation in unsaturated sandy soil using geoelectrical measurements at Gabal el Asfar farm, northeast Cairo, Egypt. Environ Earth Sci 66:749–761

Habbari K, Tifnouti A, Bitton G, Mandil A (2000) Geohelminthic infections associated with raw wastewater reuse for agricultural purposes in Beni-Mellal, Morocco. Parasitol Inte 48:249–254

Hamdan A, Abdel Rady A (2011) Vulnerability of the groundwater in the Quaternary aquifer at El Shalal-Kema area, Aswan, Egypt. Arab J Geosci 6:337–358. https://doi.org/10.1007/s12517-011-0363-y

Harter T (2003) Groundwater quality and groundwater pollution. University of Califonia, Davis & Kearney Agricultural Center

Hefny K, Farid MS, Hussein M (1992) Groundwater assessment in Egypt. Int J Water Resour Dev 8:126–134

Helmer R, Hespanhol I, Organization WH (1997) Water pollution control: a guide to the use of water quality management principles. CRC Press

Holding Company for Water and Wastewater (2020) treatment plants in Upper Egypt. Avilable on https://www.hcww.com.eg/

Hurley T, Sadiq R, Mazumder A (2012) Adaptation and evaluation of the Canadian council of ministers of the environment water quality index (CCME WQI) for use as an effective tool to characterize drinking source water quality. Water Res 46:3544–3552

INECO (Institutional and Economic Instruments for Sustainable Water Management in the Mediterranean Basin) (2009) Institutional framework and decision making practices for water management in Egypt. International Consultants Egypt

Ismail A, Anderson NL, Rogers JD (2005) Hydrogeophysical investigation at Luxor, southern Egypt. J Environ Eng Geophys 10:35–49

Ismaila E, El-Rawy M (2018) Assessment of groundwater quality in West Sohag. Egypt. Desalin Water Treat 123:101–108

Khan AA, Paterson R, Khan H (2004) Modification and application of the Canadian council of ministers of the environment water quality index (CCME WQI) for the communication of drinking water quality data in Newfoundland and Labrador. Water Qual Res J Can 39:285–293

Kloiber SM, Brezonik PL, Bauer ME (2002) Application of Landsat imagery to regional-scale assessments of lake clarity. Water Res 36:4330–4340

Lawrence FW, Upchurch SB (1982) Identification of recharge areas using geochemical factor analysis. Groundwater 20:680–687

Liu C-W, Lin K-H, Kuo Y-M (2003) Application of factor analysis in the assessment of groundwater quality in a blackfoot disease area in Taiwan. Sci Total Environ 313:77–89

Lumb A, Halliwell D, Sharma T (2006) Application of CCME water quality index to monitor water quality: a case study of the Mackenzie River basin, Canada. Environ Monit Assess 113:411–429

Macalister C, Pavelic P, Tindimugaya C, Ayenew T, Ibrahim ME, Meguid MA (2013) Overview of groundwater in the Nile River Basin. In: The Nile River basin. Routledge, pp 208–233

Masciopinto C (2013) Management of aquifer recharge in Lebanon by removing seawater intrusion from coastal aquifers. J Environ Manage 130:306–312

Masoud AA (2014) Groundwater quality assessment of the shallow aquifers west of the Nile Delta (Egypt) using multivariate statistical and geostatistical techniques. J Afri Earth Sci 95:123–137

Megahed HA, Farrag AE-HA (2019) Groundwater potentiality and evaluation in the Egyptian Nile Valley: case study from Assiut Governorate using hydrochemical, bacteriological approach, and GIS techniques. Bull Nat Res Centre 43:48

Missimer TM, Drewes JE, Amy G et al (2012) Restoration of wadi aquifers by artificial recharge with treated waste water. Ground Water 50(4):514–527

Moneim AAA, Fernández-Álvarez JP, El Ella EMA, Masoud AM (2016) Groundwater management at West El-Minia desert area, Egypt using numerical modeling. J Geosci Environ Prot 4:66

Morsi M, AF, E ES (2018) Assessment of groundwater resources for irrigation purposes in Assiut governorate, upper Egypt. Int J Res-Granthaalayah 6:248–273. https://doi.org/10.5281/zenodo.1248172

Mostafa M, Soussa H (2006) Monitoring of Lake Nasser using remote sensing and GIS techniques. In: ISPRS commission VII Mid-term symposium "Remote sensing: from pixels to processes", Enschede, the Netherlands, pp 8–11

Naithani R, Pande I (2012) Physicochemical assessment of water quality with respect to Remote sensing and GIS techniques. Adv Pure Appl Chem 1:1–6

Nashed A, Sproul A, Leslie G (2014) Water resources and the potential of brackish groundwater extraction in Egypt: a review. J Supply: Res Technol—AQUA 63:399–428

Obeidat MM, Awawdeh M, Al-Mughaid H (2013) Impact of a domestic wastewater treatment plant on groundwater pollution, north Jordan. Rev Mex De Cienc Geológicas 30:371–384

Omran E-SE, El Razek AAA (2012) Mapping and screening risk assessment of heavy metals concentrations in soils of the Bahr El-Baker Region, Egypt. J Soil Sci Environ Manag 3:182–195

Osman R, Ferrari E, Mcdonald S (2016) Water scarcity and irrigation efficiency in Egypt. Water Econ Policy 2:1650009

Page D, Bekele E, Vanderzalm J, Sidhu J (2018) Managed aquifer recharge (MAR) in sustainable urban water management. Water 10(3):239

Patela JY, Vaghanib MV (2015) Correlation study for assessment of water quality and its parameters of par river Valsad, Gujarat, India. IJIERE 2:150–156

Rainwater FH, Thatcher LL(1960) Methods for collection and analysis of water samples. US Government Printing Office

Rashed M, Awad S, Salam M, Smidt E (1995) Monitoring of groundwater in Gabal el Asfar wastewater irrigated area (Greater Cairo). Water Sci Technol 32:163–169

Research Institute for Groundwater (RIGW) (1992) Hydrogeological map of Egypt (Scale 1:500000) Nile Delta, First Edition, 1992

Salim MG (2012) Selection of groundwater sites in Egypt, using geographic information systems, for desalination by solar energy in order to reduce greenhouse gases. J Adv Res 3:11–19

Selim S, Hamdan A, Rady A (2014) Groundwater rising as environmental problem, causes and solutions: case study from aswan city, upper Egypt. Open J. Geol. 4:324–341. https://doi.org/10.4236/ojg.2014.47025

Sharaky A, Atta S, El Hassanein A, Khallaf K (2007) Hydrogeochemistry of groundwater in the Western Nile Delta aquifers, Egypt. In: 2nd International conference on the geology of tethys, vol. 1, pp 1–23

Sharma P, Meher PK, Kumar A, Gautam YP, Mishra KP (2014) Changes in water quality index of ganges river at different locations in Allahabad. Sustain Water Qual Ecol 3:67–76

Shuxia Yu JS, Zhao J, Guo H (2003) Factor analysis and dynamics of water quality of the Songhua river, northeast china. Water Air Soil Pollut 144:159–169

Simsek C, Gunduz O (2007) IWQ index: a GIS-integrated technique to assess irrigation water quality. Environ Monitor Assess 128:277–300

Staudenrausch H, Hochschild V, Helmschrot J et al (2000) Application of remote sensing and GIS for integrated water resources management in Southern Africa. Photogrammetrie Fernerkundung Geoinf 331–342

Taha A, El-Mahmoudi A, El-Haddad I (2004) Pollution sources and related environmental impacts in the new communities southeast Nile Delta, Egypt. Emirates J Eng Res 9:35–49

El Tahlawi M, Farrag A, Ahmed S (2008) Groundwater of Egypt: an environmental overview. Environ Geol 55:639–52

Tiwari AK, Singh PK, Mahato MK (2014) GIS-based evaluation of water quality index of ground water resources in West Bokaro Coalfield, India. Curr World Environ 9:843

Trpevska MT, Dodeva S (2009) Environmental impact assessment of the waste water treatment plant for city of Skopje. In: International symposium on water management and hydraulic engineering Ohrid/Macedonia, pp 1–5

Usali N, Ismail MH (2010) Use of remote sensing and GIS in monitoring water quality. J Sustain Develop 3:228

Wahaab RA, Badawy M (2004) Water quality assessment of the River Nile system: an overview. Biomed Environ Sci 17:87–100

Waly T, Elnaim EA, Omran M, El Nashar B (1987) Effect of sewage water on chemical properties and heavy metals content of El Gabal El Asfar sandy soils. Biol Wastes 22:275–284

Winter TC (1998) Ground water and surface water: a single resource. DIANE Publishing Inc.

Zaporozec A (2002) Groundwater contamination inventory: a methodological guide. HP-VI, series on groundwater, 2. IHP-VI, series on groundwater, 2

Zekri S, Ahmed M, Gaffour N et al (2013) Managed aquifer recharge using quaternary-treated wastewater: an economic perspective. Int J Water Resour Dev 30(2):246–261

Hydrochemistry and Hydrogeology Aspects of Alluvial Aquifer in Aswan City, Egypt

Ali A. A. Obeid, Ahmed Abdelrady, and Mustafa El-Rawy

Abstract This chapter highlights the groundwater characteristics of Aswan aquifer. To assess the water quality, samples were collected from the surface water bodies and groundwater at different locations. The quality of the groundwater was found to be suitable for drinking water according to the World Health Organization regulations. But a noticeable increase in the organic matter concentration was observed. It can deteriorate the quality of the water and limit its uses. To help further studying the system's dynamic, the groundwater system was modelled as well using MODFLOW. The interaction between the Nile River and the aquifer is the primary input and output of the system. Water is entering the aquifer from Aswan Dam Lake in the south and flows towards the north till it exfiltrates to the river again. In addition to the water quality, some areas experienced a rise in groundwater table problems as a result of shutting down some of the extraction wells, which can lead to damage in the infrastructure. Several solutions have been proposed to tackle down both problems. Bank infiltration technique can be useful in assuring sustainable water supply in good quality while coping with the changes in the surface water. However, it is only affecting the water table near the area of the wells and not the whole model domain. As the water table only dropped by 1.5 m after nine years at Kima area.

A. A. A. Obeid
Department of Water Resources and Ecosystems, IHE Delft Institute for Water Education, Westvest 7, 2611 AX Delft, The Netherlands
e-mail: aob001@un-ihe.org; ali.a.obeid@outlook.com

A. Abdelrady (✉)
Department of Water Management, Delft University of Technology, 2600, GA Delft, The Netherlands
e-mail: A.R.A.Mahmoud@tudelft.nl; a.mahmoud@un-ihe.org

Department of Water Science and Engineering, IHE Delft Institute for Water Education, Westvest 7, 2611 AX Delft, The Netherlands

M. El-Rawy
Civil Engineering Department, Faculty of Engineering, Minia University, Minia 61111, Egypt
e-mail: mustafa.elrawy@mu.edu.eg; mustafa.elrawy@su.edu.sa

Civil Engineering Department, College of Engineering, Shaqra University, Ar Riyadh 11911, Saudi Arabia

A. M. Negm and M. El-Rawy (eds.), *Sustainability of Groundwater in the Nile Valley, Egypt*, Earth and Environmental Sciences Library,
https://doi.org/10.1007/978-3-031-12676-5_6

Therefore, an increase in the extraction wells needs to be combined with the bank filtration technique to help to lower the water table and serve good water quality. For example, doubling the current extraction rates can lead to an average drawdown of 5 m. Moreover, the efficiency of the water distribution networks needs to be enhanced to minimize the leakage to the aquifer. Further studies are needed to maintain the stability of the soil during the water table lowering process. A coupled analysis of groundwater models and soil models need to be conducted to estimate the amount of subsidence that might occurs.

Keywords Hydrochemistry · Hydrogeology · Groundwater model · Bank infiltration · Alluvial aquifer · Aswan · Egypt

1 Introduction

People strive in arid countries to grow crops and other water-consuming products. Arid and semi-arid society faces increasingly serious water-management problems (Al-Maktoumi et al. 2016, 2018, 2020; Abd-Elaty et al. 2019; Ouhamdouch et al. 2020). Under these climate conditions, limited water resources for crop production, industrial and domestic use present a critical factor that affects development. Such issues are vital in most arid regions. Additionally, there are extreme and highly variable hydrological conditions in those areas. The management of water resources in such areas is, therefore, a vital issue (Razack et al. 2019; El-Rawy et al. 2019; Basheer and Alezabawy 2020).

Egypt is one of the driest countries in the world (Ragab and Abdelrady 2020). The per capita share of renewable water in Egypt in 2017 was about 570 cubic meters per year, and this value is predicted to decline to 500 cubic meters annually by 2025. Egypt is thus listed among water scarce countries (Luo et al. 2020). Water resources in Egypt are limited by the Nile River, which has a fixed share of 55.5 BCM/year according to the Nile Water Agreement in 1959. Groundwater is considered to be Egypt's second water resource and is a vital source for domestic and other uses, irrigation, especially in desert areas. The abstractions to groundwater account for around 7–8 BCM/year (Salman et al. 2019; El-Rawy et al. 2020, 2021a, b).

Groundwater is one of the primary sources of drinking water supply in many rural areas (Owamah 2020) and in Aswan city as well. The exploitation of groundwater of Aswan aquifer is carried out by Aswan Water and Sanitation Company where there is continuous extraction for industrial and domestic purposes, mostly in Kima and El Shallal areas (Meneisy and AI Deep 2020). The average quantity of groundwater withdrawal from this aquifer is 34560 m^3 per day from (40) pumping wells drilled for this purpose at El Shallal area (12.44 million m^3 per year). At Kima area, the average quantity of groundwater withdrawal from this aquifer for industrial purposes is 36744 m^3 per day (13.227 million m^3 per year) from (38) wells. With a total average quantity of groundwater in the study area is 71304 m3 per day (25.67 million m^3 per year) (Abdelrady et al. 2020).

In the last decade, the region characteristics have been changed. The urbanized processes have been increased. There is no safe wastewater disposal in this region. Therefore the aquifer has been contaminated. From another point of view, the main recharge source of this aquifer is the Nile River. Due to the uncontrolled water withdrawal from this aquifer, the interaction between the river and the aquifer characteristics has been changed. In 2009, Aswan Water Company had been pushed to close these wells. That leads to an increase in the static water level of the aquifer, which threaded the urban areas (Hamdan and Abdelrady 2013). In 2013, the company was obligated to re-operate the wells of the contaminated aquifer to reduce the level of the groundwater and to save access for drinking water for the citizens (Selim et al. 2014).

The rising of the water table is a serious problem that threatens the infrastructure of cities around the world (Sefry and Sen 2006; Lu et al. 2020; Yihdego et al. 2017). In this research, the hydro-chemical and hydrological characteristics of the Aswan aquifer was investigated (i) to examine the quality of the pumped water, (ii) to define the suitability of the abstracted water for drinking and agricultural uses, and (iii) to propose different solutions to decline the groundwater table and avoid the consequences of increasing its level.

The primary recharge sources of the Aswan aquifer are the River Nile, septic tanks, fish ponds as well as the agricultural disposal. The water quality characteristics have been changed rapidly in the last years. Therefore, the main objective of this research was to study the hydrological characteristics of the aquifer and the hydrochemical characteristics as well as to define the suitability of the water for drinking and agricultural uses.

2 Study Area

Aswan Town is located in the southern part of Egypt, and it occupies an area of about 77 km^2. The available water resources for the town are mainly derived from the River Nile. The southern part of the town comprises an industrial area. The groundwater is mainly for the Egyptian chemical industrial uses, and partially for domestic use. Figure 1 illustrates the study area location.

The study area is located in the arid region and characterizes by less rainfall and high temperature. Almost there are no rainfall events within the area, except some of the thunderstorms which happen occasionally. Temperature records show that January is usually the coldest month, generally from November to April; the climate is delightful with cold nights and clear, bright days (8–30°C). High-temperature records prevail from May till October (24–41°C) (Elramly 1973; Said 1962).

In terms of surface water, the River Nile is the only available source of water within the area. Along the stretch of the Nile in the study area, the Aswan dam is located. The lake behind the dam is the primary source of groundwater recharge. On other locations upstream the dam the aquifer is flowing to the river, which shows

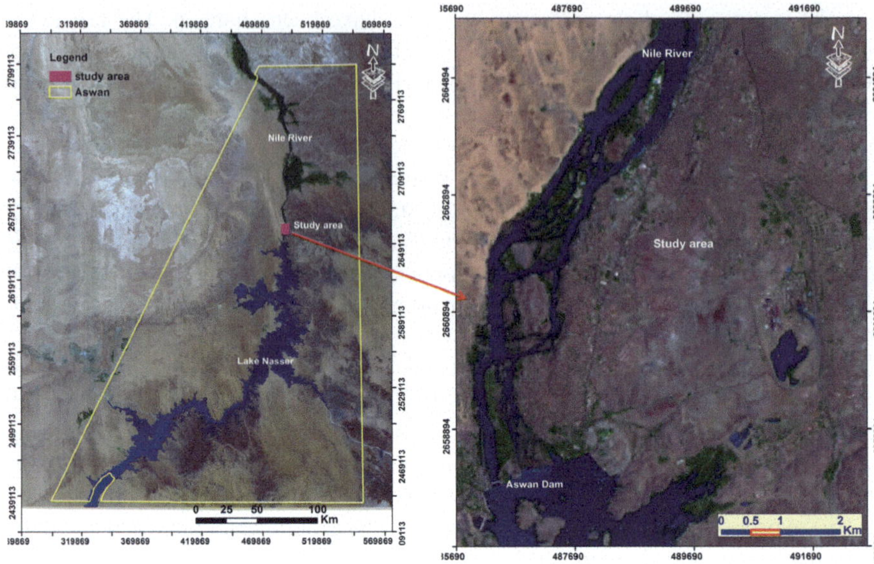

Fig. 1 Location of the study area in Aswan Governorate

the dynamics of the interaction between the river Nile and Aswan aquifer. The water level is + 85 m and + 106 m in the river and in the reservoir respectively.

3 Data Collection and Methodology

3.1 Geomorphology

Geomorphologically, the Aswan Town area can be divided into three main units: Aswan plain, highland areas, and the River Nile channel (Hamdan and Abdelrady 2013). Aswan plain was the older river Nile channel and currently occupied by the Aswan town. The old channel is now followed by two main roads Kima and El Khazan. The width of the plain is range from 3.5 km at the center of the two and narrowing to less than 1 km along EL Khazan road. The plain is almost flat with a general slope from South to North (Hamdan 2005). There are three highlands within the study area. The largest one is surrounding the town from the Eastern side. It is rugged and mountainous and is characterized by escarpment and cliffs. There are few wadis and gorges that incise this area in the east–west direction, e.g. W.AbuAgag, W.ElKeimab, and W.UmBuweirat. The other two highlands are smaller and occupy the central and the western parts of the study area. The central highland extends from GabalTagug, northwards to the Karor southwards, and has an abrupt slope in the direction of the River Nile (Hamdan and Abdelrady 2013). The River Nile Course forms the western limit of Aswan town. Aswan Dam (the old dam) with its water reservoir is situated to the South of Aswan Town.

3.2 Geology

The generalized geologic formations in the study area as shown in the studies of (Barber and Carr 1981), (Water Master Plan 1981) and (Elramly 1973), and as follows:

- Quaternary: Alluvial and Aeolian sediments.
- Cambrian – Cretaceous: Nubian sandstone formations.
- Precambrian: basement rocks

The Precambrian rocks are mainly exposed in the southern part of the study area and represent the oldest rocks. These rocks are crystalline, and constitute a complex of igneous and metamorphic rocks (mainly granites and schist).

The Nubian Sandstone unconformably overlay the basement rocks and has an age ranging from the Cambrian to the cretaceous (Issawi and Jux 1982). They are well exposed in the northern part of the study area. Attia (1954) divided the so-called Nubian sandstone formation, in Aswan into three groups. The lower group is characterized by the basal conglomerates, while the middle group by the presence of an Oolitic iron ore layer, and the upper group by the occurrence of variegated clays and has a total thickness ranges between 20 and 85 m.

The Quaternary sediments of the Aswan aquifer is mainly composed of mud and Aeolian sediments of the recent deposits and sands, gravels, and clays of the Pleistocene time and underlain by a thick bed of clays from the Pliocene epoch. These sediments are well exposed in the area through which Kima and El Khazan roads run, and this proves that the old River Nile course followed the roads of the present time (RIGW 1988).

The Quaternary gravels are mainly consisted of coarse well-rounded pebbles with a diameter varied between 15 to 20 cm. The composition and the high degree of rounding of the pebbles show that these gravels have originated from the igneous masses, and have been shaped by water action. They doubtless remain of the Nile branch which previously ran through the valley. A small portion of the surface is encompassed by recent deposits and it is mainly composed of alluvial and Aeolian deposits. The alluvial deposits, sand, and mud from the cultivated land in the northern part of the study area. The Aeolian deposits are represented by blown-sand accumulations which originated from the disintegration of the Nubian sandstone and the basement rocks (Hamdan 2005).

Thirty six wells have tapped the quaternary sediments. The lithological composition of these sediments has been investigated in the available wells, where the following points of interest have been noticed.

The Quaternary sediments are mainly represented by unconsolidated materials of sands and gravels materials of sands and gravels with a few lenses of clay. Gravels dominate in the area north of El Shallal and are present mainly as lenses, with thickness ranges between 11 and 64 m. The thickness of these sediments is not uniform and ranges between 82 and 114 m. But it shows a general increase from the south to the north (Hamdan 2005; Hamdan and Abdelrady 2013). The contact between the quaternary sediments and the underlying rocks is not clear, but in the majority of the drilled wells; the quaternary sediments are underlain by a clay layer that may belong to the Pliocene clays or the upper group of the Nubian Sandstone formation.

3.3 Hydrogeology

The quaternary sediment layer, which composes of unconsolidated deposits of gravel, sand, and clay, is defined as the aquifer. It is unconfined from the top and isolated with a layer of clay from the bottom. The thickness of the aquifer ranges between 82 and 114 m with a general increase from the south to the north. The width of the aquifer is generally small, from 2 km in Kima area to 3.5 km in the town center area (Basheer and Alezabawy 2020; Hamdan and Abdelrady 2013).

Pumping tests have been carried to determine the hydrogeological characteristics of the aquifer. From these tests, it was found that the hydraulic conductivity ranges from 0.0001 to 0.0004 m/s. Also, the values of the storage coefficient found to be lies within 0.0002 and 0.0006. Using Darcy's law, the safe yield of the aquifer was estimated to be between 18,482 and 55,645 m^3/day. In general, this quaternary aquifer is capable of Supplying high yields and sustaining considerable water supply (Basheer and Alezabawy 2020).

The exploitation of the aquifer started in 1960, mainly with wells drilled by percussion machines. Overall, there are 25 wells in use, mostly in Kima and El Shallal areas; where there is continuous withdrawal for industrial and domestic uses (Hamdan and Abdelrady 2013).

Groundwater level data was collected from these wells for consecutive five years (1986–1991). A contour map was established to illustrate the distribution of the groundwater heads within the study. The water levels generally follow the topography of the area, and ranges from 100 m southward at El Shallal area to + 80 m northward, north of town center. Also, the direction of the groundwater movement can be observed; the water flows from the south to the north. The primary source of recharge in the aquifer is the water leakage from the Aswan dam reservoir and the Nile (Abdelrady et al. 2020; Selim et al. 2014).

3.4 Water Chemistry

Water samples were collected from the surface water systems and two groundwater sites to determine the quality characteristics of the raw and pumped waters. The first groundwater field (Shallal field) located at the south of the study area at a latitude of 24°02′22″ N and longitude of 32°54′32″ E and a distance of 1.2 km of the Aswan Dam Lake. The second field (Kima field) located at 24°03′22″ N and longitude of 32°54′43″ E. The samples were collected at a monthly basis at the period (from January 2013 to December 2017).

The portable instruments (HACH, USA) were used to determine the physical characteristics of the pumped water, including (pH, temperature, turbidity, and electric conductivity) in situ. The analytical measurements were undertaken at the laboratories of the Egyptian Holding Company for water and wastewater. The physical properties of the groundwater and infiltrated water were measured in the field. The pH varied between 7.51 and 8.28 in the pumped water. A slight increase in the electric conductivity was observed at Shallal field compared to the surface water system. The electric conductivity ranged between 220–240 μS/cm for surface water, 340–410 μS/cm for Shallal area, and 430–560 μS/cm Kima area, respectively. Consequently, the total dissolved solids concentrations varied between 147–161, 228–275, 288–375 mg/L for surface water, Shallal wells, and Kima wells, respectively, which refers to the meteorological origin of the pumped water.

The major cations (Ca, Mg, Na, and K), anions (Cl, SO_4, CO_3, and HCO_3) were determined following the methods described by Baird et al. (2017). The results revealed that the pumped water has a slight hardness character; the total hardness ranged between 32–281 mg/L at Shallal region. This value increases slightly in the northern region of the aquifer with the increase of the concentrations of calcium and magnesium carbonates (Abdelrady et al. 2020). The nitrogen species (ammonia and nitrate) and phosphorus were measured spectrophotometrically (APHA 2012). The analytical data clarified that the concentrations of ammonia, nitrate, and phosphate in the pumped water are much higher than their concentrations in the surface water systems. The average concentrations of ammonia, nitrate and phosphate of the surface water systems were < 0.2, 2.5 ± 0.5 and 0.05 ± 0.02 mg/L, these values were increased to 0.6 ± 0.2, 2.8 ± 0.3 and 0.1 ± 0.02 mg/L in the pumped water of Shallal wells, and to 0.8 ± 0.2, 4.3 ± 0.8 and 0.33 ± 0.1 mg/L in the pumped water of Kima wells, respectively. This increase in the concentrations of these nutrients elements is mainly attributed to the inflow of contaminated water from the agricultural areas and septic tanks in the area toward the ambient groundwater (Buskirk et al. 2020).

The inductive coupled plasma ICP-OES (Optima 8300 from Perkin Elmer Company, USA) was used to quantify the concentrations of metals in the pumped water following the method described in APHA (2012). The results demonstrated that the concentrations of most of these toxic metals in the pumped groundwater are lower than the detection limits of the analytical instruments. However, an increase in the concentrations of Fe, Mn, and Al was observed that might ascribe to the dissolution of these metals from the soil into the groundwater during the lateral groundwater flow process (Yang et al. 2020). The concentrations of Fe, Mn, and Al reached to 33 ± 11, 24 ± 8 and 980 ± 120 μg/L in the pumped water of Shallal wells and 380 ± 40, 470 ± 70, and 580 ± 20 μg/L in the pumped water of Kima wells, respectively.

The organic content of the samples was determined using a total organic carbon (TOC) analyzer (TOC-VCPN (TN), Shimadzu, Japan). To determine the aromaticity of the pumped water, the absorbance at the ultraviolet range UV_{254} (wavelength $\lambda = 254$ nm) was determined using a UV/Vis spectrophotometer (UV-2501PC Shimadzu). Then the specific ultraviolet absorbance $SUVA_{254}$ was calculated by dividing the UV_{254} absorbance [m] by the concentration of organic matter [mg/L], and used as an indicator to the aromaticity of the water (Abdelrady et al. 2018). The analytical data illustrated that there is an increase in the concentration of organic matter in the pumped water compared to the surface water systems. The concentration of organic matter ranged between 2.7–4.5 mg/L in the surface water systems and increased to 4.2–4.9 mg/L at Shallal area and 5.1–7.3 mg/L at Kima area. The $SUVA_{254}$ data revealed that the groundwater has higher aromaticity characteristics than surface water systems, which mainly due to the release of organic matter (specifically humic compounds) from the soil into the groundwater during the groundwater flow.

These results pointed out the suitability of the groundwater for drinking uses; all the measured chemical parameters are within limits proposed by the (WHO 2011). However, the increase of organic matter could increase the potentiality of forming trihalomethanes carcinogenic compounds during the treatment processes, and therefore post-treatment process to eliminate such compounds might be needed (Kumari and Gupta 2020). In the same regard, the suitability of groundwater for irrigation uses was investigated by (Hamdan and Abdelrady 2013), and it was reported that the groundwater has low values of salinity and sodium adsorption ratio. Thus it is suitable for irrigation of all soil types.

4 Numerical Groundwater Model

4.1 Model Setup

The groundwater system within the study area has been modelled, calibrated and validated by Abdelrady et al. (2020) using MODFLOW. The model extent falls within the longitude of 32°53′ and 32°56′ E and latitude of 24°01′30″ and 24°04′30″N, with a length of 9.1 km and a width of 7 km. The grid has a spatial discretization of 100×100 m^2 in the horizontal direction. This resulted in 91 rows and 70 columns. While in the vertical direction the only one model layer was used with varying thickness. The top elevation of the model was taken from the Digital Elevation Model (DEM), while the bottom of the model was calculated by subtracting the thickness of the model gained from the geological survey of the area. To include the changes in hydrological elements, such as the river water level, the model was run for a transient state for ten years. The temporal discretization adopted has stress periods of 1-day length in the first 5 years and 10 days in the rest of the years.

The model was bounded with a no-flow boundary to the Northern, Eastern, and portion of the Southern parts of the model. The Eastern and Northern parts were considered and a no-flow due to the basement rock, there was no lateral flow in the

Northern side, and mostly the water was flowing to the river. While to the south, there are two features, the presence of basement rocks and the River Nile Boundary. The western side was bounded with the Nile River and the Aswan Dam Lake. The bottom of the model was underlaid by a basement rock and considered as a no-flow boundary. However, the top of the model was under the influence of the water table as it is an unconfined aquifer. The boundaries and the layout of the model can be seen in Fig. 2.

The hydraulic conductivities and storage coefficient values were obtained from the calibration process using the given ranges from the pumping tests, as indicated

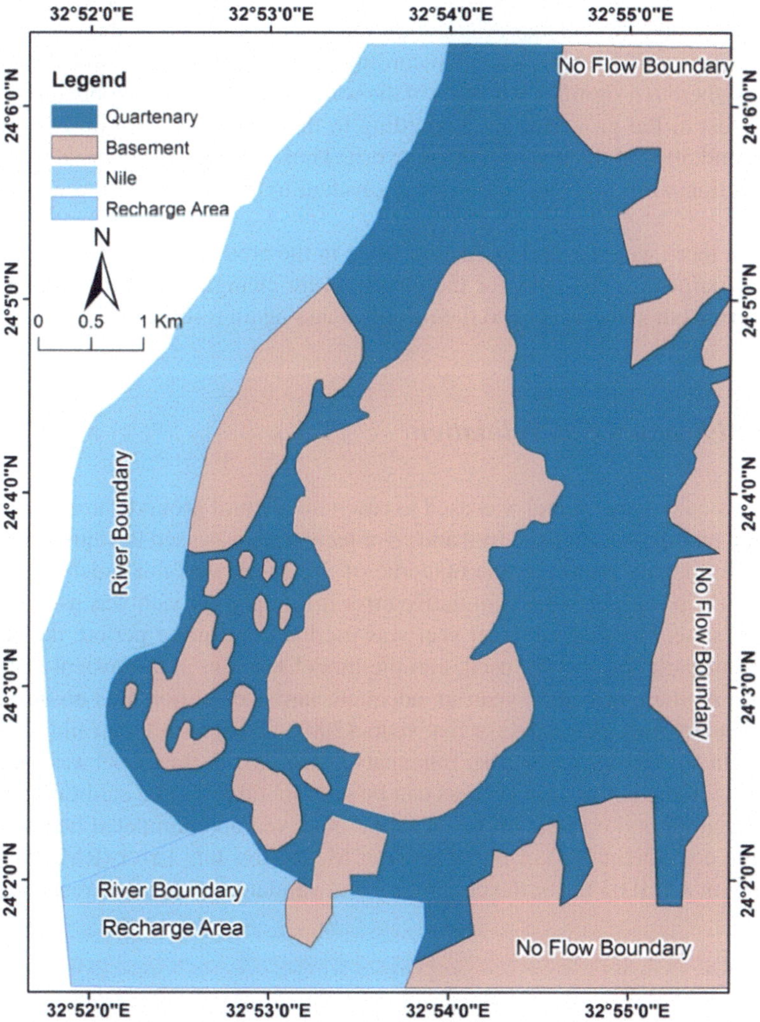

Fig. 2 Boundary condition and the layout of the model (Abdelrady et al. 2020)

earlier. The interaction between the Nile River and the Aswan Dam Lake on one side and the aquifer on the other side was simulated using the river package in the model. The values for the river stage and bottom were obtained from the Egyptian Ministry of Water Resources and Irrigation (MWRI). However, the river conductivity values were derived from the pumping tests. The data for the river stage was on a daily base. The river and the lake were the main sources of recharge to the aquifer as in this area the rainfall is scarce. The rates of evapotranspiration from the soils and the evaporation from the surface water bodies were obtained from the European Centre for Medium-Range Weather Forecasts online source and then averaged over a monthly rate. However, the evapotranspiration has no significance in our case as the water table is not shallow enough, and the soil is mostly dry. Sensitivity analysis further confirmed this for the evaporation rates, and it had no acute effects on the values of the groundwater system. According to the Aswan Water and Wastewater Company, there is a significant portion of the water from the water supply distribution network lost to the groundwater. According to their data, this portion is estimated to be around 50% of the water in the network. Thus, to incorporate these losses, we used injection wells with injection rate equivalent to the portion of the water being lost.

At last, there were several extraction wells in the area used for water supply and irrigation purposes. The rates for these wells were changing, as some of the wells were operated in some years and then stopped and again re-operated.

4.2 Calibration and Validation

At first, a steady-state model was used to reach the natural groundwater distribution before the pumping starts. The trial and error technique was used to adjust the hydrological parameters. However, the outcome of this mode was not satisfying, and to eliminate the possibility of a cumulative error different approach was used to reach the initial state. One hydrological year was used as a warming period, the data for all different parameters were used, and the model was run in a transient state with pumping. At the end of this year, an adequate initial condition was obtained, and then the model was calibrated for five years (2009–2013) with the available hydrological data set. Finally, for another four years (2014–2017), the model was validated. The results of the calibration process can be seen in Table 1. The calibration results indicate that a good correlation between the observed and simulated heads, with a coloration coefficient (R^2) of 0.9. The Root Mean Absolute Error (RMAE) values were 0.49 m and 0.65 m for the calibration and validation processes, respectively.

Table 1 Statistics of calibration and validation processes

	Calibration		Validation	
	Field data	Model data	Field data	Model data
Average	105.27	105.14	106.42	105.92
Median	105.92	105.72	106.41	105.68
Max	108.30	109.13	110.14	109.89
Min	100.10	100.31	104.45	104.56
RMSE	0.49		0.65	

4.3 Groundwater Head and Water Budget

The distribution of groundwater head can be found in Fig. 3 for the period 2008–2017. Generally, the water flows from south to north in parallel to the Nile flow direction. The water infiltrates to the aquifer at the lake and flows to the north, where it exfiltrates to the river again as this interaction represents the essential components in the aquifer water balance. The other components in the water system are the extractions and the injection wells. The water level has increased by 1 – 3 m between 2009 and 2013 as a result of shutting down some of the extraction wells. Then after 2013, some of these wells again started to re-operate, and this has decreased the water level by 0.5 m at the end of the simulation period. As a result of the water table rising, a lake was started to appear in the area of Kima.

The main cause of groundwater table rise as mentioned earlier is the reduction of the extraction rates as shown in Fig. 4. After 2008, many wells have been stopped, and the reduction in extraction rates was very high. This has influenced the soil

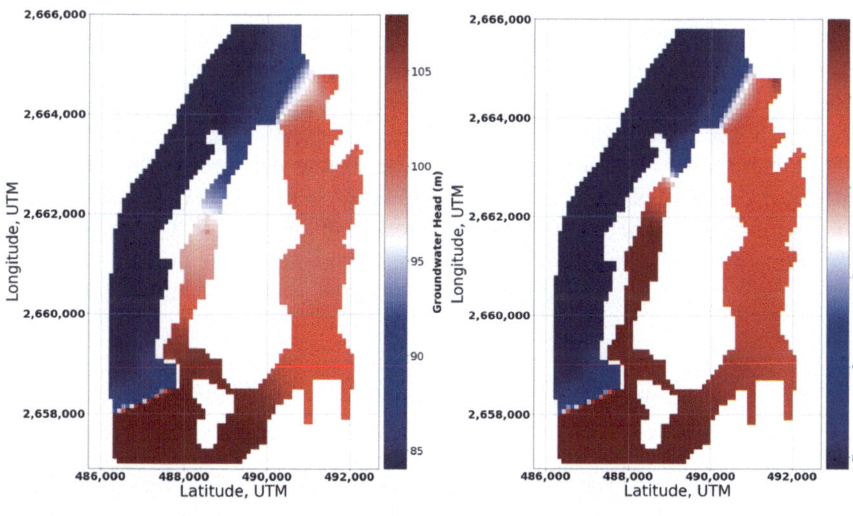

Fig. 3 Groundwater head contour map for the years **a** 2008 **b** 2013

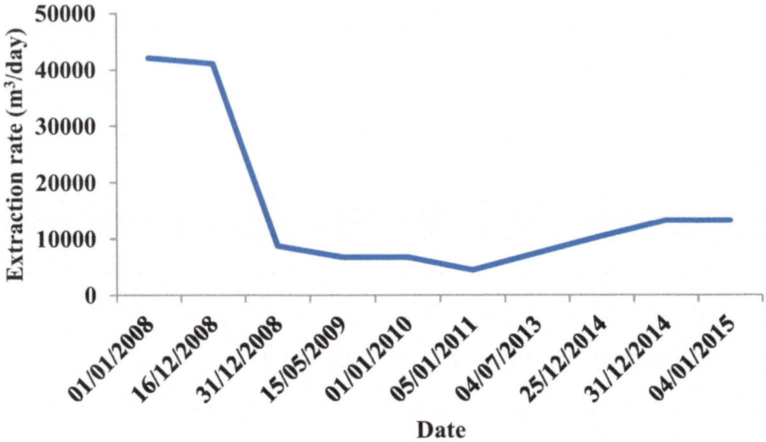

Fig. 4 The change in extraction rates with time (Abdelrady et al. 2020)

Table 2 Water budget components (m^3) for the study area

	Jun-08		Dec-08		Dec-13	
	In	Out	In	Out	In	Out
Nile River	573,528.6	562,213.3	622,873.6	627,249.8	596,426.3	600,406.7
Wells	5900	36,860	5900	35,860	5900	4520
ET	0	466.4	0	174.1	0	188.7
Total	579,428.6	599,539.7	628,773.6	663,283.9	602,326.3	605,115.4
Change in storage	−20,217.5		−34,501.6		−2791.56	

property; the higher extraction rates resulted in relatively higher conductivity than the surroundings. After the shutdown, the water started to accumulate in these areas and formed the lake.

Looking at the water balance in Table 2, we can see that between the summer and the winter there is no much difference. The river water level is changing, and this affects the amount of water exchanged. However, between 2008 and 2013, there are many differences. The water extracted is less than the water lost from the water supply network in 2013. In addition to that, the change in storage in 2008 is higher, and this means more water is leaving the storage compared to the other year.

5 Groundwater Table Rise Solutions

The infrastructure is severely affected by the increase in water level. Building's basements became full of water. This water has low quality as well, which has further effects on buildings sustainability. Therefore, several scenarios will be investigated to see how to improve both water quality and groundwater rise problems. These solutions can include but not limited, the use of bank infiltration techniques for water supply, enhance the efficiency of the water supply network to minimize the water leakage to the aquifer, and extract more water from the aquifer by the existing wells. Pumped bank filtrate of decent quality can be utilized to support the region's drinking water network, whilst groundwater wells can be used for agricultural and other domestic purposes.

5.1 Bank Infiltration

The benefits of using bank infiltration techniques can be summarized in delivering good water quilty throughout the year (Patenaude et al. 2020). The influence of any changes in the water level in the Nile River is not significant. Abdelrady et al. (2020) has studied two proposed sites for bank infiltration by applying different scenarios. Based on the study, using 10 to 15 wells with a capacity of 35,000 m^3/day near the Aswan Dam Lake can lead to acceptable results in terms of travel time, drawdown, and bank infiltration share percentage. Therefore, 10 wells with extraction rates of 3500 m^3/day each are used here.

This solution has shown its functionality in providing sustainable water quality for drinking water supply, but on the other hand, the groundwater head has only decreased by 0.5 m (on average) after the first year as in Fig. 5. The significant drawdown happened only at the location of the wells near the river. After 9 years the water has only reduced by 1.5 m (on average) which still not enough.

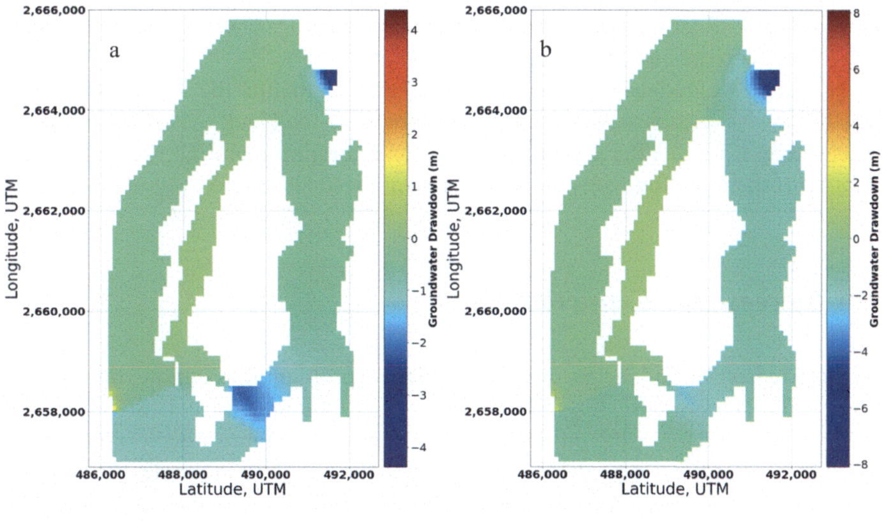

Fig. 5 Groundwater drawdowns as a result of bank infiltration after **a** 1 year **b** 9 years

Table 3 Water budget (m^3) for the study area in 2017, 2018 and 2022

	Dec-17		Dec-18		Dec-27	
	In	Out	In	Out	In	Out
Nile River	637,191	622,726	738,930	743,181	739,956	718,896
Wells	5300	13,320	5300	49,620	5300	49,620
ET	0	189.36	0	195.3	0	192.58
Total	642,491	636,235	744,230	792,996	745,256	768,708
Change in storage	6256.14		−48,696		−23,428	

This can also clearly seen by the water balance of the aquifer. In Table 3 we can see that as a result of the bank infiltration wells more water is coming from the river after the first year, at the same time the change of storage is negatively increased which means more water is now taken from the storage but this only at the areas near the Nile River. Also after 9 years, in 2027, the change in storage is reduced, and more water is entering the system from the river, as the share of the bank infiltration system is increasing and the portion of the aquifer storage is decreasing. Thus, the reduction in water level at Kima area will be small, in 9 years only reduced by 1.5 m. The system is going towards an equilibrium state as the water extracted by the wells will be mostly covered by the river.

5.2 *Reduce the Water Losses from the Water Supply Network*

Almost half of the water in the water supply network is lost to the groundwater; this amount of water is estimated to be around 5300 m^3/day. However, eliminating water losses will not affect the water table significantly, see Fig. 6 as the water will enter from the river to the aquifer and restore the balance.

The water level reduced on average by 0.05 and 0.4 m after the 1 year and 9 years respectively. Nevertheless, at the area highlighted by the black oval the water table decreased by 1.5–2 m after 1 year and by 3–5 m after nine years. This is because the majority of the injection wells used to represent these losses are located there.

5.3 *Increase the Extraction Rates*

The main reason behind the groundwater table rise problem was shutting down some of the wells. Therefore, in this scenario, the extraction rate of the existing extraction wells was doubled to see the influence in the water table. Figure 7 depicts the effects on water levels, after 1 year the water table reduced by 1.5–2 m, while after 9 years it reduced by 5 m (blue color) at the desired location. This method is very useful in lowering the water table. Therefore, either increasing the extraction rates of the

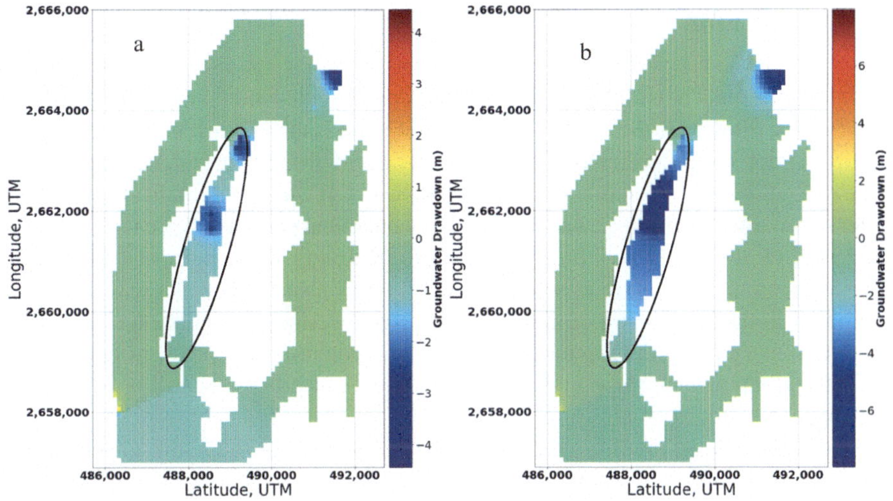

Fig. 6 Groundwater drawdown as a result of eliminating water losses from the water supply network after **a** 1 year **b** 9 years

existing wells, bringing back to service some of the other wells, or a combination of both is necessary to solve the issue. However, the desired drawdown magnitude, based on the soil properties, will govern the increase in pumping rates.

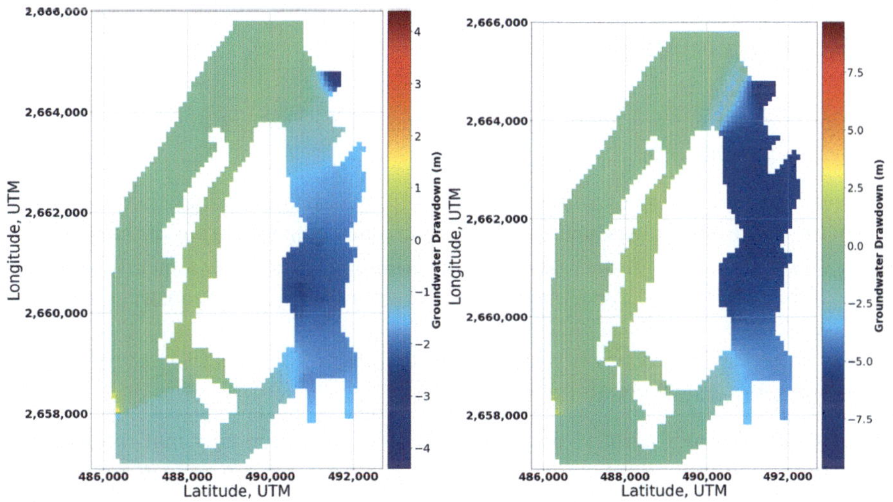

Fig. 7 Groundwater drawdown as a result of doubling the extraction rates after **a** 1 year, **b** 9 years

6 Conclusions

Aswan aquifer has outstanding potential and can provide water supply for different water users. However, the quality of the water is an issue. To investigate the water quality, samples were collected from the surface and groundwater systems to determine the quality characteristics of the aquifer's water. It was apparent that the pumped water is suitable for drinking water uses; all the chemical parameters are within limits suggested by the world health organization for drinking water purposes. However, an increase in the organic matter concentration of the abstracted water was observed compared to the surface water systems that might ascribe to the dissolution of soil organic matter during the groundwater flow. The potential risk of forming carcinogenic trihalomethane compounds during the treatment process is still needed to be investigated.

Besides the water quality problems, there is a problem with the groundwater table rise in the area of Kima industry. The groundwater system was modelled using MODFLOW to help to interpret the system dynamics. Bank infiltration technique can be used to provide sustainable water supply with outstanding quality. It can also help in lowering the water table by small magnitude. Shutting down some of the extraction wells led to a rise in the water table, so by re-operating these wells or increasing the current pumping rates, the water level will be decreased. In addition to that, the water supply authorities need to increase water distribution efficiency to minimize water losses.

7 Recommendations

In order to avoid the degradation of the groundwater quality in the study area, it is highly recommended:

- Continues monitoring of groundwater quality to evaluate any degradation of water quality for different uses and to reduce the risk of contamination of the aquifer.
- As reported that the primary source of the pollution in the study area is the outflow from the septic tanks; consequently, it is a required remedy to construct a sewage network.
- The chlorination process is necessary to disinfect the groundwater at the tanks before being distributed to the buildings.

The rise in the water table is one of the most serious problems that threaten the infrastructure of the city of Aswan. To overcome this problem, it is recommended to:

- Continuous operation of existing groundwater wells with a minimum capacity of 1440 cubic meters per hour
- A number of wells must be drilled and operated for irrigation purposes

- The loss of drinking water supply networks which feed the underground reservoir must be minimized.
- Natural treatment methods, such as river bank filtration, could be used to produce high quality drinking water and reduce the level of groundwater.
- Moreover, further studies are necessary to be conducted by coupling the soil subsidence model with the groundwater model to regulate the proposed solutions and make sure that the water table's lowering has no significant effects on the soil and will not lead to future instabilities of the infrastructure.

References

Abd-Elaty I, Sallam GA, Straface S, Scozzari A (2019) Effects of climate change on the design of subsurface drainage systems in coastal aquifers in arid/semi-arid regions: case study of the Nile delta. Sci Total Environ 672:283–295

Abdelrady A, Sharma S, Sefelnasr A, Kennedy M (2018) The fate of dissolved organic matter (DOM) during bank filtration under different environmental conditions: batch and column studies. Water 10:1730

Abdelrady A, Sharma S, Sefelnasr A, El-Rawy M, Kennedy M (2020) Analysis of the performance of bank filtration for water supply in arid climates: case study in Egypt. Water 12:1816

Al-Maktoumi A, El-Rawy M, Zekri S (2016) Management options for a multipurpose coastal aquifer in Oman. Arab J of Geosci 9:1–14

Al-Maktoumi A, Zekri S, El-Rawy M, Abdalla O, Al-Wardy M, Al-Rawas G, Charabi Y (2018) Assessment of the impact of climate change on coastal aquifers in Oman. Arab J Geosci 11:1–14

Al-Maktoumi A, Zekri S, El-Rawy M, Abdalla O, Al-Abri R, Triki C, Bazargan-Lari M R (2020) Aquifer storage and recovery, and managed aquifer recharge of reclaimed water for management of coastal aquifers. Desalination Water Treat 176:67–77

Al-Sefry SA, Şen Z (2006) Groundwater rise problem and risk evaluation in major cities of arid lands–Jedddah case in kingdom of Saudi Arabia. Water Resour Manage 20:91–108

APHA (2012) Standard methods for the examination of water and wastewater. In: Water Environment F (ed) American Public Health A, American Water Works A. American public health association, Washington

Attia MI (1954) Deposits in the Nile Valley and the delta geological survey. Egypt, Cairo

Baird RB, Eaton AD, Rice EW, Bridgewater L (2017) Standard methods for the examination of water and wastewater. American public health association, Washington, DC

Barber W, Carr D (1981) Water management capabilities of the alluvial aquifer system of the Nile Valley upper Egypt. Technical report

Basheer AA, Alezabawy AK (2020) Geophysical and hydrogeochemical investigations of Nubian sandstone aquifer, South East Sinai, Egypt: Evaluation of groundwater distribution and quality in arid region. J Afr Earth Sc 169:103862

Buskirk RE, Malzone JM, Borowski WS, Cornelison J (2020) The impact of small-scale land cover and groundwater interactions on base flow solute and nutrient export in a small agricultural stream. Environ Monit Assess 192:1–23

Elramly I (1973) Final report on geomorphology, hydrogeology, planning for groundwater resources and land reclamation in Lake Nasser region and its environs

El-Rawy M, Al-Maktoumi A, Zekri S, Abdalla O, Al-Abri R (2019) Hydrological and economic feasibility of mitigating a stressed coastal aquifer using managed aquifer recharge: a case study of Jamma aquifer, Oman. J Arid Land 11(1):148–159

El-Rawy M, Abdalla F, El Alfy M (2020) Water resources in Egypt. In: Hamimi Z, El-Barkooky A, Martínez Frías J, Fritz H, Abd El-Rahman Y (eds) The geology of Egypt, 687–711

El-Rawy M, Moghazy HE, Eltarabily MG (2021a) Impacts of decreasing Nile flow on the Nile Valley aquifer in El-Minia Governorate, Egypt. Alexandria Eng J 60:2179–2192

El-Rawy M, Makhloof AA, Hashem M D, Eltarabily M G (2021b). Groundwater management of quaternary aquifer of the nile valley under different recharge and discharge scenarios: a case study Assiut Governorate, Egypt. Ain Shams Engineering University (accepted for publication)

Hamdan A (2005) Evaluation of the Nile valley aquifer from Aswan to Assiut using geophysical and hydrogeological methods. South Valley University, Ph.D

Hamdan A, Abdelrady A (2013) Vulnerability of the groundwater in the quaternary aquifer at El Shalal-Kema area, Aswan Egypt. Arab J Geosci 6:337–358

Issawi B, Jux U (1982) Contributions to the stratigraphy of the Paleozoic rocks in Egypt. Geol Surv Egypt, Cairo

Kumari M, Gupta SK (2020) Water quality assessment, statistical analysis and kinetics of trihalomethanes formation in drinking water supplies-a complete batch study. Environ Eng Manag J (EEMJ) 19

Lu D, Li X, Du X, Lin Q, Gong Q (2020) Numerical simulation and analysis on the mechanical responses of the urban existing subway tunnel during the rising groundwater. Tunn Undergr Space Technol 98:103297

Luo P, Sun Y, Wang S, Wang S, Lyu J, Zhou M, Nakagami K, Takara K, Nover D (2020) Historical assessment and future sustainability challenges of Egyptian water resources management. J Cleaner Prod 263:121154

Meneisy AM, Al Deep M (2020) Investigation of groundwater potential using magnetic and satellite image data at Wadi El Amal Aswan Egypt. Egypt J Remote Sens Space Sci. https://doi.org/10.1016/j.ejrs.2020.06.006

Ouhamdouch S, Bahir M, Ouazar D (2020) Climate change impact assessment on a carbonate aquifer under semi-arid climate: example of the Cenomanian-Turonian aquifer within Meskala-Ouazzi region (Essaouira Basin, Morocco). Arab J Geosci 13:173

Owamah H (2020) A comprehensive assessment of groundwater quality for drinking purpose in a Nigerian rural Niger delta community. Groundw Sustain Dev 10:100286

Patenaude M, Baudron P, Labelle L, Masse-Dufresne J (2020) Evaluating bank-filtration occurrence in the province of Quebec (Canada) with a GIS Approach. Water 12:662

Ragab A, Abdelrady A (2020) Impact of green roofs on energy demand for cooling in egyptian buildings. Sustainability 12:5729

Razack M, Jalludin M, Houmed-Gaba A (2019) Simulation of climate change impact on a coastal aquifer under arid climate. The Tadjourah Aquifer (Republic of Djibouti, Horn of Africa). Water 11:2347

RIGW (1988) Hydrogeological map of Egypt, 1st edn. Research institute for groundwater, Giza

Said R (1962) The geology of Egypt. Elsevier, Amsterdam, p 377

Salman SA, Arauzo M, Elnazer AA (2019) Groundwater quality and vulnerability assessment in west Luxor Governorate Egypt. Groundwater Sustain Dev 8:271–280. https://doi.org/10.1016/j.gsd.2018.11.009

Selim SA, Hamdan AM, Abdelrady A (2014) Groundwater rising as environmental problem, causes and solutions: case study from Aswan city, Upper Egypt. Open J Geol 4:324–341. https://doi.org/10.4236/ojg.2014.47025

Water Master Plan (1981) Groundwater technical report 4 UNDP/EGY/73/024. Ministry of Public Works and Water Resources, Cairo

WHO (2011) Guidelines for drinking-water quality. World Health Organization, Geneva, pp 303–304

Yang Q, Yin H, He X, Chen F, Ali A, Mehta Y, Yan B (2020) Environmental impacts of reclaimed asphalt pavement on leaching of metals into groundwater. Transp Res Part D: Transp Environ 85:102415

Yihdego Y, Danis C, Paffard A (2017) Why is the groundwater level rising? A case study using HARTT to simulate groundwater level dynamic. Water Environ Res 89:2142–2152

Hydrochemical Analysis of Groundwater in the Area Northwest of El-Sadat City, West Nile Delta, Egypt

Amal Othman, Hosni Ghazala, and Ismael M. Ibraheem

Abstract The Nile River represents the basic surface water in Egypt as the area beside it is the most fertile land in Egypt. The Nile River's surface has become insufficient in recent years to the Egyptian requirements because of the overpopulation and the recent developing projects' problems. Therefore, searching for new possible reclamation areas turned into a critical need. Intensive efforts have been made to set large agricultural projects and new urban communities in the desert. The study region is located northwest of El-Sadat city (West Nile Delta) covering the area between longitudes 30° 16′ 48″ and 30° 30′ 40″ E and latitudes 30° 21′ 36″ and 30° 31′ 12″ N. It occupies an area of 330 km^2. Because of the presence of such a vital location, a lot of work has been put into developing it. El-Sadat city is one of Egypt's most important new industrial cities. A lot of work has been planned into developing such desired area due to the presence of agricultural soil potentialities, gentle relief, and groundwater resources. The main purpose of the present study is to evaluate the hydrochemical settings of the groundwater and defeat the groundwater shortage problem in this area. The water samples gathered from 57 different drilled water boreholes were tested for the major cations [Mg^{2+}, Ca^{2+}, Na^+, K^+], major anions [$CO_3{}^{2-}$, $HCO_3{}^-$, $SO_4{}^{2-}$, Cl^-], Mn and Fe trace elements, in addition to electrical conductivity (EC) and total dissolved solids (TDS). Moreover, the suitability of 57 groundwater boreholes for drinking and irrigation purposes was assessed by comparing with the known standard guidelines' values. Water quality indices such sodium content (SC), sodium adsorption ratio (SAR), residual sodium carbonate

A. Othman (✉) · H. Ghazala
Department of Geology, Faculty of Science, Mansoura University, Mansoura 35116, Egypt
e-mail: osman83@mans.edu.eg

H. Ghazala
e-mail: Ghazala@mans.edu.eg

I. M. Ibraheem
Institute of Geophysics and Meteorology, University of Cologne, Pohligstr. 3, 50969 Cologne, Germany
e-mail: ismael.ibraheem@geo.uni-koeln.de

A. M. Negm and M. El-Rawy (eds.), *Sustainability of Groundwater in the Nile Valley, Egypt*, Earth and Environmental Sciences Library,
https://doi.org/10.1007/978-3-031-12676-5_7

(RSC), permeability index (PI), Kelly's ratio (KI), chloride classification, magnesium hazard (MH), and chloro-alkaline indices were also taken into consideration while assessing the quality of groundwater for agricultural objectives.

Keywords West Nile Delta · El-Sadat city · Water quality · Hydrochemical analysis

1 Overview

The Nile River, as Egypt's principal surface water, aids in boosting the fertility of the land beside the Nile River and around its two branches. These days, the Nile River's surface water got lacking to the Egyptian demands due to the overpopulation issue. The Nile Basin has subjected to several changes in discharge, geomorphology as well as hydrography, since its inception (Gaber et al. 2020). Several efforts have been made to start new urban settlements and enormous agricultural activities in the desert fringe and many researches concern with this issue (Attwa and Ali 2018; El Osta et al. 2018; Gad et al. 2020; Hegazy et al. 2020; McCool 2019; Mohamed 2020; Mosaad and Basheer 2020; Salem et al. 2018, 2019; Salem and Osman 2017).

The main objective of this research is to highlight some information on the quality of the groundwater and its availability and appropriateness for irrigation and human utilities in the region northwest of El-Sadat city in the West Nile Delta. Such current research could be a great informative tool to help constructing new projects concern with establishing new communities and land reclamation purposes. This can help the Egyptian government to implement the desired development strategy for 2030. Groundwater hydrochemistry has a considerable significance concerning the groundwater's suitability for different uses. It helps to assess the hydrochemical processes in charge of the areal changes in the groundwater chemistry (Heath 1983; Miller 1991; Kimblin 1995; Mayo and Loucks 1995; Hudson and Golding 1997; Todd and Mays 2005; Fetter 2014; Şen 2015). As there are several interactions between the earth layers and its surroundings of the atmospheric gasses caused by the man-made materials, no pure water will exist. Therefore, there is an urgent need for groundwater analysis. The hydrochemical analysis deals with the groundwater's chemical composition and its source. In addition, it aims to discover the relation between the rock constitution and the groundwater passing through (Zaporozec 1972). As a result, both the quality and the geological context can be inferred from the groundwater chemical composition (Dauda and Habib 2015). Variations in the geochemistry of the groundwater are commonly used indicators for the aquifer composition variations, because groundwater's geochemistry is a follower of rock mineral composition (Rogers 1989). Ion dissolution results from water circulation in the surrounding soils and rocks. The hydrochemical analysis exhibits to what extent if waters would be suitable for human demands, industrial, and agriculture purposes. This will be carried out through determination of the water type from its constitution, different diagrams, moreover, calculating several indicators helping in taking decision of its suitability for different purposes.

2 Methodology

During many campaigns, water samples were taken directly from boreholes, which are widely utilised for irrigation and drinking. They were collected in pre-cleaned one-liter plastic bottles after a certain pumping duration (exceeding 30 minutes) in order to overcome the contamination and stagnant groundwater problems. The used bottles were washed out numerous times with sampling water, then they were filled and sealed off quickly to avoid being exposed to the air. For identification, each bottle was named and thereafter saved in an icebox container. Within less than three days, the obtained samples were sent to laboratory. A total of fifty seven water samples were collected from different drilled boreholes in the area (Fig. 1). Water samples were subjected to a series of tests to determine the presence of the major anions [CO_3^{2-}, Cl^-, HCO_3^-, SO_4^{2-}], major cations [Mg^{2+}, Ca^{2+}, Na^+, K^+], Fe and Mn trace elements, total dissolved solids, and furthermore electrical conductivity at the Soil Fertility Tests and Fertilizers Quality Control Laboratory, Faculty of Agriculture, Mansoura University, Egypt. Hydrochemical analysis mainly evaluates the different ions concentrations. There are several computer software's that are very useful in the water analysis, especially for large volumes of data with small percentages of error and it would exhibit these data in different manners such as graphs, maps, and tables. Both Rockware AqQA (RockWare Inc. 2004) and Rockwork software have been used to carry out the evaluation of the groundwater data analysis. Graphical presentation of the output of these programs include the most common aqueous geochemistry plots and diagrams, such as Piper and Schoeller charts. These diagrams have a good role in deducing the hydrogeological facies and facilitating data analysis in the investigated region.

3 Geological Settings

Geological and structural settings of the West Nile Delta including the area of interest has been studied by many researchers such as Shata (1953, 1955, 1959, 1961), Zaghloul (1976), Abdel Baki (1983), Shedid (1989), Harms and Wray (1990), Sadek et al. (1990), Abdel Aal et al. (1994, 2001), Sarhan and Hemdan (1994), Sarhan et al. (1996), Garfunkel (1998), Guiraud and Bosworth (1999), Dawoud et al. (2005), ElGalladi et al. (2009), Abd El-Kawy et al. (2011), El Kashouty and El Sabbagh (2011) and Ibraheem et al. (2018).

The surface geology of the West Nile Delta area is represented by Oligocene, Miocene, Pliocene, and Quaternary sediments. Oligocene sediments consist mainly of sandstone, sand, and gravel. These sediments exist at the southwest part of the Nile Delta. Pliocene, Miocene, besides Quaternary deposits consist of sand and sandstones including intercalations of clay and limestone. These deposits are predominant at western and southern parts at Wadi El-Natrun, El Ralat, and Wadi El Farigh depressions, El Washika, Dahr El Tashasha, and Gabel El Hadid (Ibraheem 2009; Ibraheem

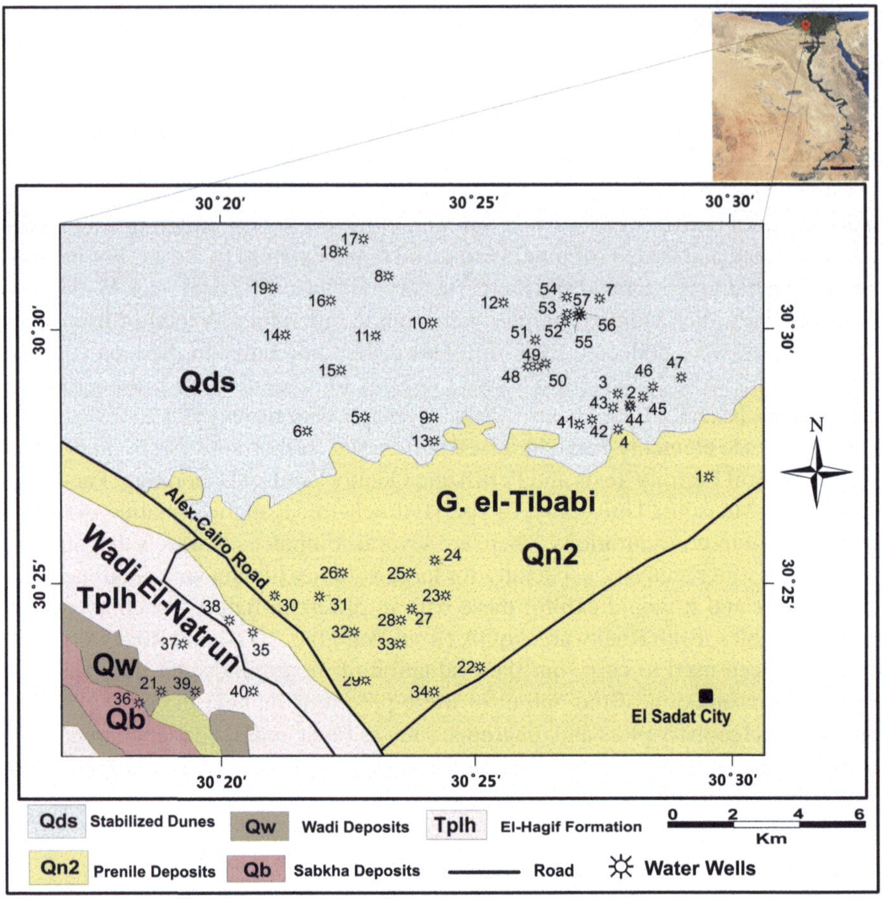

Fig. 1 Groundwater samples' locations posted on the geologic map of the area of interest

et al. 2016). Generally, the studied area is mainly covered by Prenile deposits and stabilized dunes (Fig. 1). The subsurface stratigraphy includes sediments of Quaternary, Pliocene, Miocene, Oligocene, Eocene, Cretaceous, Jurassic, and Triassic ages (Fig. 2). A thick sequence of Late Cretaceous to Quaternary sediments was recorded in the West Nile Delta area (Sharaky et al. 2007).

4 Hydrogeological Settings

Many studies have been carried out on the hydrogeochemical and hydrogeological settings at the area of the West Nile Delta such as: Abdel Baki (1983), Ahmed (1999), Al-Kilany (2001), Embaby (2003), Atta et al. (2005), Dawoud et al. (2005), Ahmed

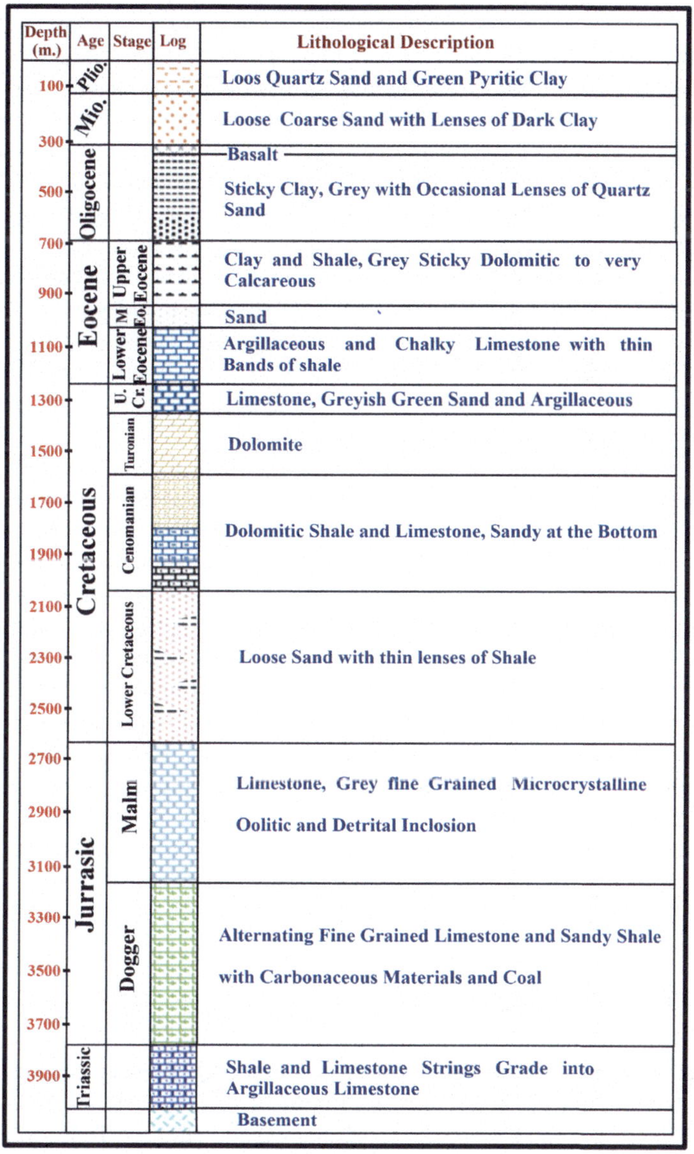

Fig. 2 Stratigraphic column of Sahara Wadi El-Natrun well, Western Desert, Egypt (modified after Ibraheem and El-Qady 2019)

et al. (2011), Massoud et al. (2014), Salem and El-Bayumy (2016), Ibraheem and El-Qady (2017), Abd-Elhamid et al. (2019), Othman et al. (2019), Salem et al. (2019), and Hegazy et al. (2020). They have studied the area of El-Sadat industrial city which is situated to the south of our investigated area and they came to the conclusion that the groundwater in that location is not uniform, with salinity and ionic composition varying. With groundwater movement, the salinity of the groundwater rises gradually from northeastern to southwestern directions. They mentioned also that, as mineralization decreases with depth, this indicates limitations of water penetration to great depths. They claimed that shallow groundwater has high nitrate concentration caused by agriculture procedures and groundwater at northeastern and southwestern parts of industrial areas has high heavy metals concentrations because of the industrial activities. Five different chemical facies ($Na–HCO_3$, $Ca–HCO_3$, $Na–SO_4$, Ca–Cl and Na–Cl) have been recorded at this area indicting process of ion exchange and interactions between the groundwater and the formations.

Salem and El-Bayumy (2016) carried out a study on the groundwater's hydrochemistry of the area east to Wadi El-Natrun. They concluded that most of the groundwater samples are convenient for irrigation purposes except some samples having high EC values. The samples show low to medium salinity and low sodium absorption ratio values. The main groundwater types are: $Na–HCO_3$, Na–Cl, $Ca–HCO_3$, Mg–Cl and $Mg–HCO_3$. Sodium/Chloride/Carbonate water types are dominant at that area. Also, El Osta et al. (2018) discussed the groundwater flow and Vulnerability at the district of Wadi El-Natrun depression and its surroundings using numerical simulation and concluded that the present groundwater over-abstraction has resulted in a general head drawdown of 3 m in 2015 and will reach 40 m in 2050. Moreover, they mentioned that the western parts of their study area have a high vulnerability rate (>110).

Generally, the West Nile Delta area has four main groundwater aquifers: Oligocene, Miocene, Pliocene and Quaternary aquifers.

4.1 *Quaternary Aquifer*

The first main aquifer in the West Nile Delta is the Quaternary aquifer which is located at the Coastal Plain, the Rolling Plains of the Nile Delta, and Nile Delta floodplain (Geirnaert and Laeven 1992; Mohamed and Hua 2010). It represents the substantial water-bearing zone in the region along east Abu Rawash–Khatatba–El-Sadat–Alexandria where it changes from 100 to 500 m (in thickness) towards the coast. It is composed of sandy and gravely layers. Clay strata split the aquifer into smaller confined sub-aquifers.

Quaternary aquifer is subdivided into Recent aquifer and Pleistocene aquifer. Recent aquifer (sand with some intercalations of aeolian deposits) is predominant at the depression of Wadi El-Natrun (Ibraheem 2009). The main groundwater unit of the Pleistocene deposits is Sahl El-Tahrer which is remarked by the presence of Pliocene sediments at its base (Menco 1990). The Nile Delta groundwater is the main recharger for the Pleistocene aquifer at El-Sadat area. There are other water source supplies by seepages from the irrigation canals and/or cultivated lands, and strong rainstorms. Pleistocene deltaic deposits compose the main Nile Delta aquifer. These deposits consist of unconsolidated coarse sands and gravels with the existence of lenses, and it covers a thick clay layer of Pliocene age (Sherif 1999).

4.2 Pliocene Aquifer

Omara and Sanad (1975) mentioned that Pliocene aquifer is made up of multi layers of sand and clay. It is subdivided into two parts: the lower part consists of sandy horizon with thickness varying from 1 to 10 m, while the upper part is made up of loose sand and sandstone with a thickness of 15 m. Between the two parts, there is a thick clay bed, so that the groundwater at the lower part is found under confined condition. The Upper part is found in a semi-confined condition due to the existence of layers of sandy clay above it.

4.3 Miocene Aquifer

Miocene (Moghra Fm.) aquifer is the second main aquifer in the West Nile Delta. Dawoud et al. (2005) mentioned this aquifer as well developed at Wadi El Farigh and has a variable thickness (50–250 m). This aquifer is mainly sand, sandstone, and clay interbeds with silicified wood and vertebrate remains (Said 1962).

4.4 Oligocene Aquifer

Oligocene aquifer is found at the western parts of Cairo. It consists of gravel, sand, thin bands of limestone and clay interbeds (Salem and El-Bayumy 2016). Both rainwater and paleowater are the main recharge sources for this aquifer (Abdel Baki 1983; El Abd 2005).

5 Results

5.1 Numerical Water Quality Indicators

Water classification based on different combinations of the major ions can provide water quality indicators representing guidelines for assessing the groundwater

Table 1 Classification of water according to EC, TDS and TH

Parameter	Classification		References
TDS (mg/L)	0–1000	Fresh	Fetter (2014)
	1000–10,000	Brackish	
	10,000–100,000	Saline	
	>100,000	Brine	
EC (μS/cm)	<250	Excellent	Wilcox (1955) and Şen (2015)
	250–750	Good	
	750–2000	Permissible	
	2000–3000	Doubtful	
	>3000	Unsuitable	
Hardness (mg/L)	0–75	Soft	Şen (2015)
	75–150	Moderately hard	
	150–300	Hard	
	>300	Very hard	

quality. The classification of groundwater according to total dissolved solids (TDS), electrical conductivity (EC), and total hardness (TH) is given in Table 1.

5.1.1 Electrical Conductivity (EC) and Total Dissolved Solids (TDS)

Electrical conductivity and total dissolved solids are the most commonly calculated physical parameters. TDS is known as the total quantity of the solids especially the inorganic one existing in the water. High concentrations of TDS would affect negatively both soil and crops by reducing the groundwater's suitability for the human utilities and irrigation supply. TDS levels in water samples are significantly associated with the rock weathering process and many other factors (Jacks 1973; Rao 2002; Bartarya 1993).

Electrical conductivity is defined as the water ability to pass the electrical current and it helps to study the groundwater's quality. High EC values may result in a gastrointestinal irritation for consumers (World Health Organization [WHO] 2011). Groundwater having high EC levels are unsuitable for both domestic and irrigation utilities.

Figure 3a, b exhibit the distribution maps of TDS and EC of the surveyed area. They exhibit a zone of high EC and TDS levels at the northwestern sector of the study area. The TDS and EC of the water samples have a clear linear relationship shown in Fig. 4. The following empirical formula illustrates the relationship between TDS and EC (Walton 1989):

$$\text{TDS}\ (\text{mg/L}) = \delta \text{EC}\ (\mu\text{s/cm}) \tag{1}$$

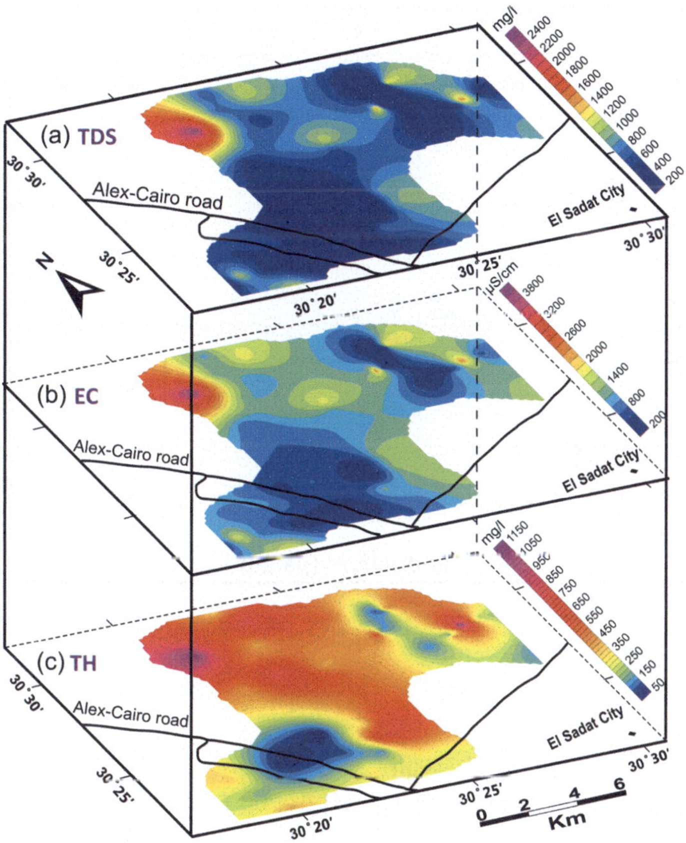

Fig. 3 A map showing the spatial distribution for **a** TDS, **b** EC, and **c** TH

The parameter δ was computed for the water samples in the area of interest as it equals 0.625, therefore the formula can be expressed by Eq. 2:

$$TDS = 0.625\,EC \quad (2)$$

5.1.2 Total Hardness (TH)

Hardness is taken into account as a crucial property for domestic and industrial classification of water quality. It depends mainly on the presence of magnesium (Mg^{2+}) and calcium (Ca^{2+}) and it is estimated in milligram per liter (Şen 2015). The following equation expresses the total hardness in terms of the calcium carbonate equivalent (Todd and Mays 2005).

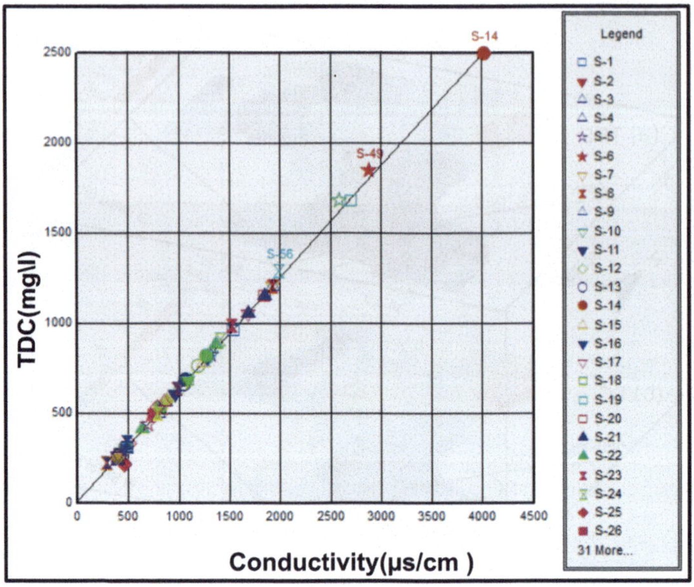

Fig. 4 Linear cross plot between TDS and EC for the gathered water samples in the study area

$$\mathrm{TH} = 2.5\left[\mathrm{Ca}^{2+}\right] + 4.1\left[\mathrm{Mg}^{2+}\right] \quad (3)$$

as TH, $[Mg^{2+}]$ and $[Ca^{2+}]$, are in mg/L.

For domestic purposes, if hardness exceeds 200 mg/L, then it would affect the water taste and may form a dirty layer of the groundwater (scum formation) (Burbery and Vincent 2009). The total hardness has been calculated for the water samples (Fig. 3c). High TH levels can be seen at the surveyed area except at the southwestern part in addition to other places at the eastern part of the region.

5.2 Groundwater Hydrochemical Analysis

Major ions existing in the groundwater samples are controlled by several factors such as climate conditions, rock nature and the movement ability of water. Both variable nature and the human activity are obvious through the differences in the hydrochemical parameters (Karanth 1991). Surface and subsurface physicochemical conditions usually affect the ion distribution (Aghazadeh and Mogaddam 2010).

Water chemical parameters include different ions concentrations: major anions (Cl^-, $SO_4{}^{2-}$, $HCO_3{}^-$ and $CO_3{}^{2-}$) and major cations (Na^+, K^+, Ca^{2+}, and Mg^{2+}). A summarize of the maximum, minimum, average, and standard deviation for different chemical parameters obtained in our study is shown in Table 2. Figures 5 and 6 represent a graphical representation for exhibiting the different concentrations of both cations and anions.

Table 2 Physicochemical parameter statistics

Parameter	Minimum value	Maximum value	Average value	Standard deviation
Na^+	2.9	329.4	117.9	90.71
K^+	0	9	6.2	2.15
Mg^{2+}	0	122.7	24.5	18.54
Ca^{2+}	0	418	113.5	88
HCO_3^-	7.2	585.6	227.5	113.3
CO_3^{2-}	0	27.7	2.38	5.83
Cl^-	20	630.9	159	144.5
SO_4^{2-}	0	930	228	180.8
EC	300	4032	1127	734.6
TDS	216.2	2500	709	458.9

All previous parameters are measured in mg/L unless EC in μS/cm

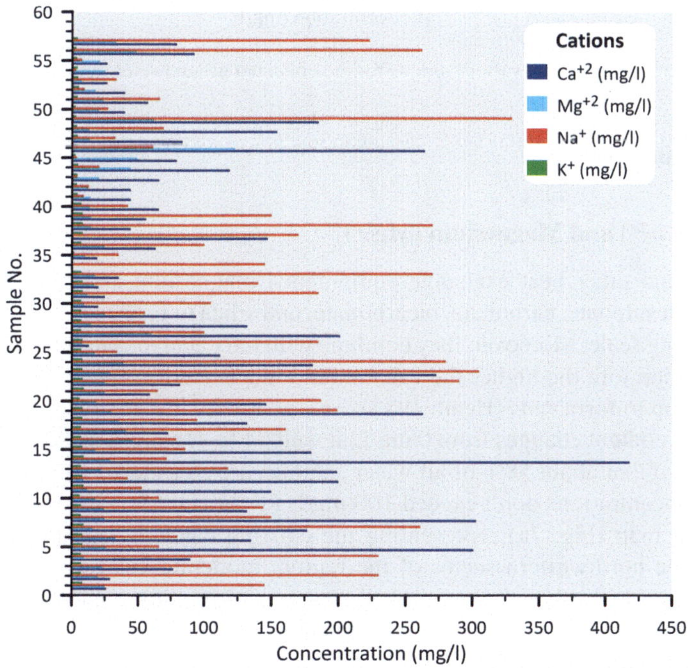

Fig. 5 The concentrations of various cations for the gathered groundwater samples

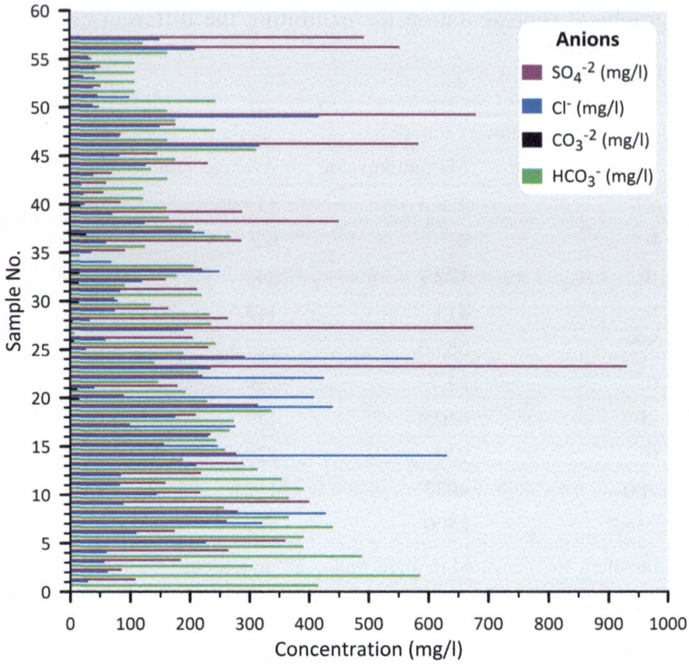

Fig. 6 The concentrations of various anions for the collected groundwater samples

5.2.1 Major Cations

Calcium (Ca^{2+}) and Magnesium (Mg^{2+})

In boilers and other heat exchange equipments, magnesium and calcium cations interact with sulphate, carbonate, bicarbonate, and silica to produce a heat-retarding, pipe-clogging scale. Moreover, they combine with fatty acid ions in soaps to generate soapsuds, therefore the higher the calcium and magnesium content, the greater the need for soap to form suds (Heath 1983).

Calcium content changes from 0 mg/L at well 34 to 418 mg/L at well 14. It could be concluded that about 58% of all water samples are considered natural waters, as calcium concentrations don't exceed 100 mg/L (Todd and Mays 2005). The spatial distribution map (Fig. 7a) representing the existing calcium cation exhibits high values at the northwestern sector of the region, moderate values in the centre of the area towards northwest-southeast direction, and low levels at the eastern and southwestern parts of the study region. This map shows a good harmonization with the total hardness distribution map (Fig. 3c).

High content of magnesium cation has laxative effect especially on the supply's new users (Todd and Mays 2005). It also has a side effect on the soil causing alkaline environment and hence lowering the crops product. Magnesium levels change from

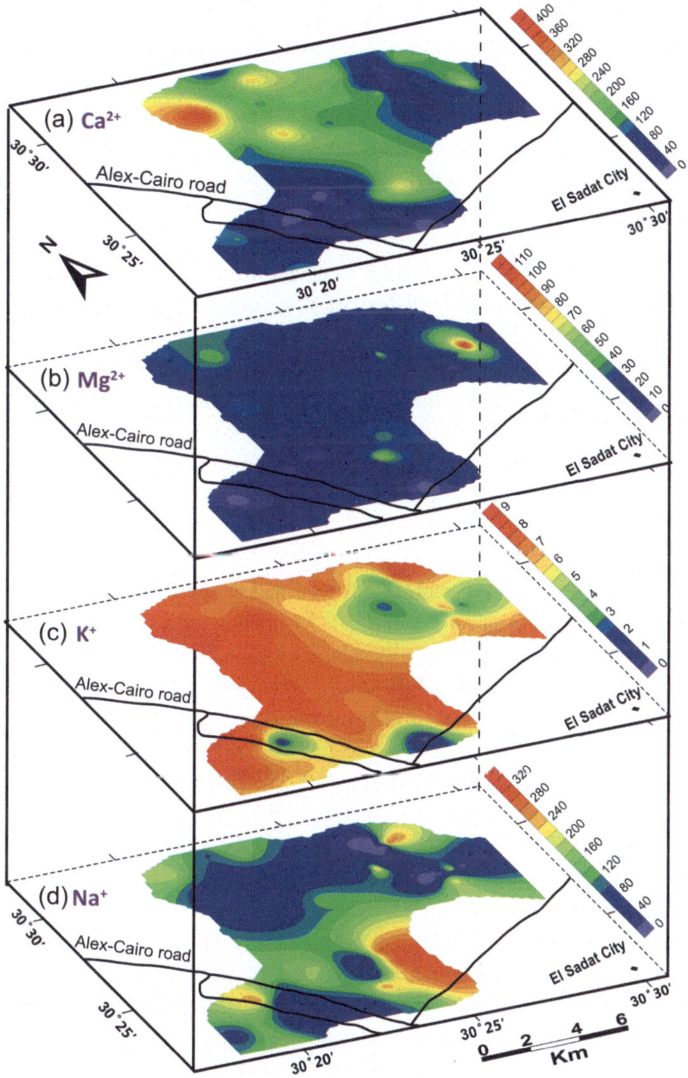

Fig. 7 The major cations' spatial distribution maps

0 mg/L at well 34 to 122.7 mg/L at well 46 and has 30.6 mg/L as an average value. The distribution map of Mg^{2+} shows high Mg^{2+} concentration zone at the northeast portion (Fig. 7b). Based on Todd and Mays (2005), all water samples are considered to have natural magnesium concentration except three groundwater samples (at wells 27, 46, and 49). These samples have Mg^{2+} concentration exceeding 50 mg/L.

Groundwater equilibrium state is well related to the existence of Ca^{2+} and Mg^{2+} cations in water (Negm and Armanuos 2017). Concentrations of these two cations in natural waters differ over a wide range relying on the history of the water sample including the geochemical characteristics and the geological formations (Şen 2015). Mg^{2+} and Ca^{2+} are very similar in several aspects and they could be dissolved from different soils and rocks (Salomons and Forstner 1984). Mg^{2+} and Ca^{2+} cations distribution maps indicate that the majority of the research area has low concentrations of both magnesium and calcium except some small zones.

Potassium (K^+)

Most natural waters have K^+ levels less than 10 mg/L (Chow 1964). K^+ cation is similar to Na^+ in many aspects but differs in other manners affecting its existence in water and its effect on other utilities (Şen 2015). All of the samples in the research area have K^+ concentration below 10 mg/L where K^+ concentration changes between 0 and 9 mg/L with an average value of 6.23 mg/L. The aerial distribution map of potassium level (Fig. 7c) reveals normal K^+ levels all over the study area. These normal K^+ levels are probably due to the nature of the sediments constituting the aquifer as potassium salts are insoluble in several types of rocks (Sharaky et al. 2017).

Sodium (Na^+)

Sodium is a wide common alkali metal in the natural water that is linked to the salinity of the groundwater (Ghoraba and Khan 2013). Existence of high concentration of Na^+ cation would be resulted from several factors such as rock weathering and ion exchange reactions (Stallard and Edmond 1983; Stimson et al. 2001). Chemical analysis of the sodium relative to the total cations is important in case of cation exchange process in sediments (Zaporozec 1972). Under normal cases, groundwater includes Na^+ and Cl^-. In the current study; Na^+ concentration differs from 2.9 mg/L as a minimum value at borehole no. 54 to 329.4 mg/L at borehole no. 49 with an averaged value of 116.11 mg/L. There are eleven samples of the groundwater having concentrations exceeding 200 mg/L, while the rest have acceptable values. Hence, it could be considered as natural water and is suitable for domestic purposes. Figure 7d shows the spatial distribution map of Na^+. This map highlights four high Na^+ zones in the northeast, southeast, northwest, and western sectors of the area under investigation. This may be a result from the fertilizers used for agricultural purposes. In the presence of suspended particles, salt and potassium concentrations greater than 50 mg/L create foaming, which increases scale formation and corrosion in boilers (Todd and Mays 2005). High concentration of sodium would also cause cardiac problems, hypertension, and other medical difficulties. Depending on the presence of calcium and magnesium concentrations in water, sodium would be harmful to certain irrigated crops (Heath 1983).

5.2.2 Major Anions

Bicarbonate (HCO_3^-) and Carbonate (CO_3^{2-})

Concentrations of carbonate (CO_3^{2-}) and bicarbonate (HCO_3^-) are related to the value of the PH and they could be described as the total alkalinity (Fetter 2014). They control the water capacity to neutralized strong acids (Heath 1983). Concerning HCO_3^-, the minimum concentration is 7.2 mg/L, while the highest value is 585.6 mg/L with a value of 225.3 mg/L on average. Groundwater samples in the research region have normal HCO_3^- levels (less than 500 mg/L), except the sample no. 2 which has high value (585.6 mg/L). The spatial distribution map of HCO_3^- (Fig. 8a) shows two relatively high-level zones in the eastern parts, while moderate values are found at the northwestern parts, and low levels are encountered at the southern zones of the study area, but mostly in permissible levels. Bicarbonates of magnesium and calcium dissolve in the water heater to form scale and emit corrosive carbon dioxide gas (Heath 1983).

Percolation process for the waters enriched in CO_3^{2-} resulting from the dissolution of the carbonate rocks with the aid of atmosphere increase the bicarbonate concentration and led to the existence of Ca–HCO_3 groundwater type (Edmunds et al. 1987; Appelo and Postma 1996). CO_3^{2-} is found in the water samples with a value of 2.34 mg/L on average, with the highest value of 27.7 mg/L at borehole no. 37, whereas the minimum value is zero mg/L. Only six water samples have CO_3^{2-} concentration more than 10 mg/L, and the rest of water samples in the area have normal levels and suggest natural waters suitable for domestic purposes. The spatial distribution map of CO_3^{2-} is well presented in Fig. 8b that appears to have two high CO_3^{2-} level sectors in the southwestern parts of the research area.

Chloride (Cl^-)

Groundwater content of chloride initiates from natural and anthropogenic origins (Sayyed and Bhosle 2011). Gradually, Cl^- concentration could be high because of leaching of the saline residues from the sediments (Patil et al. 2020). Moreover, it might be related to included NaCl fluid and by atmospheric deposition. Human activities also result in the Cl^- ions existence in the groundwater (Rogers 1989). The threshold value of Cl^- in groundwater is 250 mg/L (WHO 2011), therefore samples exceeding this limit would have salty taste when chloride combines with sodium. Water would be described as corrosive water in case of high concentration of chloride (Heath 1983). Concentration of Cl^- is very important in case of agricultural activities as it is mostly absorbed by plants. Cl^- is a good guide for groundwater chemical processes. Low Cl^- levels indicate freshwater entrance into the aquifer, while high Cl^- concentrations might be resulted from different process of saline water intrusion, solid salts solutions from the rocks, and evapotranspiration (Şen 2015).

Chloride anion is found with different concentrations in the obtained water samples, as it changes from 20 mg/L at well no. 42 to 630.92 mg/L at borehole no. 14, and has average value of 157.312 mg/L. Cl^- aerial distribution map (Fig. 8c) shows several zones with high Cl^- levels at different parts of the region; northeast,

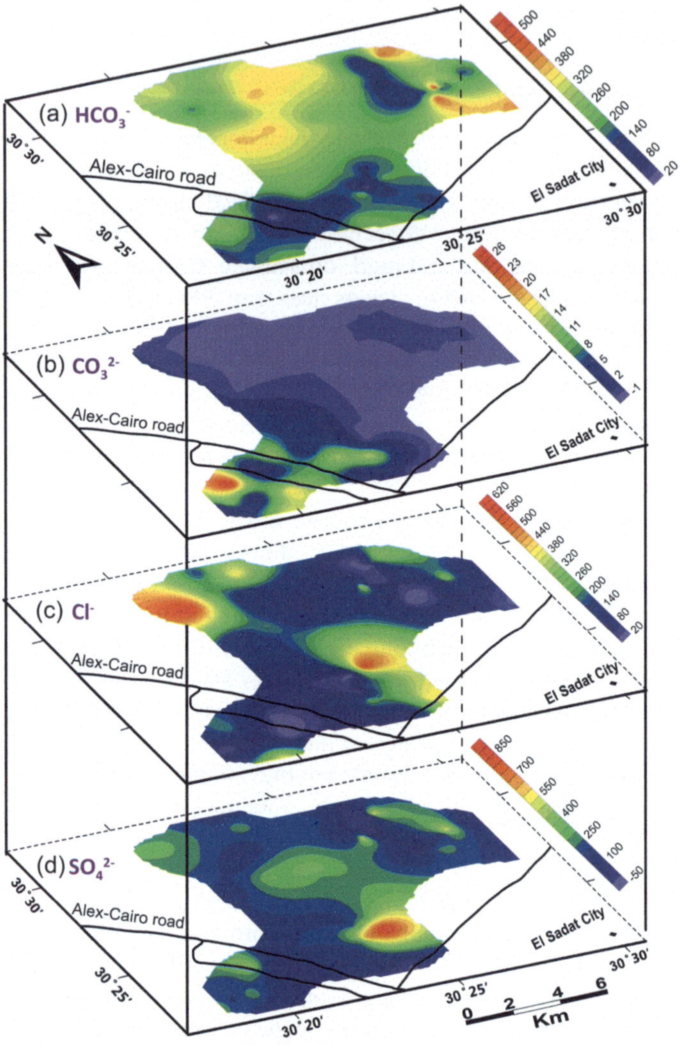

Fig. 8 The major anions' spatial distribution maps

northwest, southeast and western parts, whereas the central part has low Cl^- concentrations. There is a good consensus between the spatial distribution maps of Na^+ and Cl^-. High levels of Cl^- exceeding 350 mg/L would be harmful to the agricultural crops (Hopkins et al. 2007), and this was observed only in 6 groundwater samples (8, 19, 20, 22, 24, and 49).

Sulfate (SO_4^{2-})

The SO_4^{2-} levels usually increase with the salinity of the water samples. It could result from the interaction between water and gypsum presents in the Pleistocene sediments (Atta et al. 2005). An existence of SO_4^{2-} could be caused by pyrite oxidation, gypsum dissolution and/or by barite, especially in sedimentary bedrock areas (Rogers 1989). The maximum SO_4^{2-} level in fresh groundwater is 1360 m/L (Şen 2015), therefore all water samples have acceptable concentrations of SO_4^{2-} anions. SO_4^{2-} content fluctuates from zero mg/L at borehole no. 34 to 930 mg/L at borehole no. 23 as a maximum value. In accordance with (Heath 1983), presence of sulfate with certain concentrations (300–400 mg/L) results in a bitter taste, while at higher concentrations (600–1000 mg/L) it provides a laxative effect. So, we can deduce that only 12.2% of all samples gives a laxative effect and the rest of samples show bitter taste. Figure 8d presents the SO_4^{2-} distribution map in the study area. It reveals only one high SO_4^{2-} zone in the southern part, while almost the whole area has permissible levels of SO_4^{2-}.

5.2.3 Trace Elements

Secondary concentrations of trace elements were identified in the studied groundwater samples. The acceptable drinking water limits are 0.1 and 0.3 mg/L for manganese and iron, respectively (WHO 2011). The concentration of manganese ranges from 0.05 and 0.96 mg/L. The map representing the spatial distribution of manganese clarifies exceeding the permissible guideline value in the surveyed area unless the central and the western parts (Fig. 9a). Iron level differs from 0.01 mg/L at borehole no. 15 to 1.12 mg/L at borehole no. 12 with a value of 0.361 mg/L on average. Iron spatial distribution map shows two high iron level zones in the study region (Fig. 9b). The existence of iron and manganese is an indicator that the groundwater quality is reduced (Burbery and Vincent 2009). Although these ions (Fe^{2+} and Mn^{2+}) are useful for stain laundry, they are harmful in food processing, ice manufacturing, bleaching, brewing, dyeing, and different industry process (Heath 1983).

5.3 *Water Quality Graphical Representation*

Water quality could be analyzed by graphical representation of different ions existing in water. This method of analysis is quick, easy, effective, and informative way to understand water quality and to decide groundwater's convenience for different intents (drinking, domestic, agricultural, or industrial), so it enhances and manages drilling new wells. The graphical representation includes Schoeller, Piper's tri-linear and Durov diagrams.

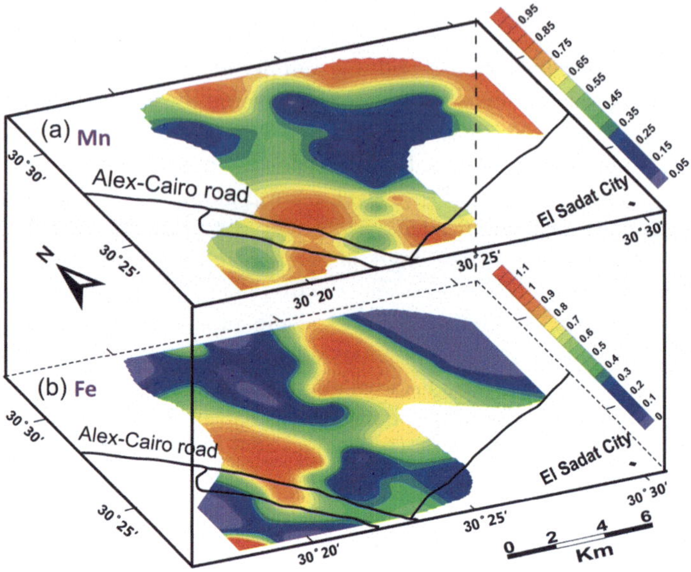

Fig. 9 Spatial distribuation maps of concentrations of trace elements of **a** Mn^{2+} and **b** Fe^{2+} at the research area

5.3.1 Schoeller Diagram

Schoeller diagram is a semi-logarithmic graphic with eight vertical logarithmic scales that are evenly spaced representing several ions concentrations in equivalents per million (Schoeller 1967). Applying the logarithmic scale for this diagram is to distribute the appearance of the concentrations as it amplifies low concentrations and decreases high levels. All ions concentrations are plotted as points for each sample then a straight line tied them. This diagram reveals the absolute concentration of each ion and the difference between samples. This diagram evaluates the saturation degree of groundwater with both of calcium carbonate and calcium sulfate (Şen 2015). Moreover, it could describe the average chemical composition of the samples (Ghoraba and Khan 2013). Schoeller diagram represented in Fig. 10 has been sketched for all gathered groundwater samples in the region of the investigation using Aq.QA® software. It reveals that the general trend of ions in mg/L exposes $Ca^{2+} > Na^{+} > Mg^{2+} > K^{+}$ and $HCO_3^{-} > Cl^{-} > SO_4^{2-} > CO_3^{2-}$.

5.3.2 Piper's Tri-Linear Diagram

Piper diagram is another graphical representation which achieves the combined effect of major constitutes of water controlling its quality in a single diagram (Şen 2015). It explains different variables related to the major cations and anions and reveals the main chemical components of the waters. It also shows both similarities and

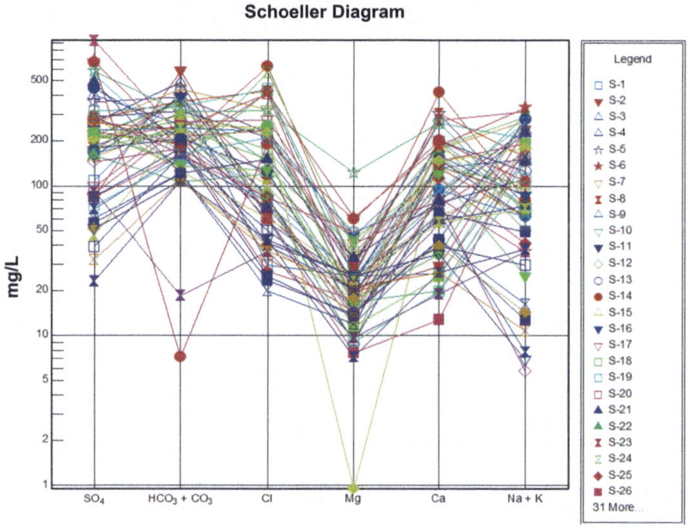

Fig. 10 Schoeller diagram representing all water samples in the study area

differences between samples of water (Piper 1944, 1953). Piper diagram represents a powerful technique in chemistry analysis of the groundwater (Dauda and Habib 2015). It is made up of a rhombus and two triangles. Each triangle represents three ion concentration percentages. The rhombus form depicts the composition of water in terms of cations and anions (Fetter 2014). Piper's tri-linear diagram is a good analytical and representative way to illustrate the classification of the main chemical components and hydrochemical facies of the waters within the area. Aq.QA® computer software has been used to establish the Piper diagram for the collected water samples (Fig. 11).

The triangle corresponding cations reflects a percentage of 42.12% of the existing water samples is of calcium type, moreover, sodium and potassium type represents 33.33%, while 24.56% of the samples showed no sign of a dominating type. Furthermore, the triangle demonstrating anions suggests that 45.61% of the water samples did not display any dominant element, 22.82% represents a type of bicarbonate, 19.29% reflects sulfate type, and chloride type has a percentage of 12.28%. Furthermore, 66.7% of the water samples are on the left side of the diamond demonstrating the alkaline ($Ca^{2+} + Mg^{2+}$) predominance over alkalies earth ($Na^{+} + K^{+}$), whereas the right side of the rhombus is filled with 33.3% of the samples and suggests that alkalies prevail over alkaline earth. Also, groundwater's strong acids (${SO_4}^{2-} + Cl^{-}$) representing 77.2% exceed week acids (${CO_3}^{2-} + {HCO_3}^{-}$) whereas the left over samples (22.8%) demonstrate that strong acids are overcome by weak acids. Besides, mixed water type is found in 43.86% of groundwater samples, sodium-chloride type corresponds 26.33% of the samples, 15.78% show magnesium-bicarbonate type, while calcium-chloride type is represented by 10.52%, and lastly 3.53% are of sodium-bicarbonate type. Figure 12 shows the geographical distribution for the different

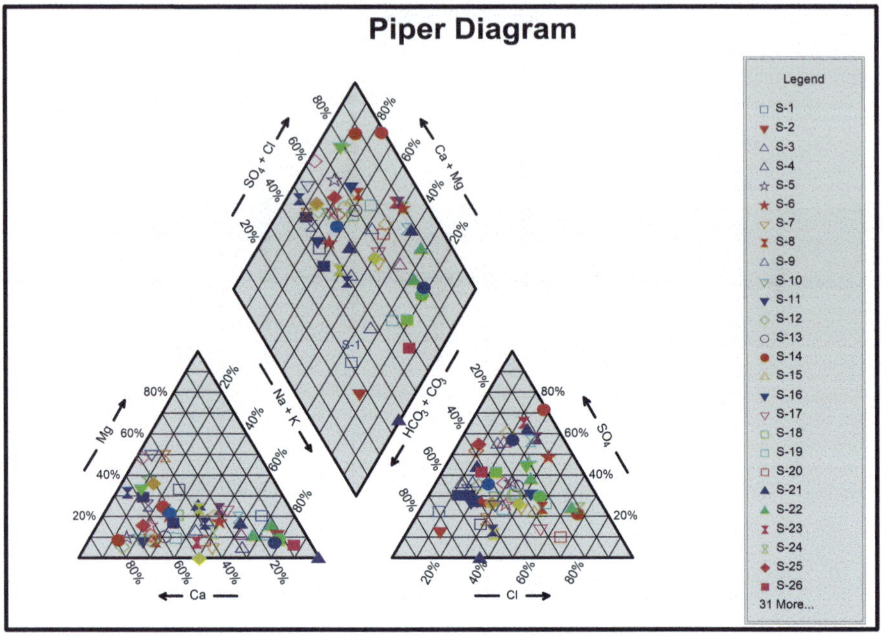

Fig. 11 Piper diagram represents the groundwater samples' hydrogeological facies at the area of interest

hydrochemical types according to Piper tri-linear scheme in the study region. It reveals that the mixed water type is located at northern, central and southwestern parts of the area. While, CaCl type is found at northern locations and this is related with relatively high concentrations of calcium cations which may be dissolved from different rocks and soils. On the other hand, NaCl type is dominant at the southern parts of the area. This could be caused by agriculture activities effect and/or ion exchange reactions. Moreover, $MgHCO_3$ water type characterizes the northeastern parts of the research region corresponding to high concentrations of Magnesium cations.

5.3.3 Durov Diagram

Another hydrochemical graphical representation could be carried out by Durov diagram which has been constructed by Durov (1948). Generally, it relies on the percentages of each cation and anion in milliequivalent. It is made of two ternary charts. Both anions and cations are plotted in separate triangles, then by extending points with lines to intersect, the resultant point in the central rectangle reflects the water type. Durov diagram plays a crucial function in determining the concentration, the origin of the groundwater chemical ingredients, and the total dissolved solids. A code has been provided by Al-Bassam et al. (1997) to process data using Durov

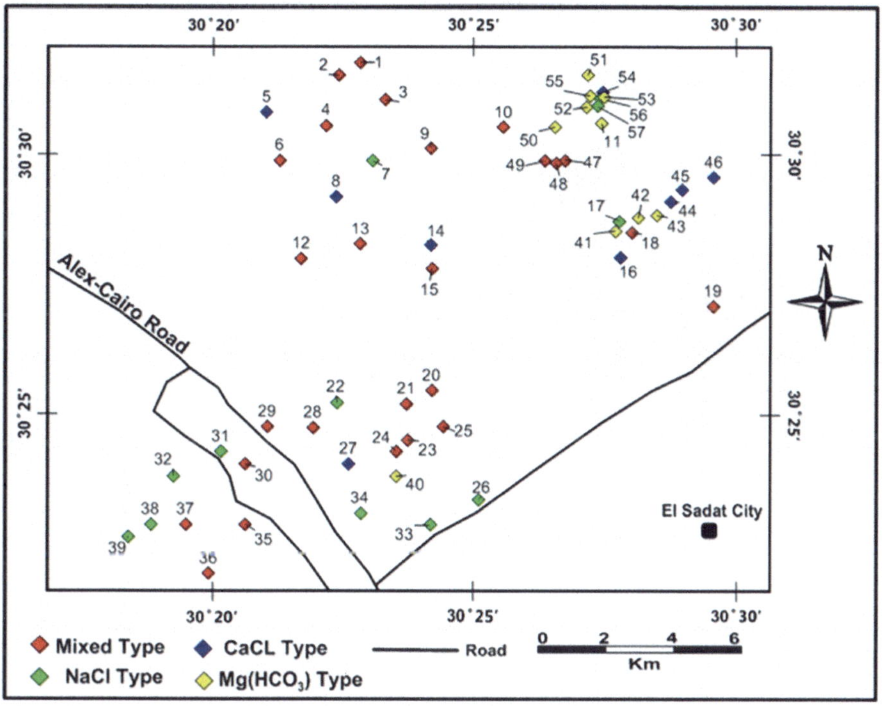

Fig. 12 The spatial distribution map representing the different hydrochemical water types revealed from Piper's tri-linear diagram

diagram. This diagram may give more information about the hydrochemical facies of the groundwater by defining its type or provide some useful geochemical processes for more understanding and analyzing the quality of the groundwater (Ghoraba and Khan 2013). Analyzing Durov diagram according to groundwater classification based on this diagram (Lloyd and Heathcote 1985) reveals that more than 61% of water samples are in a simple dissolution or mixing phase with no dominant anion or cation. Durov diagram exhibits that the mixed water type in the current region are found at field 5 (61% of all groundwater samples) along the mixing or dissolution line. The trend of this line may be related to the recent fresh water recharge indicating little dissolution or mixing with no dominant cation or anion (based on Lloyd and Heathcote 1985). While about 18% of all water samples are at field 2 that suggests the water type is characterized by Ca and HCO_3 with the presence of Na. These waters are affected by ion exchange. Moreover, a percent of 12% of the groundwater samples is located at field 8 reveals that these water types are dominated with Cl and Na and assumes that the groundwater is affected with the process of reverse ion exchange of Na–Cl waters. Furthermore, the rest of samples are dominant with HCO_3 and Ca, SO_4, Cl and Na, and SO_4 and Na (Fig. 13).

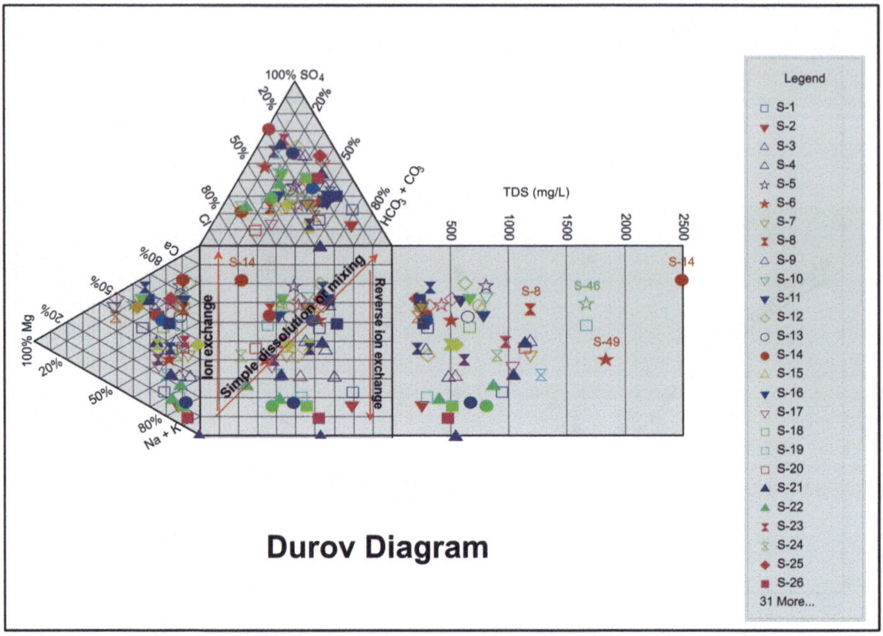

Fig. 13 The gathered groundwater samples are depicted in a Durov diagram drawn using Aq.QA® computer software (RockWare Inc. 2004)

6 Suitability of Groundwater for Irrigation Purposes

Groundwater suitability for irrigation purposes relies on the TDS concentrations which in turn have a high affect on both productivity and quality of crops. Groundwater quality could be decided based on different indicators like sodium content (SC), permeability index (PI), sodium adsorption ratio (SAR), residual sodium carbonate (RSC), Kelly's ratio (KI), magnesium hazard (MH), chloro-alkaline Indices (CAI), chloride classification, and corrosively Ratio (CR), as shown in Tables 3 and 4. Figure 14 represents bar chart plots revealing several parameters of water quality in the investigated area.

6.1 Sodium Content (SC)

Sodium ion concentration has a negative impact on the crop yield, which is reduced when Na^+ enters the vacuum areas in the soil causing a decrease in permeability (Şen 2015). Therefore, SC represents a very significant index. SC could be obtained by expression (Eq. 4).

Table 3 Classification of water quality

Parameter	Classification		Reference
SC (%)	<20	Excellent	Wilcox (1955)
	20–40	Good	
	40–60	Permissible	
	60–80	Doubtful	
	>80	Unsuitable	
SAR (epm)	<10	Excellent	Todd and Mays (2005) and Şen (2015)
	10–18	Good	
	18–26	Permissible	
	>26	Doubtful	
RSC (meq/L)	<1.25	Safe for drinking	Eaton (1950) and Richards (1954)
	1.25–2.50	Marginal as an irrigation source	
	>2.5	Unsuitable for irrigation without amendment	
MH	<50%	Suitable	Negm and Armanuos (2017)
	>50%	Unsuitable	
PI	>75%	Class-I	Doneen (1962)
	25–75%	Class-II	
	<25%	Class-III	

$$SC\ (\%) = 100 * \frac{\left(Na^{+} + K^{+}\right)}{(\mathrm{Ca}^{2+} + \mathrm{Mg}^{2+} + \mathrm{Na}^{+} + \mathrm{K}^{+})} \tag{4}$$

6.2 *Sodium Adsorption Ratio (SAR)*

Both water salinity and SAR are good indicators for groundwater suitability for irrigation. This is depending on sodium concentration that impacts the soil and hence the permeability. SAR evaluates the hazard of Na^+ to agricultural crops and is obtained by Eq. 5, after converting mg/L unit into meq/L unit (Todd and Mays 2005).

$$SAR = \frac{Na^{+}}{\sqrt{\frac{Ca^{2+}+Mg^{2+}}{2}}} \tag{5}$$

It reflects the ability of the soil to absorb Na^+. At high values of SAR, the soil must be adapted to avoid long term disturbances. SAR calculation is important due to Na^+ displacement phenomenon with both Ca^{2+} and Mg^{2+} in the same soil resulting

Table 4 Water quality classification depending on a variety of indicators

Sample number	EC	TDS	SC	PI	SAR	RSC	Hardness	CAI	KI	CR	MH
1	1548	960	65.81	92	4.85	3.42	167.5	−6.66906	1.86	0.52	61.57
2	419	260	76.8	100.18	8.07	6.52	152.45	−4.62177	3.25	0.46	52.84
3	484	300	45.14	65.83	2.69	−0.77	286.8	−1.89131	0.79	1.26	33.23
4	806	500	65.93	84.07	6.17	2.74	262.39	−4.80312	1.9	1.023	14.44
5	1323	820	15.15	26.88	0.98	−10.82	859.1	0.527539	0.17	2.65	12.58
6	839	520	27.94	46.3	1.6	−3.1	472.8	−0.16668	0.37	1.28	28.94
7	1935	1200	53.27	65.24	5.04	−2.96	506.87	−0.25535	1.12	2.59	12.87
8	1935	1200	28	37.77	2.22	−11.23	860	0.454261	0.38	3.9	12.08
9	1935	1200	42.75	55.83	3.02	−4.42	429.63	−1.52922	0.73	2.92	23.47
10	1323	820	20.68	36.76	1.19	−5.6	578.8	0.268107	0.25	1.75	12.93
11	968	600	23.04	44.88	1.12	−2.07	422.56	−0.06658	0.27	1.06	11.41
12	1032	640	15.9	33.69	0.79	−4.94	536.49	0.16531	0.17	1.41	6.9
13	1065	660	31.07	43.71	2.11	−6.66	587.7	0.123019	0.44	2.89	15.55
14	4032	2500	12.6	18.47	0.92	−20.24	1163.9	0.814388	0.13	10.23	10.36
15	1258	780	27.34	40.77	1.63	−6.24	522.98	0.442393	0.36	3.145	14.14
16	1290	800	23.47	35.72	1.42	−7.78	586.1	0.454105	0.29	3.6	32.52
17	1694	1050	50.4	64.61	3.72	−2.71	350.6	0.093731	0.99	3.05	48.34
18	1097	680	31.17	45.58	1.89	−5.09	476.37	0.139142	0.43	2.6	31.06
19	2710	1680	36.18	45.32	3.1	−10.1	779.27	0.298076	0.55	4.45	15.83
20	1855	1150	48.37	59.33	3.87	−4.56	440.85	0.292737	0.92	4.31	17.43
21	1855	1150	37.83	58.64	1.78	−1.46	249.59	−1.57224	0.56	1.56	41.28

(continued)

Table 4 (continued)

Sample number	EC	TDS	SC	PI	SAR	RSC	Hardness	CAI	KI	CR	MH
22	1375	880	67.47	75.36	7.17	−3.78	307.34	−0.05054	2.04	9.16	33.65
23	1540	985.6	45.62	51.63	4.68	−12.55	800.58	−0.99662	0.83	8.48	14.95
24	1420	908.8	50.16	55.94	4.91	−10	612.4	0.249859	0.99	12.86	26.41
25	470	216.2	19.06	38.02	0.79	−3.94	350.85	−1.13901	0.21	1.97	20.45
26	764	488.9	86.57	107.68	10.06	2.92	63.49	−3.79013	6.29	1.64	49.99
27	470	300.8	23.13	24.19	1.59	−14.93	748.5	0.169778	0.29	189.19	33.21
28	518	331.5	24.79	42.74	1.21	−3.8	389.04	−1.71418	0.31	1.76	15.38
29	477	305	27.49	47.41	1.3	−2.88	337.34	−0.19424	0.35	1.69	30.67
30	481	307	73.06	95.78	4.88	0.95	36.89	−1.06646	2.61	1.8	43.42
31	824	527.8	78.95	97.05	7.67	1.46	109.23	−2.42152	3.66	2.18	43.63
32	650	416	72.39	86.16	5.56	−0.4	118.88	−0.82695	2.54	4.16	58.32
33	1280	819	78.84	90.02	9.28	0.26	158.85	−0.66116	3.67	4.31	45
34	874	559	100	129.24	6.3	3.4	0	−2.63594	3.15	0.64	50
35	990	633.6	46.89	62.24	1.62	−1.43	37.68	−0.49298	0.86	10.38	46.22
36	490	313.6	46.55	60.53	2.7	−3.15	258.37	−1.58594	0.84	4.24	39.22
37	888	568.3	50.55	64.02	3.83	−2.11	371.44	−0.17637	0.99	2.54	1.08
38	1070	684.8	77.84	89.72	9	0	169.04	−2.32945	3.45	4.36	35.28
39	872	558	62.71	77.6	4.61	−1.3	199.04	−2.25532	1.63	3.39	29.99
40	430	275.2	31.31	55.43	1.23	−1.16	219.1	−1.79493	0.42	0.88	25.45
41	400	245.8	12.52	47.96	0.29	−1.36	166.99	0.284261	0.11	1.09	34.48
42	400	245.8	16.71	52.39	0.44	−1.2	159.2	−0.11811	0.17	1.07	31.24

(continued)

Table 4 (continued)

Sample number	EC	TDS	SC	PI	SAR	RSC	Hardness	CAI	KI	CR	MH
43	500	336.6	5.09	34.86	0.1	−2.33	248.23	0.766878	0.03	1.18	33.88
44	1100	678.4	9.46	22.67	0.4	−7.38	476.99	0.724609	0.09	4.57	38.1
45	800	529.3	2.51	22.46	0.07	−5.17	399.72	0.913316	0.02	2.26	50.86
46	2600	1676.8	10.6	18.87	0.78	−18.32	1163.07	0.694855	0.11	4.99	43.64
47	700	434.6	22.25	45.46	0.87	−2.62	263.12	0.368665	0.27	1.96	20.83
48	1200	761.6	26.04	42.41	1.43	−4.8	439.12	0.281233	0.34	2.48	12.5
49	2900	1843.2	50	55.79	5.34	−11.51	715.79	−0.20286	0.99	10.96	35.82
50	500	313	25.41	58.25	0.87	−0.98	181	0.136008	0.32	1.08	45.69
51	800	498.6	38.37	63.56	1.86	−0.8	238.53	−0.04059	0.6	1.27	40.42
52	400	255.4	13.35	45.01	0.31	−1.68	171.57	0.536268	0.12	1.55	42.68
53	300	222.7	44.81	82.53	1.47	−0.14	95.52	−0.25141	0.75	1.27	31.24
54	400	260.5	5.61	36.75	0.09	−2.07	190.5	0.816558	0.03	1.62	54.2
55	300	222.7	11.71	48.71	0.26	−1.29	152.1	0.613588	0.1	1.23	56.99
56	2000	1288.4	57.03	65.02	5.48	−5.97	428.62	−0.89961	1.32	7.69	46.5
57	1700	1055	59.29	67.65	5.28	−4.72	333.71	−1.269	1.44	8.39	41.05
Minimum value	**300**	**216.2**	**2.51**	**18.47**	**0.07**	**−20.24**	**63.49**	**−6.67**	**0.018**	**0.46**	**1.08**
Maximum value	**4032**	**2500**	**100**	**129.24**	**10.06**	**6.52**	**1163.9**	**0.913**	**6.29**	**189.19**	**58.32**
Average value	**1127.83**	**709.08**	**39.53**	**57.24**	**3.02**	**−3.91**	**391.15**	**−0.66**	**1**	**6.61**	**31.26**
Standard deviation	**734.6**	**458.9**	**23.56**	**23.67**	**2.6**	**6.42**	**258.94**	**1.52**	**1.21**	**24.56**	**14.95**

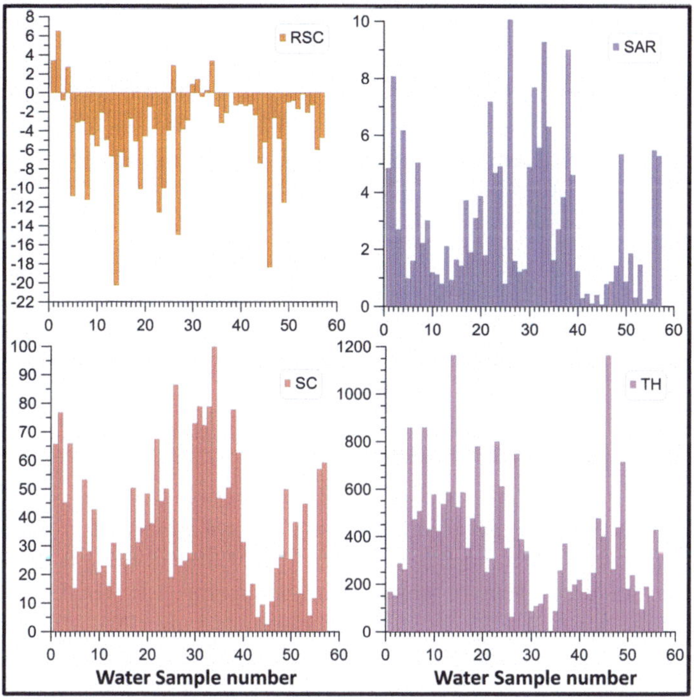

Fig. 14 Bar chart plots represent different water quality parameters

in lowering the soil ability to construct stable aggregates and soil structure loss (Şen 2015).

6.3 *Residual Sodium Carbonate (RSC)*

RSC evaluates the risk effect on the groundwater caused by excess concentration of both bicarbonate and carbonate (Raju 2007). High content of HCO_3^- results in the precipitation of both Ca^{2+} and Mg^{2+} as CO_3^{2-}, therefore soil would be enriched with Na^+. RSC influences the electric conductivity, alkalinity, and SAR (Naseem et al. 2010). RSC values would be estimated by Eq. 6 (Şen 2015):

$$RSC = \left(CO_3^{2-} + HCO_3^-\right) - \left(Ca^{2+} + Mg^{2+}\right) \tag{6}$$

all concentrations are in meq/L.

6.4 Permeability Index (PI)

Permeability index (PI) is important to deduce the groundwater appropriateness for irrigation aims. Water content of calcium, sodium, bicarbonates and magnesium influences by long term use the permeability of the soil. Doneen (1962) has calculated PI value using Eq. 7 where ions are in meq/L.

$$PI = 100 * \frac{Na^{+} + \sqrt{HCO_3^{-}}}{Ca^{2+} + Mg^{2+} + Na^{+}} \tag{7}$$

6.5 EC and SAR Relationship

EC together with SAR are good indicators for the availability of groundwater for cultivation purposes. SAR and EC data could be exhibited using a USSL (1954) diagram to decide the validity of water for irrigation (Fig. 15). The USSL diagram shows EC on the horizontal versus SAR on the vertical axes. USSL diagram reveals that about 93% of the water samples show a wide range of both salinity and alkalinity having the following types C2S1, C3S1 and C3S2. These types reveals that the groundwater types are good to excellent depending on the sodium hazard class and are good to doubtful depending on the salinity hazard, while the rest of samples: S14, S19, S46 and S49 are not suitable for irrigation because they are classified as C4S3 and C4S4 (see Fig. 16 to see the locations of these samples).

6.6 Magnesium Hazard (MH)

Generally, both calcium and magnesium ions are found in equilibrium state in groundwater (Negm and Armanuos 2017). High levels of Mg^{2+} in water cause converting the quality of soils to alkaline and reducing the crops. Szabolcs and Darab (1964) suggested an index expresses the water suitability for irrigation purposes, this parameter is known as magnesium hazard (MH) which is given by Eq. 8:

$$MH = Mg^{2+}/\left(Ca^{2+} + Mg^{2+}\right) \times 100 \tag{8}$$

where Mg^{2+} and Ca^{2+} concentrations are in meq/L.

When MH exceeds 50%, then groundwater would be harmful and unsuitable for irrigation aims. Generally, groundwater samples are classified based on MH ratio (Table 3) to be suitable for irrigation goals (89.48%), whereas only 10.52% of the samples (boreholes 1, 2, 32, 45, 54, and 55) are classified as unsuitable for the agricultural purposes (refre to Table 4).

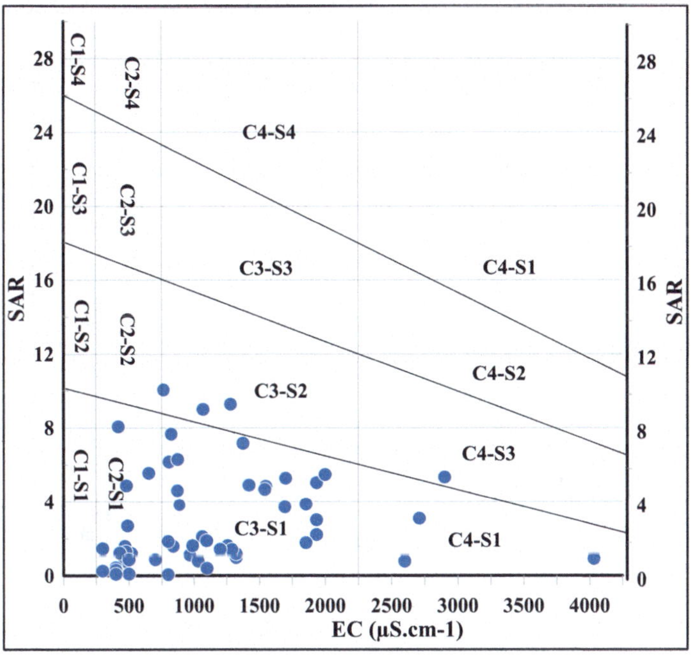

Fig. 15 SAR-EC water classification diagram for irrigation purposes

6.7 *Kelly's Ratio (KI)*

Kelly's ratio is also a parameter that indicates the validity of groundwater for agricultural activities. This ratio is expressed by the following formula (Negm and Armanuos 2017):

$$KI = Na^{+}/\left(Ca^{2+} + Mg^{2+}\right) \quad (9)$$

where Ca^{2+}, Na^{+}, and Mg^{2+} concentrations are in meq/L.

According to Narsimha et al. (2013) and according to the calculated KI values, groundwater would be grouped into suitable water for irrigation purposes (KI < 1) representing 73.7% of the groundwater samples, while 26.3% of the samples are unsuitable water (KI > 1).

6.8 *Chloride Classification*

Chloride concentration in groundwater is affected by agricultural activities, and human and animal waste sources because Cl^{-} is well transported through the soil (Stallard and Edmond 1981). Stuyfzand (1989) has classified groundwater based

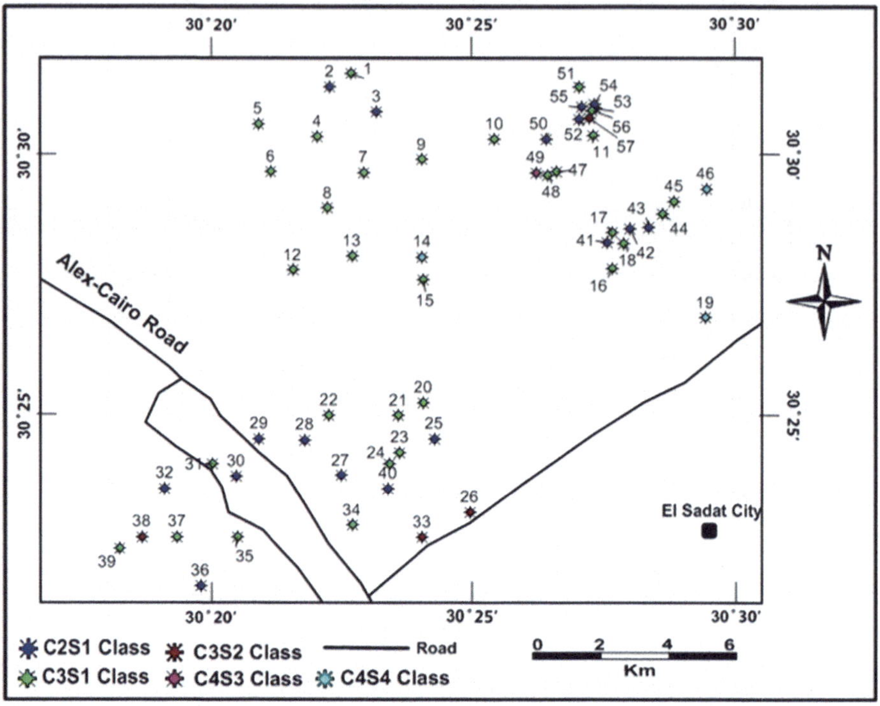

Fig. 16 Spatial distribution map for the types revealed from SAR-EC water classification diagram

on Cl^- concentration in meq/L into: hypersaline, salt, brackish-salt, brackish, fresh-brackish, fresh, very fresh and extremely fresh. Thus, our groundwater samples are grouped into: very fresh water (61.4%), fresh-brackish water (22.8%), and brackish water (15.8%).

6.9 *Chloro-Alkaline Indices (CAI)*

This index denotes the aquifer's ion exchange phase which occurs between the soil and the groundwater. CAI is expressed by Eq. 10 (Schoeller 1967).

$$\text{CAI} = \frac{Cl^- - \left(Na^+ + K^+\right)}{Cl^-} \tag{10}$$

Based on CAI values; negative values refer to ion exchange phase between K^+ and Na^+ in groundwater and Mg^{2+} and Ca^{2+} in the rock (Negm and Armanuos 2017). By calculating the CAI values, it could be concluded that 57.9% of water samples has negative CAI values (Table 4) and this is emphasized by Durov diagram.

6.10 Corrosively Ratio (CR)

Corrosively ratio (CR) is an index reflecting the safety for groundwater transportation in metallic pipes (Negm and Armanuos 2017). If this ratio is less than 1, this means that it is safe to transport groundwater in any type of pipes, otherwise if it is higher than 1, this refers to the corrosive nature of the groundwater and it would be not safe to transport water through metal pipes. This index is given by Eq. 11:

$$CR = \left(\frac{Cl^-}{35.5} + 2\frac{SO_4{}^{2-}}{96}\right)/2\left(\frac{HCO_3{}^- + CO_3{}^{2-}}{100}\right) \tag{11}$$

All anions' concentrations are in meq/L unit.

In accordance with the calculated CR values (Table 4), it would be deduced that the CR values of almost water samples are greater than 1, so it is unsafe to transport this water in metal pipes for long distances.

7 Results and Discussion

Hydrochemistry has an incredible significance in quality assurance and characterization of groundwater, and subsequently, choice if this water is appropriate for various demands, for example, drinking, household, cultivation, and industrial purposes. Depending on the groundwater analysis of 57 collected samples from several drilled boreholes in the surveyed region, important information about the concentrations of both major anions and cations, Fe–Mn trace elements, furthermore TDS and EC are obtained. Groundwater reasonableness to use for drinking and irrigation was assessed depending on the realized standard values of the guidelines.

Several numerical indicators including EC, TDS and TH have been studied. EC concentrations range between 300 and 4032 μS/cm with the value of 1127.83 μS/cm on average. As a result, groundwater in the examined region could be divided into permissible water (59.65%) and good water (33.33%) categories, with just 3 water samples categorised as doubtful (boreholes 19, 46, 49). Only one sample from borehole 14 was declared unsuitable for drinking. Concerning TDS level, when TDS level in the water samples is beneath 600 mg/L, the groundwater would be suitable for drinking while the threshold limit is 1000 mg/L (WHO 2011). TDS values change between 216 and 2500 mg/L with an average value of 709 mg/L. Most of the water samples collected from the study area are categorized into freshwater (79.31%, their TDS < 1000 mg/L), except 12 samples represent brackish water (21.05%). Furthermore, it is worth mentioning that groundwater having TDS values below 600 mg/L represents 50.88% of the existing samples.

Distribution maps of TDS, EC and TH exhibit a high EC and TDS level zone at the northwestern part. It is also obvious that there are moderate concentration zones in the central (extending NW–SE) and eastern regions at the studied area, while the lowest levels, on the other hand, are clustered towards the southwest. This might be resulted from different reasons such as recharge decrease from the Rosetta branch and/or increase discharge process for agricultural drainage. Moreover, EC has a strong relation with TDS, TH, Ca^{2+}, Cl^{-}, Na^{+} and intermediate linkage with HCO_3^{-}, SO_4^{2-}, Mg^{2+} indicating that such cations and anions have an identical source and are implicated in the reactions of ion exchange (Rao 2002). Furthermore, TDS has high correlations with Ca^{2+}, Mg^{2+}, and Na^{+}. As EC could be estimated directly in the field; hence, TDS level can be computed by using the obtained formula (Eq. 2). Hence, with great benefits such as saving time, money, and efforts, an economic study of water quality can be carried out using this equation.

Groundwater's total hardness reveals that the maximum concentration is 1163 mg/L at well 46 where there are some high contents of Mg^{2+} and Ca^{2+} (264 and 122.7 mg/L; respectively), while the minimum value is 63 mg/L at well 26 this corresponds low concentrations of Ca^{2+} (12.8 mg/L) and Mg^{2+} (7.68 mg/L). The average hardness value is 391 mg/L. According to hardness levels, groundwater is grouped into soft (3.51%), moderately hard (8.77%), hard (33.33%), and very hard (54.39%) waters (Tables 1 and 4). TH is highly related to Mg^{2+}, Ca^{2+}. This analysis suggests that the groundwater in the study area is rich with Mg^{2+} and Ca^{2+} salts. From the previous numerical indicators, it could be extracted that most of the groundwater is suggested to be fresh to brackish water and suitable for drinking and domestic utilities.

Groundwater quality for agricultural purposes was evaluated based on several indicators such SC, PI, SAR, RSC, MH, Kelly's ratio, and chloride classification. These indicators suggest suitable and safe groundwater for agricultural activities (Tables 3 and 4). SC is a very significant index for agriculture purposes. Based on Wilcox (1955) classification, groundwater is categorized into good to excellent water (54.39% of the samples), permissible water (24.56% of water samples), while 17.54% is doubtful water for irrigation, but 3.51% of the samples (found at boreholes no. 26 and 34) are not suited for irrigation as SC exceeds 80%. The estimated groundwater SAR values change from 0.07 to 10.06 epm. So, it is worth to conclude that the groundwater in the research region is ideal for irrigation aims for all types of soils with very low sodium hazards.

Based on RSC values, generally groundwater in the study area is suitable for both drinking and irrigation purposes. 91.13% of the water has RSC less than1.25 while 8.77% exceeds the acceptable value (2.5 mg/L) at five boreholes (1, 4, 2, 26, and 34) that suggests the unsuitability of the water for irrigation purposes (Eaton 1950; Richards 1954). Another classification of groundwater is depending on PI values. This classification is subdivided into three classes: I, II, and III. Naseem et al. (2010) mentioned classes I and II when the highest permeability is 75% or more, then groundwater is good for irrigation, furthermore class III is obtained when the highest permeability is 25% and this causes the unsuitability of the groundwater for irrigation. Calculated values of PI reveal that classes I and II account 91.23% of the samples, indicating the suitability of the groundwater for irrigation. However, 8.77% of the water samples override the PI limits and are described as unsuitable water. This only applies to five groundwater samples at boreholes 14, 27, 44, 45, and 46.

Both electrical conductivity and the sodium concentration (%) are commonly used to describe the groundwater quality for agricultural needs (Yidana et al. 2010). Representing EC versus SAR values on the USSL diagram reveals that 93% of groundwater samples are of the following types C2S1, C3S1 and C3S2, which indicates that these waters are good to excellent based on the sodium hazard class and these waters also are good to doubtful depending on the salinity hazard. Moreover, about 7% of the samples (S14, S19, S46 and S49) are not suitable for irrigation purposes as they are classified as C4S3 and C4S4.

8 Conclusions

Results obtained from the hydrochemical interpretation show a clear harmony deduced from EC, TDS, and TH distribution maps. In the northwest part, it appears to have high values of EC, TDS, and TH. Moreover, cations distribution maps present high values at the same area. Therefore, this part is somehow not suitable for domestic purposes but could be used for agriculture utilities with some restrictions. The rest regions of the area have moderate to low values of EC, TDS, TH, cations, and anions levels. Hence, at these areas, waters are good for the different uses. In majority of the boreholes, the EC and TDS readings indicate fresh to brackish water, while the hardness indicates hard to extremely hard water. The hydrochemical outcomes were correlated with the levels of well-known standard guidelines. Water types in the area are classed as follows based on the amounts of major ions: Ca–HCO_3 (22.81%), Ca–SO_4 (17.54%), Na–SO_4 (15.79%), Na–Cl (14.03%), Ca–Cl (14.03%), Na–HCO_3 (10.53%), Mg–HCO_3 (3.51%), and Mg–SO_4 (1.75%). The research area's groundwater is characterised by a high concentration of alkaline earth and strong acids. Also, the cations and anions have mostly the same source and are implicated in the ion exchange reactions. Different water quality indicators such as SC, SAR, MH, PI, and RSC recommend safe and proper groundwater for agricultural objectives.

9 Recommendations

The southern, central, and eastern parts of the study area are thought to be the best locations for future water supply boreholes. However, prior to any process of groundwater exploitation in such favorable district, excessive research should be considered to control the basis for groundwater abstraction in order to avoid potential difficulties related with excessive groundwater abstraction. This involves e.g. groundwater age dating using isotopes, recognizing and distinguishing areas of recharge, and simulating the impact of various groundwater abstraction scenarios. Moreover, it is important to detect the recharge areas to protect them and avoid pollution problem of the aquifer.

References

Abd El-Kawy OA, Rød JK, Ismail HA, Suliman AS (2011) Land use and land cover change detection in the western Nile delta of Egypt using remote sensing data. Appl Geogr 31:483–494
Abdel Aal A, Price R, Vaitl J, Shrallow J (1994) Tectonic evolution of the Nile delta, its impact on sedimentation and hydrocarbon potential. EGPC 12:19–34
Abdel Aal A, El Barkooky A, Gerrits M, Meyer HJ, Schwander M, Zaki H (2001) Tectonic evolution of the eastern Mediterranean basin and its significance for the hydrocarbon prospectivity of the Nile delta deepwater area. GeoArabia 6:363–384
Abdel Baki AA (1983) Hydrogeological and hydrogeochemical studies in the area west of Rosetta branch and south of El Nasr canal. Ph.D. thesis, Faculty of Science, Ain Shams University, Cairo, Egypt, p 156
Abd-Elhamid H, Abdelaty I, Sherif M (2019) Evaluation of potential impact of Grand Ethiopian Renaissance Dam on seawater intrusion in the Nile delta aquifer. Int J Environ Sci Technol 16(5):2321–2332
Aghazadeh N, Mogaddam AA (2010) Assessment of groundwater quality and its suitability for drinking and agricultural uses in the Oshnavieh area, northwest of Iran. J Environ Prot 1:30
Ahmed S (1999) Hydrological and isotope assessment of groundwater in Wadi El-Natrun and Sadat city, Egypt. Ain Shams University, Cairo, p 237
Ahmed MA, Samie SA, El-Maghrabi HM (2011) Recharge and contamination sources of shallow and deep groundwater of Pleistocene aquifer in El-Sadat industrial city: isotope and hydrochemical approaches. Environ Earth Sci 62(4):751–768
Al-Bassam AM, Awad HS, Al-Alawi JA (1997) Durov plot: a computer program for processing and plotting hydrochemical data. Ground Water 35(2):362–367
Al-Kilany S (2001) A new geoelectrical studies on groundwater aquifer, El-Sadat area and its vicinities Egypt. Dissertation, Menoufiya University, Egypt
Appelo CAJ, Postma D (1996) Geochemistry groundwater and pollution. Balkema, Rotterdam
Atta S, Sharaky A, El Hassanein A, Khallaf K (2005) Salinization of the groundwater in the coastal shallow aquifer, northwestern Nile delta, Egypt. In: First international conference on the geology of Tethys, Cairo University, pp 151–166
Attwa M, Ali H (2018) Resistivity characterization of aquifer in coastal semiarid areas: an approach for hydrogeological evaluation. In: Groundwater in the Nile delta. Springer, Cham, pp 213–233
Bartarya S (1993) Hydrochemistry and rock weathering in a sub-tropical lesser Himalayan river basin in Kumaun, India. J Hydrol 146:149–174
Burbery L, Vincent C (2009) Hydrochemistry of south Canterbury tertiary aquifers. Environment Canterbury. Christchurch, New Zealand
Chow VT (1964) Handbook of Applied Hydrology. Sec. 11: Evapotranspiration. Mc-Graw Hill Co., New York, pp 11–38
Dauda M, Habib G (2015) Graphical techniques of presentation of hydro-chemical data. J Environ Earth Sci 5:65–76
Dawoud MA, Darwish MM, El-Kady MM (2005) GIS-based groundwater management model for western Nile delta. Water Resour Manage 19(5):585–604
Doneen LD (1962) The influence of crop and soil on percolating water. In: Proceeding 1961 biennial conference on groundwater recharge, pp 156–163
Durov S (1948) Natural waters and graphic representation of their composition. Dokl Akad Nauk SSSR 59:87–90
Eaton FM (1950) Significance of carbonates in irrigation waters. Soil Sci 69(2):123–134
Edmunds W, Cook J, Darling W, Kinniburgh D, Miles D, Bath A, Morgan-Jones M, Andrews J (1987) Baseline geochemical conditions in the Chalk aquifer, Berkshire, UK: a basis for groundwater quality management. Appl Geochem 2:251–274
El Abd E (2005) The geological impact on the water bearing formations in the area southwest Nile delta, Egypt. Unpublished PhD thesis, Faculty of Science, Menufiya University, Egypt

El Kashouty M, El Sabbagh A (2011) Distribution and immobilization of heavy metals in Pliocene aquifer sediments in Wadi El Natrun depression, Western Desert. Arab J Geosci 4:1019–1039

El Osta M, Hussein H, Tomas K (2018) Numerical simulation of groundwater flow and vulnerability in Wadi El-Natrun depression and vicinities, west Nile delta, Egypt. J Geol Soc India 92(2):235–247

ElGalladi A, Ghazala HH, Ibraheem IM, Ibrahim E-K (2009) Tectonic interpretation of gravity field data of west of Nile delta region, Egypt. NRIAG J Astron Geophys Egypt Spec Issue 487–508

Embaby A (2003) Environmental evolution for geomorphological situation in relation to the water and soil resources of the region north of the Sadat city, west Nile delta, Egypt. Unpublished Ph.D. thesis, Faculty of Science, Mansoura University, Egypt, vol 60, p 323

Fetter CW (2014) Applied hydrogeology, 4th edn. Pearson, UK, p 604

Gaber A, Mohamed AK, ElGalladi A, Abdelkareem M, Beshr AM, Koch M (2020) Mapping the groundwater potentiality of West Qena Area, Egypt, using integrated remote sensing and hydro-geophysical techniques. Remote Sens 12(10):1559

Gad M, El-Hendawy S, Al-Suhaibani N, Tahir MU, Mubushar M, Elsayed S (2020) Combining hydrogeochemical characterization and a hyperspectral reflectance tool for assessing quality and suitability of two groundwater resources for irrigation in Egypt. Water 12(8):2169

Garfunkel Z (1998) Constrains on the origin and history of the eastern Mediterranean basin. Tectonophysics 298:5–35

Geirnaert W, Laeven M (1992) Composition and history of ground water in the western Nile delta. J Hydrol 138:169–189

Ghoraba SM, Khan AD (2013) Hydrochemistry and groundwater quality assessment in Balochistan province, Pakistan. Int J Res Rev Appl Sci 17(2):185–199

Guiraud R, Bosworth W (1999) Phanerozoic geodynamic evolution of northeastern Africa and the northwestern Arabian platform. Tectonophysics 315:73–104

Harms JC, Wray JL (1990) Nile delta. In: Said R (ed) The geology of Egypt. Balkema, Rotterdam, Netherlands

Heath RC (1983) Basic groundwater hydrology. US geological survey water supply paper 2220. US Geological Survey. https://doi.org/10.3133/wsp2220

Hegazy D, Abotalib AZ, El-Bastaweesy M, El-Said MA, Melegy A, Garamoon H (2020) Geo-environmental impacts of hydrogeological setting and anthropogenic activities on water quality in the Quaternary aquifer southeast of the Nile delta, Egypt. J Afr Earth Sci 172:103947

Hopkins BG, Horneck DA, Stevens RG, Ellsworth JW, Sullivan DM (2007) Managing irrigation water quality for crop production in the Pacific Northwest. [Covallis, Or.]: Oregon State University Extension Service

Hudson RO, Golding DL (1997) Controls on groundwater chemistry in subalpine catchments in the southern interior of British Columbia. J Hydrol 201(1–4):1–20

Ibraheem IM (2009) Geophysical potential field studies for developmental purposes at El-Nubariya–Wadi El-Natrun, west Nile delta, Egypt. Ph.D thesis, Faculty of Science, Mansoura University, Egypt, p 193

Ibraheem IM, El-Qady G (2017) Hydrogeophysical investigations at El-Nubariya–Wadi El-Natrun area, west Nile delta, Egypt. In: The handbook of environmental chemistry. Springer, Berlin, Heidelberg. https://doi.org/10.1007/698_2017_154

Ibraheem IM, El-Qady G (2019) Hydrogeophysical investigations at El-Nubariya–Wadi El-Natrun area, west Nile delta, Egypt. In: Negm A (ed) Groundwater in the Nile delta. The handbook of environmental chemistry, vol 73. Springer, Cham, pp 235–271

Ibraheem IM, El-Qadi GM, ElGalladi A (2016) Hydrogeophysical and structural investigation using VES and TDEM: a case study at El-Nubariya–Wadi El-Natrun area, west Nile delta, Egypt. NRIAG J Astron Geophys 5(1):198–215

Ibraheem IM, Elawadi EA, El-Qady GM (2018) Structural interpretation of aeromagnetic data for the Wadi El Natrun area, northwestern desert, Egypt. J Afr Earth Sci 139:14–25

Jacks G (1973) Chemistry of ground water in a district in southern India. J Hydrol 18:185–200

Karanth KR (1991) Impact of human activities on hydrogeological environment. J Geol Soc India 38(2):195–206

Kimblin RT (1995) The chemistry and origin of groundwater in Triassic sandstone and Quaternary deposits, northwest England and some UK comparisons. J Hydrol 172(1–4):293–308

Lloyd JW, Heathcote JA (1985) Natural inorganic hydrochemistry in relation to groundwater: an introduction. Oxford University Press, New York, p 296

Massoud U, Kenawy AA, Ragab ES, Abbas AM, El-Kosery HM (2014) Characterization of the groundwater aquifers at El Sadat city by joint inversion of VES and TEM data. NRIAG J Astron Geophys 3:137–149

Mayo AL, Loucks MD (1995) Solute and isotopic geochemistry and groundwater flow in the central Wasatch Range, Utah. J Hydrol 172(1–4):31–59

McCool JP (2019) Carbonates as evidence for groundwater discharge to the Nile river during the Late Pleistocene and Holocene. Geomorphology 331:4–21

Menco (1990) Miser company for engineering and agricultural projects report

Miller JA (1991) Summary of the hydrology of the southeastern coastal plain aquifer system in Mississippi, Alabama, Georgia, and South Carolina. U.S. Geol Surv Prof 1410-A:33–36

Mohamed A (2020) Gravity applications to groundwater storage variations of the Nile delta aquifer. J Appl Geophys 104177

Mohamed RF, Hua CZ (2010) Regional groundwater flow modeling in western Nile delta, Egypt. World Rural Obs 2:37–42

Mosaad S, Basheer AA (2020) Utilizing the geophysical and hydrogeological data for the assessment of the groundwater occurrences in Gallaba plain, western desert, Egypt. Pure Appl Geophys 1–22

Narsimha A, Sudarshan V, Swathi P (2013) Groundwater and its assessment for irrigation purpose in Hanmakonda area, Warangal district, Andhra Pradesh, India. Int J Res Chem Environ 3(2):196–200

Naseem S, Hamza S, Bashir E (2010) Groundwater geochemistry of Winder agricultural farms, Balochistan, Pakistan and for irrigation water quality. Eur Water 31:21–32

Negm AM, Armanuos AM (2017) GIS-based spatial distribution of groundwater quality in the western Nile delta, Egypt. In: Negm AM (ed) The Nile delta. Springer, Cham, pp 89–119

Omara SM, Sanad S (1975) Rock stratigraphy and structural features of the area between Wadi El Natrun and the Moghra depression, Western Desert, Egypt. Geol J Hannover 16:45–73

Othman A, Ibraheem IM, Ghazala H, Mesbah H, Dahlin T (2019) Hydrogeophysical and hydrochemical characteristics of Pliocene groundwater aquifer at the area northwest El Sadat city, west Nile delta, Egypt. J Afr Earth Sci 150:1–11

Patil VB, Pinto SM, Govindaraju T, Hebbalu VS, Bhat V, Kannanur LN (2020) Multivariate statistics and water quality index (WQI) approach for geochemical assessment of groundwater quality—a case study of Kanavi Halla sub-basin, Belagavi, India. Environ Geochem Health 1–18

Piper AM (1944) A graphic procedure in the geochemical interpretation of water-analyses. EOS Trans Am Geophys Union 25(6):914–928

Piper AM (1953) A graphic procedure in the geochemical interpretation of water analysis. US department of the interior, geological survey. Water Resources Division, Ground Water Branch, Washington

Raju NJ (2007) Hydrogeochemical parameters for assessment of groundwater quality in the upper Gunjanaeru River basin, Cuddapah district, Andhra Pradesh, South India. Environ Geol 52(6):1067–1074

Rao NS (2002) Geochemistry of groundwater in parts of Guntur district, Andhra Pradesh, India. Environ Geol 41(5):552–562

Richards LA (1954) Diagnosis and improvement of saline and alkali soils. Handbook, vol 60. U.S. Department of Agriculture, Washington, DC, p 160

RockWare Inc. (2004) A user's guide to RockWare® Aq·QA®, version 1.1. RockWare, Inc. Golden, Colorado, USA

Rogers RJ (1989) Geochemical comparison of ground water in areas of New England, New York, and Pennsylvania. J Groundwater 27(5):690–712

Sadek HS, Soliman SA, Abdulhadi HM (1990) A correlation between the different models of resistivity sounding data to discover a new fresh water field in El Sadat city, western desert of Egypt. In: Geophysical data inversion methods and applications. Vieweg + Teubner Verlag, Wiesbaden, pp 329–346
Said R (1962) The geology of Egypt. Elsevier, Amsterdam
Salem ZE, El-Bayumy DA (2016) Hydrogeological, petrophysical and hydrogeochemical characteristics of the groundwater aquifers east of Wadi El-Natrun, Egypt. NRIAG J Astron Geophys
Salem ZE, Osman OM (2017) Use of major ions to evaluate the hydrogeochemistry of groundwater influenced by reclamation and seawater intrusion, west Nile delta, Egypt. Environ Sci Pollut Res 24(4):3675–3704
Salem ZE, Gaame OM, Hassan TM (2018) Integrated subsurface thermal regime and hydrogeochemical data to delineate the groundwater flow system and seawater intrusion in the middle Nile delta, Egypt. In: Groundwater in the Nile delta. Springer, Cham, pp 461–486
Salem ZE, Sefelnasr AM, Hasan SS (2019) Assessment of groundwater vulnerability for pollution using DRASTIC index, young alluvial plain, western Nile delta, Egypt. Arab J Geosci 12(23):727
Salomons W, Forstner U (1984) Metals in the hydrocycle. Springer-Verlag, Berlin, p 349
Sarhan M, Hemdan K (1994) North Nile delta structural setting and trapping mechanism, Egypt. In: Proceedings of the 12th petroleum conference of EGPC, vol 12, pp 1–17
Sarhan M, Barsoum F, Bertello F, Talaat M, Nobili M (1996) The Pliocene play in the Mediterranean offshore: structural setting and growth faults controlled hydrocarbon accumulations in the Nile delta basin. A comparison with the Niger delta basin. In: 13th EGPC, Egyptian general petroleum corporation conference, vol 1, Cairo, pp 1–19
Sayyed A, Bhosle B (2011) Analysis of chloride, sodium and potassium in groundwater samples of Nanded City in Mahabharata, India. Europ J Exp Biol 1:74–82
Schoeller H (1967) Geochemistry of groundwater. An international guide for research and practice, chap 15. UNESCO, pp 1–18
Şen Z (2015) Practical and applied hydrogeology. Elsevier, p 407
Sharaky AM, Atta SA, El Hassanein AS, Khallaf KM (2007) Hydrogeochemistry of groundwater in the western Nile delta aquifers, Egypt. In: 2nd international conference on the geology of Tethys, vol 1, pp 1–23
Sharaky AM, El Hasanein AS, Atta SA, Khallaf KM (2017) Nile and groundwater interaction in the Western Nile Delta, Egypt. In: Negm AM (ed) The Nile Delta. Springer, Cham, pp 33–62
Shata A (1953) New light on the structural development of the western desert of Egypt. Bull Inst Desert Egypt 6:117–157
Shata A (1955) An introductory note on the geology of the northern portion of the western desert of Egypt. Bull Inst Desert Egypt 5(3):87–98
Shata A (1959) Geological problems related to the groundwater supply of some desert areas of Egypt. Bull Soc Geogr Egypt 32:247–262
Shata A (1961) The geology of groundwater supplies in some arable lands in the desert of Egypt. Desert Institute, Cairo, Egypt
Shedid A (1989) Geological and hydrogeological studies of El Sadat area and its vicinities. MSc thesis, Faculty of Science, El Menoufia University, Shibin El Kom, Cairo, Egypt, p 157
Sherif M (1999) Nile delta aquifer in Egypt. In: Seawater intrusion in coastal aquifers—concepts, methods and practices. Springer
Stallard R, Edmond J (1981) Geochemistry of the Amazon: 1. Precipitation chemistry and the marine contribution to the dissolved load at the time of peak discharge. J Geophys Res 86:9844–9858
Stallard R, Edmond J (1983) Geochemistry of the Amazon: 2. the influence of geology and weathering environment on the dissolved load. J Geophys Res Oceans 88:9671–9688
Stimson J, Frape S, Drimmie R, Rudolph D (2001) Isotopic and geochemical evidence of regional-scale anisotropy and interconnectivity of an alluvial fan system, Cochabamba Valley, Bolivia. Appl Geochem 16(9–10):1097–1114

Stuyfzand PJ (1989) Hydrology and water quality aspects of Rhine bank groundwater in the Netherlands. J Hydrol 106(3–4):341–363

Szabolcs I, Darab C (1964) The influence of irrigation water of high sodium carbonate content on soils. Agrokém Talajt 13:237–246

Todd DK, Mays LW (2005) Groundwater hydrology, 3rd edn. Wiley, Hoboken, NJ, p 891

USL Staff (1954) Diagnosis and improvement of saline and alkali soils. In: Agriculture handbook, vol 60, pp 83–100

Walton NRG (1989) Electrical conductivity and total dissolved solids—what is their precise relationship? Desalination 72(3):275–292

Wilcox LV (1955) Classification and use of irrigation waters. Circ. 969. US Department of Agriculture, Washington, DC, p 19

World Health Organization (WHO) (2011) Guidelines for drinking-water quality, vol 216. World Health Organization, pp 303–304

Yidana SM, Banoeng-Yakubo B, Akabzaa TM (2010) Analysis of groundwater quality using multivariate and spatial analyses in the Keta basin, Ghana. J Afr Earth Sci 58(2):220–234

Zaghloul Z (1976) Stratigraphy of the Nile delta. In: Proceedings of the seminar on Nile delta sedimentology, pp 40–49

Zaporozec A (1972) Graphical interpretation of water quality data. Groundwater 10(2):32–43

As a Water Resources Management Tool, Groundwater Quality Assessment for Irrigation in the Young Alluvial Plain of Western Nile Delta, Egypt

Zenhom E. Salem, Ahmed Sefelnasr, and Samia S. Hasan

Abstract Seventy-five groundwater samples were collected and chemically analyzed for assessing groundwater use for irrigation in the area between Rosetta Nile Branch and El Nubariya canal. The TDS, Na %, SAR, RSC, Cl, KI, PI, MH, CAI, CR and IQW were calculated. Approximately, 75% of the groundwater samples are usable having TDS less than 3000 mg/l. Concerning Na%, about 53% of the groundwater samples have good to permissible quality. The SAR values indicated that the majority of groundwater wells (82.6%) are suitable for irrigation, while, about 91% of samples have good quality based on RSC values and 31% of the samples are suitable for irrigation based on PI. The chloride content shows a changing from fresh (in the south) to salt characters (in the north). According to MH, most of the samples are suitable, while, based on KI; most of the samples are unsuitable. The CR values showed that about 83% of samples have corrosive properties. According to IQW, about 43% of samples have medium quality occupying the central part of the study area. Besides, about 35% of samples have good quality that located in the southern part, while, the low quality class appears in the northern part representing about 23% of the samples. Generally, the groundwater is good, accepted and unsuitable for irrigation in the southern, middle and northern parts of the study area, respectively.

Keywords Nile Delta · Quaternary aquifer · Irrigation water quality index · Water quality assessment · GIS

Z. E. Salem (✉) · S. S. Hasan
Geology Department, Faculty of Science, Tanta University, Tanta, Egypt
e-mail: zenhomsalem@yahoo.com

A. Sefelnasr
Geology Department, Faculty of Science, Assiut University, Asyut, Egypt

A. M. Negm and M. El-Rawy (eds.), *Sustainability of Groundwater in the Nile Valley, Egypt*, Earth and Environmental Sciences Library,
https://doi.org/10.1007/978-3-031-12676-5_8

1 Introduction

One of the basic human needs for survivability is appropriate quantity and suitable water quality, but keeping desirable water quality is a tough task for decision makers. Water quality assessment depends on numerous factors, such as physical, chemical and biological characteristics. Concluding the water quality from analyses of such variables is a difficult task (Mukate et al. 2019). To simplify this complex task, a large number of parameters need to be combined into a single composite value or model to describe water quality in an easily understandable manner without missing its scientific basis (Ponsadailakshmi et al. 2018; Wu et al. 2018).

The irrigation water quality index (IWQI) is an effective tool for determining the overall quality of water instead of traditional methods of water quality assessment. Different kinds of water quality index models have been developed by different researchers (Shabbir and Ahmad 2015; Singh et al. 2015; Arulbalaji and Gurugnanam 2017; Ewaid and Abed 2017; Sener et al. 2017; Adimalla and Li 2018; Khangembam and Kshetrimayum 2019; Mukate et al. 2019; Solangi et al. 2019). Geographic information system is applied to indicate the spatial distribution of water quality data. Evaluation of the groundwater quality in the Nile Delta was achieved by several authors, such as Armanuos et al. (2016), El Ramly (1997), Eltarabily and Negm (2018), Negm and Armanuos (2016), Salem et al. (2017a, b, 2018a, b).

This study aimed to determine the irrigation groundwater quality in the region between the Rosetta Nile branch and the El Nubariya canal using several quality parameters and IWQI. Mapping and parameters spatial distribution were made using GIS technique.

2 Study Area

The study area is a part of the western Nile Delta. It is bounded by the Rosetta Nile branch from east, the El Nubariya canal from west and south and the Mediterranean Sea in north (Fig. 1). The population of the region is about 4.43 million people, with a population density of 441 men/km^2. Various anthropogenic activities especially agricultural activities in the region depend mainly on the groundwater (Afifi and Darwish 2018). Generally, the study area slopes gently from west to east and from south to north. Western Nile Delta geology has been studied by several authors such as Shata (1955), Said (1962), Attia (1975), Abdel Baki (1983), Diab et al. (1995), Embaby (2003), Mohamed (2016), Salem and El Bayumy (2016a, b), Salem and Osman (2016), Sharaky et al. (2016), Tarabees and El-Qady (2016), Ibraheem and El-Qady (2017), Salem and Osman (2017, 2018), Sobeih et al. (2017), El Osta et al. (2018), Salem et al. (2019), and Armanuos et al. (2020).

The study area is covered entirely by sediments that which belong to Quaternary and Tertiary periods, where the Quaternary is represented, from base to top, by Mit-Ghamr Formation and Belqas Formation. Belqas Formation (of Holocene age) was

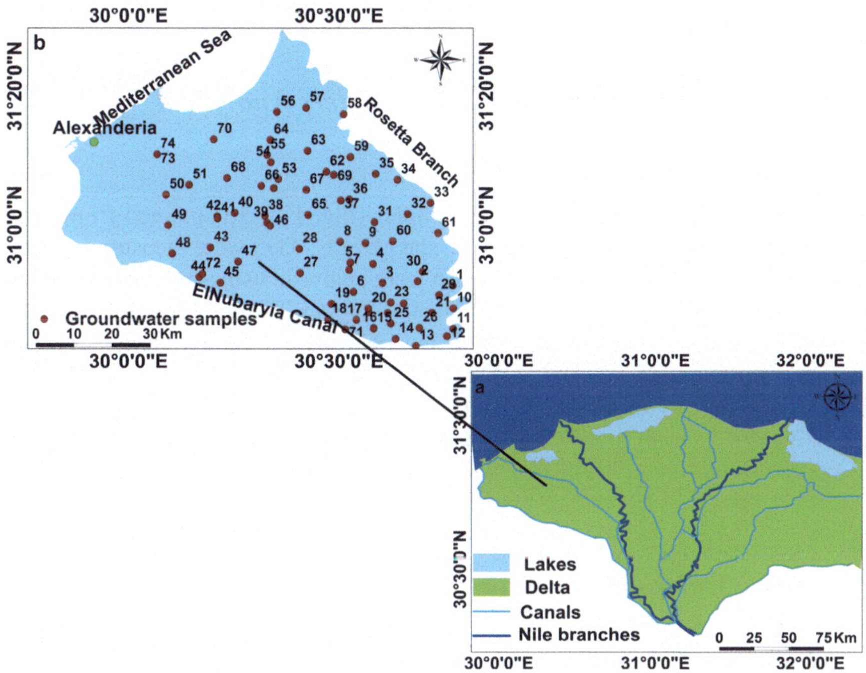

Fig. 1 Location map of the study area showing locations of the collected groundwater smaples

consisted of sand intertwined with clay-rich deposits. On the other hand, Mit Ghamr Formation (of Upper Pliocene–Pleistocene age) was consisted of sand and gravel, with interferences of clay deposits, representing the main aquifer in the Nile Delta, and overlain by Belqas Formation (Abu El-Ella 1990; Azzam 1994; Barakat 1982; Badran 1996; Schlumberger 1984).

The Quaternary aquifer is semi-confined with increasing thickness that varied from a few tens of meters at Cairo (in south) to about 1000 m on the Mediterranean shore (in north). The groundwater is recharged from the Nile water, as well as the leakage from irrigation and the surface water canals and drains (Attia 1985). The hydraulic parameters affecting the groundwater flow system are hydraulic conductivity (50–150 m/day), transmissivity (15,000–75,000 m^2/day), storage coefficient (10^{-4}–10^{-3} m/day) and porosity (25–40%), in addition to the pumping water rate that reaches 3.30 mm^3/year in different parts of the Nile Delta (Leaven1991; RIGW 2002; Salem et al. 2017c).

3 Methodology

3.1 *Groundwater Sampling and Parameters Analytical Procedure*

In the investigated area 75 groundwater samples (Fig. 1) were obtained from hand pumps and production wells during 2018. The pH, EC, TDS were measured in situ. Besides, Ca, Na, Mg, K, Al, Ba, Fe and Mn were analyzed using Inductive Coupled Plasma (ICP)—Optima 7000 DV. The Hach's Digital Titrator Model 16,900–01 was used for determining the alkalinity and chloride, while Sulfate and Nitrate anions were measured using Hach's Direct Reading Spectrophotometer–(DR/2000), following the standard analytical methods according to Hach (1990). Chemical analyses were carried out in the Center of Scientific Research and Measurements, and the Geology Department, Faculty of Science, Tanta University.

3.2 *Groundwater Quality Calculations*

3.2.1 Individual Parameters

Because the groundwater is an important source for irrigation in the western Nile Delta, assessing the groundwater suitability for irrigation needs is a necessary issue. Parameters such as TDS, Na%, SAR, RSC, Cl, KI, PI, MH, Cl, CAI, and CR (Table 1) were applied for evaluating the groundwater quality for irrigation. The parameters calculation equations are given in Table 1.

Table 1 Irrigation groundwater suitability parameters and its calculation equations

Parameter	Equation	Reference
Sodium adsorption ratio	$SAR = \frac{Na}{\sqrt{\frac{(Ca+Mg)}{2}}}$ (1)	Richards (1954)
Residual sodium carbonate	$RSC = (CO_3 + HCO_3) - (Ca + Mg)$ (2)	Richards (1954)
Sodium percent	$Na\% = \frac{(Na+K)}{(ca+Mg+Na+K)} \times 100$ (3)	Tiwari and Manzoor (1988)
Magnesium hardness	$MH = \frac{Mg}{(Ca+Mg)} * 100$ (4)	Szabolcs and Darab (1964)
Kelly's index	$KI = \frac{Na}{(Ca+Mg)}$ (5)	Kelly (1940)
Permeability index	$PI = \frac{(Na+\sqrt{HCO_3})}{Ca+Mg+Na} \times 100$ (6)	Doneen (1954)
Chloro-Alkaline index	$CAI = \frac{(Cl-(Na+K))}{Cl}$ (7)	Schoeller (1965)
Corrosivity ratio	$CR = \frac{\frac{Cl}{35.5}+2\left(\frac{SO_4}{96}\right)}{2\left(\frac{HCO_3+CO_3}{100}\right)}$ (8)	Aher and Gaikwad (2017)

3.2.2 Irrigation Water Quality Index (IWQI)

Table 2 shows the variables used for determining the IWQI, which were grouped based on the guidelines set by Ayers and Westcost (1985) and Simsek and Gunduz (2007). The IWQ index is calculated by Eq. (9) (Simsek and Gunduz 2007; Spandana et al. 2013).

$$\text{IWQI} = \sum_{i=1}^{5} G_i \tag{9}$$

where i is an incremental index and G is the value of each variable.

4 Results and Discussion

4.1 Individual Parameter

The obtained results of TDS, Na%, SAR, RSC, KI, PI, MH, Cl, CAI, and CR are summarized in Table 4 and their spatial distributions are shown in Figs. 2 and 3.

a. ***Total Dissolved Solids (TDS)***

Based on TDS values, about 75% of the samples are suitable and 25% of the samples are unfit for irrigation (Table 5). The suitable water is located in the central and southern parts, while the northern part had unsuitable groundwater (Fig. 2a).

b. ***Sodium Percent (Na%)***

Sodium percent was used to determine the suitability for irrigation (Tiwari and Manzoor 1988; Wilcox 1948). According to the computed Na%, the groundwater in the southern corner has good to permissible quality (Table 6), while, in the northern part it has a doubtful to unsuitable quality (Fig. 2b).

c. ***Sodium adsorption ratio (SAR)***

Sodium is a critical component as it represents an index of the soluble alkali/sodium influence on soils (Richards 1954). Since sodium decreases soil penetrability that has a negative impact on the seeding process. The calculated SAR values showed that most of samples (70.6%) are excellent, while 12% of samples have good quality. In addition, the groundwater in the area near the cost showed doubtful to unsuitable quality (Table 7 and Fig. 2c).

Table 2 IWQI's parameters classification (Ayers and Westcost 1985; Simsek and Gunduz 2007)

Hazard	Weight	Parameter	Range	Rating	Suitability	No.	%
Salinity hazard $*G_1 = w_1 r_1$ (10)	5	EC (μs/cm)	EC < 700	3	High	11	14.6
			700 ≤ EC ≤ 3000	2	Medium	40	53.4
			EC > 3000	1	Low	24	32
Infiltration and permeability hazard $*G_2 = w_2 r_2$ (11)	4	See Table 3 for details					
Specific ion toxicity $**G_3 = \frac{W_3}{3}\sum_{j=1}^{3} r_j$ (12)	3	SAR	SAR < 3.0	3	High	23	30.6
			3.0 ≤ SAR ≤ 9.0	2	Medium	26	34.7
			SAR > 9.0	1	Low	26	34.7
		B (mg/l)	B < 0.7	3	High		
			0.7 ≤ B ≤ 3.0	2	Medium		
			B > 3.0	1	Low		
		Cl (mg/l)	Cl < 140	3	High	15	20
			140 ≤ Cl ≤ 350	2	Medium	25	33.3
			Cl > 350	1	Low	35	46.7
Trace element toxicity $***G_4 = \frac{W_4}{N}\sum_{k=1}^{N} r_k$ (13)	2	Al (mg/l)	Al < 5.0	3	High	75	100
			5.0 ≤ Al ≤ 20.0	2	Medium	–	–
			Al > 20.0	1	Low	–	–
		Fe (mg/l)	Fe < 5.0	3	High	72	96
			5.0 ≤ Fe ≤ 20.0	2	Medium	2	2.6
			Fe > 20.0	1	Low	1	1.4
		Mn (mg/l)	Mn < 0.2	3	High	23	30.6
			0.2 ≤ Mn ≤ 10.0	2	Medium	52	69.4
			Mn > 10.0	1	Low	–	
Miscellaneous effects to sensitive cops $****G_5 = \frac{W_5}{3}\sum_{m=1}^{3} r_m$ (14)	1	NO_3–N (mg/l)	NO_3–N < 5.0	3	High	11	14.7
			5.0 ≤ NO_3-N ≤ 30.0	2	Medium	50	66.7
			NO3-N > 30.0	1	Low	14	18.6
		HCO_3 (mg/l)	HCO3 < 90	3	High	2	2.7
			90 ≤ HCO_3 ≤ 500	2	Medium	73	97.3
			HCO_3 > 500	1	Low	–	–
		pH	7.0 ≤ pH ≤ 8.0	3	High	32	42.7
			6.5 ≤ pH < 7.0 and 8.0 < pH ≤ 8.5	2	Medium	20	26.7
			pH < 6.5 or pH > 8.5	1	Low	23	30.6

*w is the group weighting, and r is the factor rating value, ** j is the incremental index, *** k is an incremental index and N is the total number of trace element available for the analysis and *** *m is an incremental index

Table 3 Classification of the infiltration and permeability hazard for irrigation water (Simsek and Gunduz 2007)

Parameters	Ranges					Rating	Suitability	No	%
SAR	< 3	3–6	6–12	12–20	> 20				
EC	> 700	> 1200	> 1900	2900	5000	3	High	49	65.3
	700–200	1200–300	1900–500	2900–1300	5000–2900	2	Medium	25	33.3
	< 200	< 300	< 500	< 1300	< 2900	1	Low	1	1.4

Table 4 The descriptive statistics of the obtained results of the water quality parameters

Parameter	Mean	Maximum	Minimum	Standard deviation
TDS	4080	43,600	190	8106
Na%	57.56	94.69	17.03	17.98
SAR	12.63	71.33	0.93	18.39
RSC	−11.61	3.82	−64.27	16.92
Cl	52.16	403.40	4	102.91
KI	2.15	17.73	0.183	2.56
PI	68.94	116.09	30.85	14.99
MH	45.92	73.94	31.98	10.59
CAI	−0.12	0.65	−1.43	0.45
CR	55.26	2383	0.66	277.36

d. ***Residual Sodium Carbonate (RSC)***

The relative abundance of sodium, alkaline earths, boron, HCO_3 and CO_3 in the soils have an additional effect on the suitability of irrigation water (Richards 1954). Table 8 indicated that about 91% of the groundwater samples have good quality, while 8% and 1.3% are doubtful and unsuitable, respectively (Fig. 2d).

e. ***Permeability Index (PI)***

Long-run use of irrigation water affects the soil penetrability, where the Ca, Mg, Na and HCO_3 in the soil influenced it. Doneen (1954) established the Permeability Index. Regarding to PI values, 69.7% (52 wells) of the collected samples represented in class II (Table 9) while 23 wells (30.6%) are related to class I and appear in the southern areas (Fig. 2e).

f. ***Chloro-Alkaline Index (CAI)***

The ion exchanging between the groundwater and aquifer matrix are expressed as CAI that is influential factor. When exchange occurs between the water Na^+ (or K^+) and the rock Ca^{+2} (or Mg^{+2}), the CAI index is positive and it is known as direct exchange process. While, the CAI values is negative, when the exchange occurs between the water Ca^{+2} (or Mg^{+2}) with the rock Na^+ (or K^+), and this is known as a reverse ion exchange (Schoeller 1965). The obtained results indicated the dominance of base-exchange (57.3%) that covers most of the study area (Table 10). Besides, cation–anion reactions (42.7%) mainly represented in the area located in the southeastern and northwestern parts (Fig. 2f).

g. ***Chloride Concentrations (Cl)***

Based on Cl ion concentrations, the groundwater of the southern areas belongs to the fresh and fresh–brackish classes, while, in the central and northern parts the groundwater varied from brackish, brackish–salt to salt class (Table 11 and Fig. 3a).

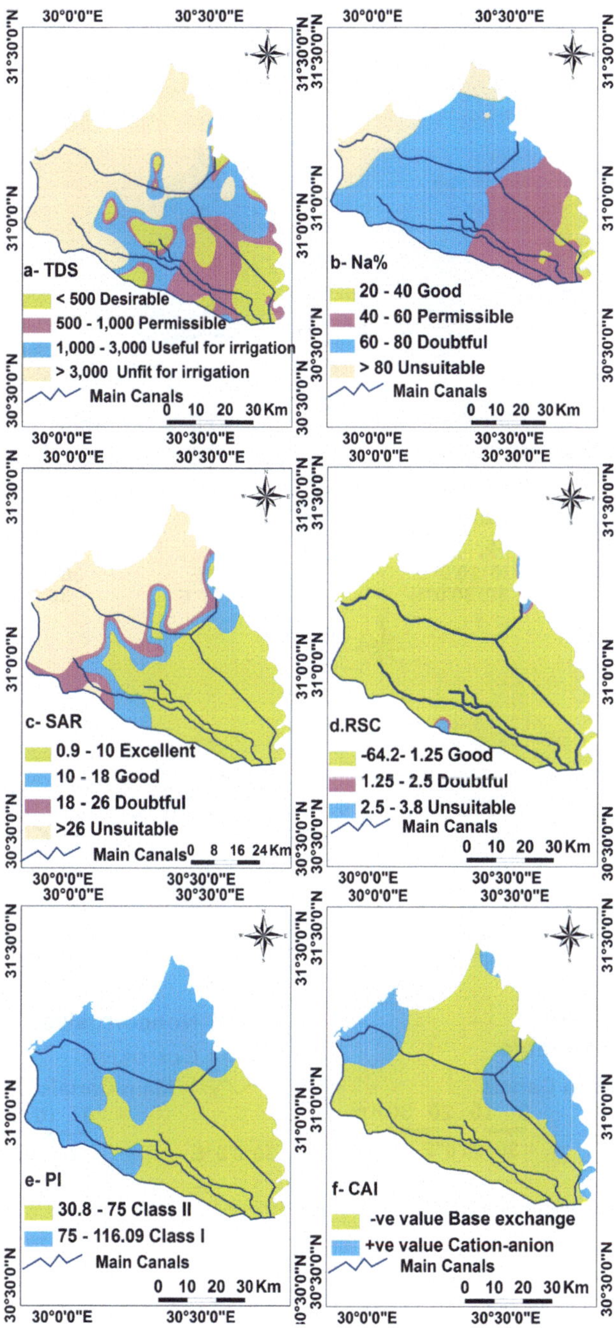

Fig. 2 Variation of the groundwater suitability for irrigation according to **a** TDS, **b** Na%, **c** SAR, **d** RSC, **e** PI, and **f** CAI

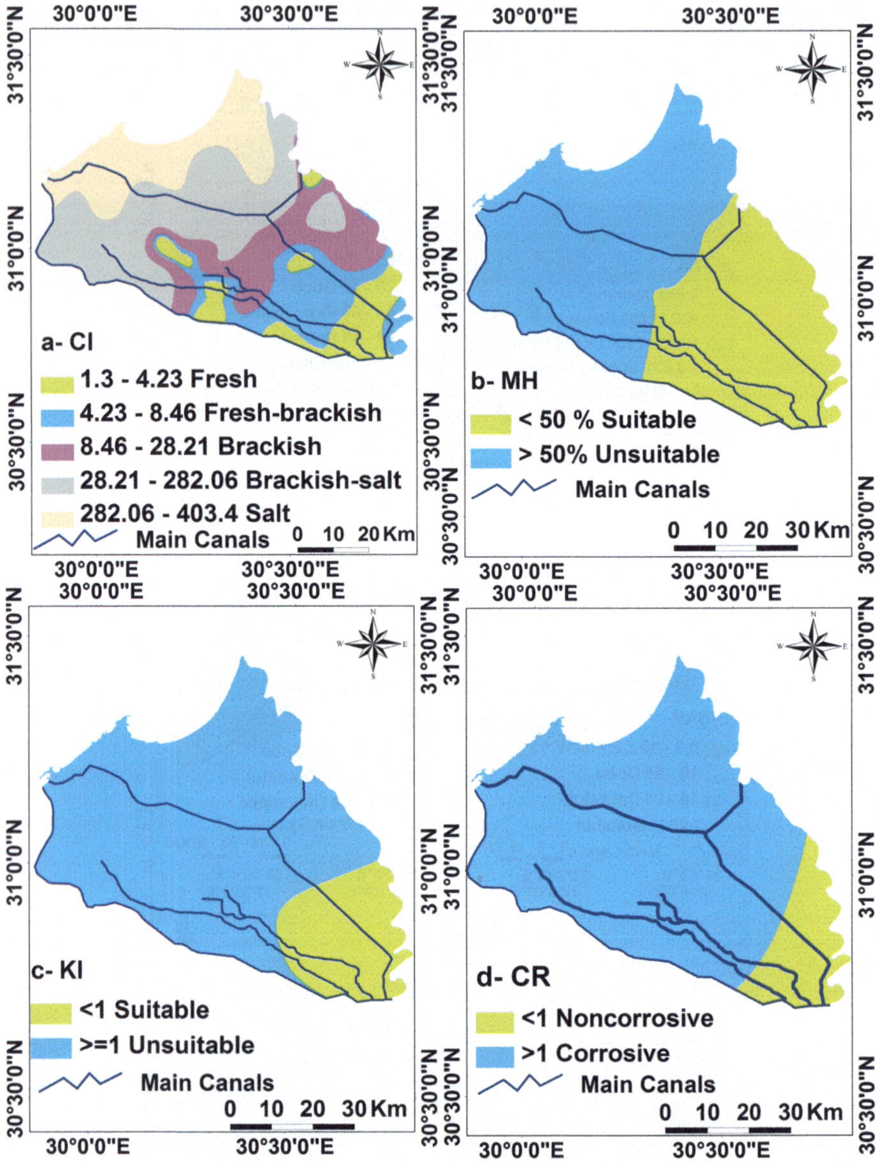

Fig. 3 Variation of the groundwater suitability for irrigation according to based on **a** Cl, **b** MH, **c** KI, and **d** CR

Table 5 Groundwater classification for irrigation according to TDS

TDS (mg/l)	Quality class	No. of samples	%
< 500	Desirable	19	25.3
500–100	Permissible	21	28
1000–3000	Useful	16	21.4
> 3000	Unfit	19	25.3

Table 6 Groundwater classification for irrigation according to Na%

Na%	Quality class	No. of samples	%
< 20	Excellent	–	–
20–40	Good	14	18.7
40–60	Permissible	29	38.7
60–80	Doubtful	20	26.6
> 80	Unsuitable	12	16

Table 7 Groundwater classification for irrigation according to SAR

SAR	Quality class	No. of samples	%
< 10	Excellent	53	70.6
10–18	Good	9	12
18–26	Doubtful	3	4
> 26	Unsuitable	10	13.4

Table 8 Groundwater classification for irrigation according to RSC

RSC	Quality class	No. of samples	%
< 1.25	Good	68	90.7
1.25–2.5	Doubtful	6	8
> 2.5	Unsuitable	1	1.3

Table 9 Groundwater classification for irrigation according to PI

PI	Quality class	No. of samples	%
> 75	Class I	23	30.6
25–75	Class II	52	69.4

Table 10 Groundwater classification for irrigation according to CAI

CAI	Quality class	No. of samples	%
Negative	Base exchange	43	57.3
Positive	Cation–anion	32	42.7

Table 11 Groundwater classification for irrigation according to Cl

Cl	Quality class	No. of samples	%
< 0.14	Extremely fresh	–	–
0.14–0.85	Very fresh	–	–
0.85–4.23	Fresh	17	22.7
4.23–8.46	Fresh–brackish	22	29.3
8.46–28.21	Brackish	15	20
28.21–282.06	Brackish–salt	15	20
282.06–564.13	Salt	6	8
> 564.13	Hypersaline	–	–

Table 12 Groundwater classification for irrigation according to MH

MH	Quality class	No. of samples	%
< 50%	Suitable	51	68
> 50%	Unsuitable	24	32

h. ***Magnesium Hardness (MH)***

Magnesium ratio (MH) was used for assessing the irrigation water quality (Szabolcs and Darab 1964). With respect to the calculated magnesium hazard, most groundwater samples (68%) are suitable for irrigation covering the southern parts of the study area, while, about 32% of samples are unsuitable that located in the northern part (Table 12 and Fig. 3b).

i. ***Kelly's Index (KI)***

Kelley index measures Na versus Mg and Ca (Kelly 1940). The KI values showed that about 59% of samples are unsuitable, while about 41.3% are suitable that located in the southern parts (Table 13 and Fig. 3c).

j. ***Corrosively Ratio (CR)***

Corrosively ratio is important to determine the water ability for corroding the metallic pipes. Safe water has CR < 1, while corrosive water has CR > 1 and therefore should not be transferred through metallic pipes. As illustrated in Table 14, 17.3% of the wells (13 wells) of the collected samples belong to noncorrosive class. 62 wells (82.7%) are of corrosive class; occupy the northern and central parts (Fig. 3d).

Table 13 Groundwater classification for irrigation according to KI

KI	Quality class	No. of samples	%
< 1	Suitable	31	41.3
≥ 1	Unsuitable	44	58.7

Table 14 Groundwater classification for irrigation according to CR

CR	Quality class	No. of samples	%
< 1	Noncorrosive	13	17.3
> 1	Corrosive	62	82.7

k. ***Salinity and Sodium Hazards Relationships***

Figures 4 and 5 show the plotting of SAR versus salinity hazards (EC), and their spatial distribution, respectively. Total ion concentrations and Na% relationship and its aerial distribution are represented in Figs. 6 and 7. These relationships deduced that most of the groundwater samples have good to permissible quality. Both of the aerial representation of groundwater quality based on salinity hazards and SAR values (Fig. 5) and spatial distribution of groundwater quality according to Wilcox Diagram (Fig. 7) illustrate that the groundwater in the southern and middle parts of the study area has excellent to permissible quality for irrigation, while, the groundwater in the northern part has doubtful to unsuitable quality.

4.2 *Irrigation Water Quality Index (IWQ)*

The calculated IWQ of the studied groundwater (Fig. 8 and Table 15) shows that 17 wells (22.6%) have low water quality for irrigation and represented in the northern parts. The central regions of the study area (32 wells, 42.8%) are fall within the medium category. The southern parts are distinguished by high water quality (26 wells, 34.6%). The correlation matrix between IWQ and various water quality factors indicates that TDS, SAR, Ca, Mg, Na, K and SO_4 show strong positive relationship with IWQI (Table 16). The RSC has a strong negative relationship with IWQ. Moderate positive correlation value is appeared between IWQ and pH, EC, Na%, PI, KI, MH, CR, Cl, Al. CAI show positive weak correlation to IWQ. While a negative weak to very weak correlations were noticed between NO_3, HCO_3, Fe and Mn. This means salinity, sodium hazards and major ions are the most effective parameters in determining the suitability of groundwater for irrigation.

5 Conclusions

The quality of the analyzed groundwater and its suitability for irrigation were assessed using ten individual parameters (TDS, Na%, SAR, RSC, Cl, PI, CAI, MH, CR, and KI), EC-SAR relationship, Wilcox Diagram and irrigation water quality index (IWQ). Accordingly, the study area could be geographically classified into three parts. In the southern part, the groundwater has TDS lower than 3000 mg/l, in addition the Na% is good to permissible, chloride content showed fresh to fresh-brackish characters, noncorrosive propensity for the agricultural device so they are safe for irrigation uses

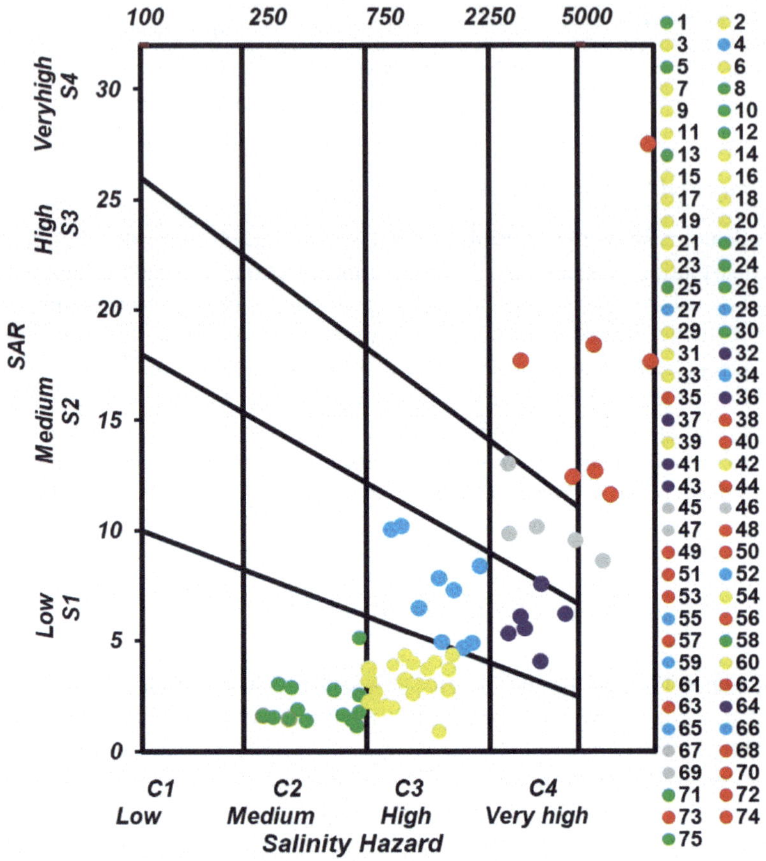

Fig. 4 Groundwater classification according to salinity hazards and SAR

and furthermore the groundwater samples are suitable based on KI and MH and PI values. The computed SAR values are excellent. While, in the central part, the TDS is useful for seeding, Cl content is of brackish characters, PI and MH values are proper for agriculture, excellent SAR values for soil, Na% values are doubtful to unsuitable and KI values are unsuitable. Referring to CR value, the determined value showed corrosive characters to metal pipes.

Based on TDS, SAR, PI, KI and MH, the groundwater of the northern part is unfit for irrigation. As well as, the chloride content showed brackish-salt characters, which is mainly due to seawater intrusion. The calculated Na% values of this region are of doubtful to unsuitable range for agriculture purposes. Likewise, the obtained results of the CR for groundwater samples in these territory showed that they are not suitable for irrigation, due to their corrosive capability for irrigation tools. In contrast.

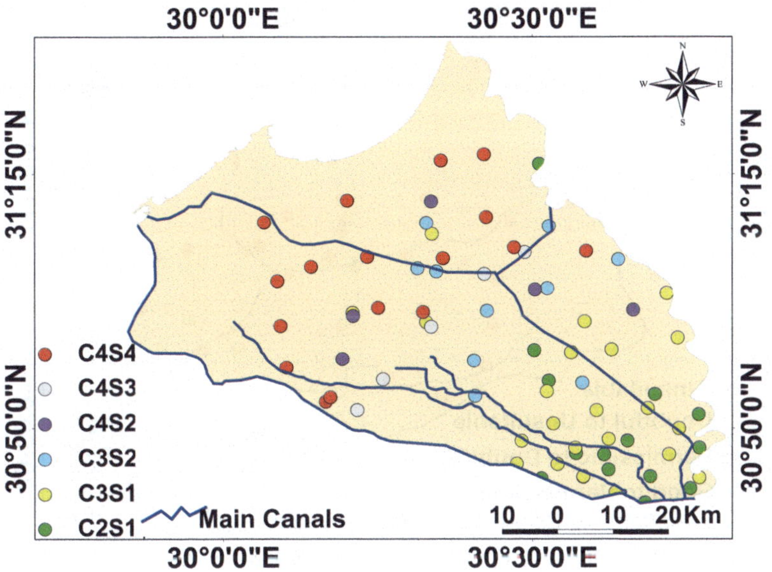

Fig. 5 Classification of groundwater samples based on salinity hazards and SAR values

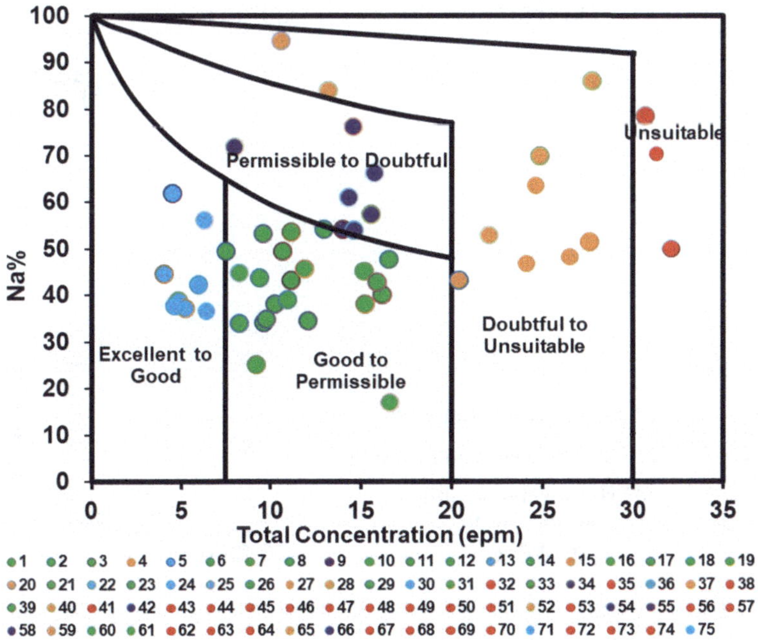

Fig. 6 Groundwater classification for irrigation use based on Wilcox diagrams

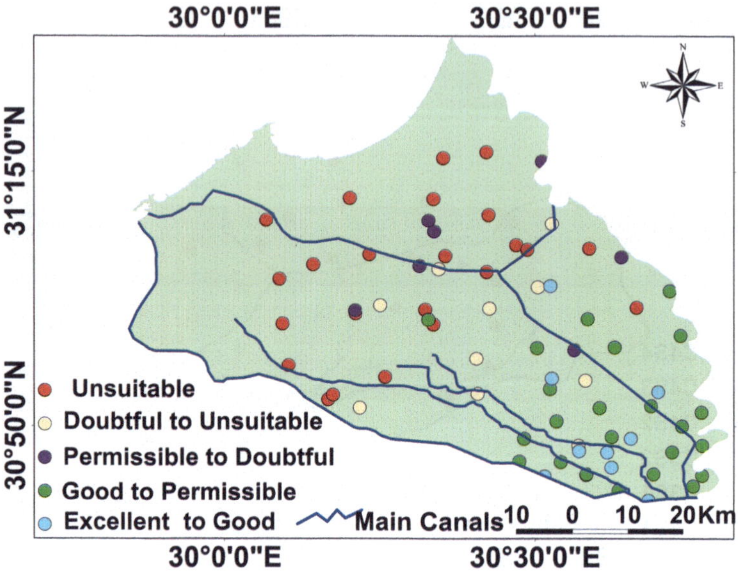

Fig. 7 Spatial distribution of groundwater samples based on Wilcox diagram

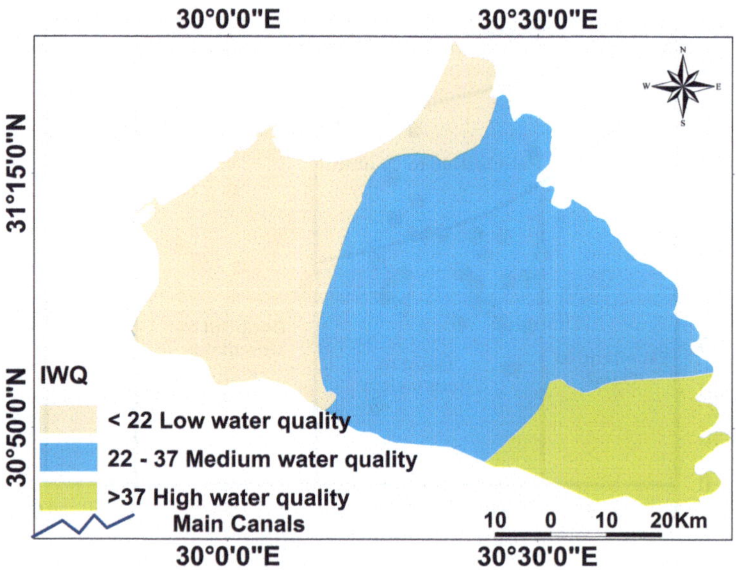

Fig. 8 Spatial distribution of groundwater quality for irrigation based on the IWQI

Table 15 Groundwater classification according to its suitability for irrigation

Type of water	Range	No. of wells	% of wells
Low water quality	< 22	17	22.6
Medium water quality	22 – 37	32	42.8
High water quality	> 37	26	34.6

Table 16 Correlation coefficients of IWQI and the determined parameters

Parameter	Correlation coefficient
TDS	0.98
EC	0.67
SAR	0.70
Na%	0.58
RSC	−0.95
PI	0.36
KI	0.50
MH	0.32
CAI	0.14
CR	0.47
Ca	0.93
Mg	0.94
Na	0.98
K	0.83
SO_4	0.77
Cl	0.67
NO_3	−0.27
HCO_3	−0.07
pH	0.62
Al	0.35
Fe	−0.21
Mn	−0.11

Concerning to RSC, almost all samples of the study region fall within the good category. With regard to CAI, the southeastern and the northwestern parts have + ve values indicating ion exchange between the water Na^+ and K^+ with the soil Mg^{+2} or Ca^{+2}. While the other regions (57.3% of total wells) have –ve values which means most of the groundwater in the study region could reduce Na and K of the soil through irrigation activities. Based on salinity hazards, SAR and the Wilcox diagram, the groundwater quality in the southern and middle parts is excellent to acceptable, while, it is doubtful to unfit in the northern region. According to IWQ, most of the groundwater samples have low to medium quality for agriculture except 17 groundwater samples that located in the northern part of the study area having poor irrigation quality.

In general, the groundwater quality is suitable in the central and southern parts of the study region. This is because these parts are not affected by seawater intrusion. Water seepage from the irrigation channel network plays an important source for freshwater feeding for groundwater in these areas.

6 Recommendations

In the Nile Delta, the groundwater is influenced by numerous environmental parameters, among them intruded seawater, anthropogenic activities, and aquifer's sediment mineralogical composition. Therefore, the studied groundwater could have different qualities depending on the geographic location and the drilling depth. The current study is considered a helpful monitoring work for the decision-makers to determine the most suitable location for drilling wells for irrigation purposes and could be also a guide for regulating the pumping rate. It is urgent for Governors to prevent drilling any wells in the northern part of the study area. Controlling the pumping rate of the wells drilled in the southern and central parts of the current area is important to avoid deterioration of the groundwater quality. A detailed hydrogeochemical study should be done for estimating the impact of different environmental factors on the groundwater.

Acknowledgements This work was financially supported by Faculty of Science, Tanta University, Egypt.

References

Abdel Bakib AMA (1983) Hydrogeological and hydrogeochemical studies in the area west of Rosetta branch and south El Nasr canal. Ph.D. Thesis, Faculty Sci

Abu El-Ella R (1990) The neogene-quaternary section in the Nile Delta, Egypt: geology and hydrocarbon potential. J Pet Geol 13(3):329–340

Adimalla N, Li P (2018) Occurrence, health risks, and geochemical mechanisms of fluoride and nitrate in groundwater of the rock dominant semi-arid region, Telangana State. India Hum Ecol Risk Assess. https://doi.org/10.1080/10807039.2018.1480353

Afifi AA, Darwish KM (2018) Detection and impact of land encroachment in El-Beheira governorate. Egypt Model Earth Syst Environ 4:517–526

Aher K, Gaikwad S (2017) Irrigation groundwater quality based on hydrochemical analysis of Nandgaon block, Nashik district in Maharashtra. Int J Adv Geosci 5(1):1

Armanuos AM, Allam A, Negm AM (2020) Assesment of groundwater vulnerability to pollution in Western Nile Delta aquifer. Egypt Int Water Technol J (IWTJ) 10:18–40

Armanuos AM, Negm A, Valeriano OCS (2016) Groundwater quality investigation using multivariate analysis-case study: Western Nile Delta Aquifer. Egypt Int J Environ Sci Develop 7(1):1–9. https://doi.org/10.7763/IJESD.2016.V7.732
Arulbalaji P, Gurugnanam B (2017) Groundwater quality assessment using geospatial and statistical tools in Salem District, Tamil Nadu. India Appl Water Sci 7:2737–2751
Attia SH (1975) Pedology and soil genesis of the quaternary deposits in the region west of the Nile Delta. Ph.D. Thesis, Faculty Sci. Ain Shams Univ. Cairo, p 288
Attia FA (1985) Management of water systems in Upper Egypt. Ph.D. Thesis, Cairo University, Cairo, Egypt. p 156
Ayers S, Westcost W (1985) Water quality for agriculture, FAO irrigation and drainage paper. Food Agriculture Organization, Rome, p 29
Azzam AS (1994)An integrated seismo-facies and seismo-tectonic study of the Nile Delta of Egypt, utilizing common-depth seismic reflection data. Ph.D. Thesis, Geology Department, Faculty Sci. Ein Shams Univ. Cairo, Egypt, p 240
Badran AM (1996) The contribution of logging analysis and seismic stratigraphy in the reservoir valuation and Lithology determination of the north western Delta Abu Qir Offshore Egypt. M. Sc. Thesis, Geophysics Department, Faculty Sci.Ein Shams Univ. Cairo, Egypt, p 112
Barakat MG, (1982) General review of the petroliferous provinces of Egypt with special emphasis on their geological setting and oil Potentialities. Cairo Univ. Cairo, Egypt, Geology Department, Faculty Sci, p 87
Diab MSh, Mahammed MA, Rizk ZS (1995) The role of geology, hydrogeology and human activities in the contamination of shallow water resources northwest of the Rosetta Nile Branch, Egypt. J Faculty Sci Unite Arab Emirates Univ 8(2):260–291
Doneen D (1954) Salinzation of soil by salts in the irrigation water. Am Geophys Union Trans 35:943–950
El Osta M, Hussein Hand Tomas K (2018) Numerical simulation of groundwater flow and vulnerability in wadi el-natrun depression and Vicinities, West Nile Delta. Egypt J Geol Soc India 92:235–247. https://doi.org/10.1007/s12594-018-0986-7
El Ramly IM (1997) Hydrogeological and water quality characteristics of the saturated zone beneath the various land uses in the Nile Delta region, Egypt. Freshwater contamination in proceedings of rabat symposium S4. IAHS Publ. no 243. pp 255–261
Eltarabily MGA, Negm AM (2018) Groundwater management for sustainable development plans for the Western Nile Delta. Negm AM (ed.), Groundwater in the Nile Delta, HdbEnv Chem. https://doi.org/10.1007/698_2018_247
Embaby AAA (2003) Environmental evaluation for geomorphological situation in relationto the water and soil resources of the region north of the Sadat City, west Nile Delta, Egypt.PhD thesis, Faculty of Science, Mansoura University, Mansoura, p 383
Ewaid SH, Abed SA (2017) Water quality index for Al-Gharraf River, southern Iraq. J Aquat Res 43:117–122
HACH (1990) Chemical procedures explained. Hach Technical Center for Applied Analytical Chemistry, Colorado
Ibraheem IM, El-Qady G (2017) Hydrogeophysical investigations at El-Nubariya-Wadi El-Natrun area, west Nile Delta, Egypt. Handb Environ Chem. https://doi.org/10.1007/698_2017_154
Kelly P (1940) Permissible composition and concentration of irrigation waters. Proc Am Soc Civ Eng 66:607–613
Khangembam S, Kshetrimayum KS (2019) Evaluation of hydrogeochemical controlling factors and water quality index of water resources of the Barak valley of Assam, Northeast India. Groundwat Sustain Develop
Leaven MT (1991) Hydrological study of the Nile Delta and adjacent Desert Areas, Egypt, with emphasis on hydrochemistry and isotope hydrology. M. Sc. thesis, Free University, Amsterdam. Published RIGW/IWACO Technical Note TN 77:01300-91–1301
Mohamed AK (2016) Application of DC resistivity method for groundwater investigation, case study at West Nile Delta. Egypt Arab J Geosci 9:11

Mukate S, Wagha V, Panaskara D, Jacobsb JA, Sawantc A (2019) Development of new integrated water quality index (IWQI) model to evaluate the drinking suitability of water. Ecolog Indicat 101:348–354

Negm AM, Armanuos AM (2016) GIS-based spatial distribution of groundwater quality in the Western Nile Delta, Egypt. Negm AM (ed.), The Nile Delta, HdbEnv Chem. https://doi.org/10.1007/698_2016_66

Ponsadailakshmi S, GanapathySankari S, Mythili Prasanna S, Madhurambal G (2018) Evaluation of water quality suitability for drinking using drinking water quality index in Nagapattinam district, Tamil Nadu in Southern India. Groundwat Sust Develop 6 (43–49)

Richards LA (1954) Diagnosis and improvement of saline and alkali soils, U. S. Department of Agriculture Handbook, 60 Washington USA. 160. https ://www.ars.usda.gov/ARSUs erFiles/20360500/hb60_pdf/hb60c omple te.pdf

RIGW (2002) Nile Delta groundwater modeling report. Res Inst Groundwat, Kanater El-Khairia, Egypt

Said R (1962) The geology of Egypt. Elsevier Publishing Company, Amsterdam, p 377

Salem ZE, Osman MO (2016) Shallow subsurface temperature in the environs of El-Nubaria canal, northwestern Nile Delta of Egypt: mplications for monitoring groundwater flow system. Environ Earth Sci 75:1241. https://doi.org/10.1007/s12665-016-6046-y

Salem ZE, El Bayumy DA (2016a) Use of the subsurface thermal regime as a groundwater-flowtracer in the semi-arid western Nile Delta, Egypt Hydrogeol J 24(4):1001–1014. https://doi.org/10.1007/s10040-016-1377-z

Salem ZE, El Bayumy DA (2016b) Hydrogeological, petrophysical and hydrogeochemical characteristics of the groundwater aquifers east of Wadi El-Natrun. Egypt NRIAG J Astron Geophys 5:106–123

Salem ZE, Osman OM (2017) Use of geoelectrical resistivity to delineate the seawater intrusion in the northwestern part of the Nile Delta, Egypt. Handb Environ Chem. https://doi.org/10.1007/698_2017_175

Salem ZE, Osman OM (2018) Use of one dimensional subsurface temperature profiles to characterize the groundwater flow system in the northwestern part of the Nile Delta, Egypt. In: The Handbook of Environmental Chemistry (Groundwater in the Nile Delta, 2019). Springer, Berlin, Heidelberg. ISBN 978-3-319-94283-4

Salem ZE, Elsaiedy G, ElNahrawy A (2017a) Hydrogeochemistry and Quality Assessment of Groundwater under Some Central Nile Delta Villages, Egypt. Negm AM (ed.) Groundwater in the Nile Delta, Hdb Env Chem. https://doi.org/10.1007/698_2017a_111

Salem ZE, Elsaiedy G, ElNahrawy A (2017b) Assessment of the groundwater quality for drinking and irrigation purposes in the Central Nile Delta Region, Egypt. Negm AM (ed.) Groundwater in the Nile Delta, Hdb Env Chem. https://doi.org/10.1007/698_2017b_137

Salem ZE, Negm AM, ElNahrawy A (2017c) Hydrogeophysical characteristics of the central Nile Delta aquifer. In: Negm AM (ed.) Groundwater in the Nile Delta, Hdb Env Chem ©Springer International Publishing AG. https://doi.org/10.1007/698_2017_75

Salem ZE, Temamy AMA, Salah MK, Kassab M (2018a) Evaluation of water resources qualities for agriculture irrigation in Abu Madi area. Northern Middle Nile Delta 74:277–316. https://doi.org/10.1007/698_2018_273

Salem ZE, El Nahrawy A, Ghobara M (2018b) Spatiotemporal hydrochemical evaluation and quality assessment of drainage water compared to canal surface water in the middle nile delta. Egypt 75:123–161. https://doi.org/10.1007/698_2018_296

Salem ZE, Sefelnasr A, Hasan SS (2019) Assessment of groundwater vulnerability for pollution using DRASTIC Index, young alluvial plain, Western Nile Delta. Egypt Arab J Geosci 12:727. https://doi.org/10.1007/s12517-019-4883-1

Schlumberger (1984) Well evaluation conference, Egypt. Schlumberger Middle East, pp 1–64

Schoeller H (1965) Hydrodynamiquelanslekarst (ecoulementedemmagusinement). Actes Colloques Doubronik, I AIHS et UNESCO pp 3–20

Sener S, Sener E, Davraz A (2017) Evaluation of water quality using water quality index (WQI) method and GIS in Aksu River (SW-Turkey). Sci Total Environ 584–585:131–144
Shabbir R, Ahmad SS (2015) Use of geographic information system and water quality index to assess groundwater quality in Rawalpindi and Islamabad. Arabian J Sci Eng 40:2033–2047
Sharaky AM, El Hasanein AS, Atta SA, Khallaf KM (2016) Nile and groundwater interaction in the Western Nile Delta, Egypt. In: Negm A. (eds) The Nile Delta. The Handbook of Environmental Chemistry, vol 55. Springer, Cham
Shata AA (1955A) n introductory note on the geology of the northern period of the western desert of Egypt. Bull Desert Res Inst Egypt 5(3):96–106
Simsek C, Gunduz O (2007) IWQ index: a GIS-integrated technique to assess irrigation water quality. Environ Monit Assess 128:277–300
Singh SK, Srivastava PK, Singh D, Han D, Gautam SK, Pandey AC (2015) Modelling groundwater quality over a humid subtropical region using numerical indices, earth observation datasets, and X-ray diffraction technique: a case study of Allahabad district, India. Environ Geochem Health 37:157–180
Sobeih MM, El-Arabi NE, Helal EY, Awad BS (2017) Management of water resources to control groundwater levels in the southern area of the western Nile delta. Egypt Water Sci 31(2):137–150. https://doi.org/10.1016/j.wsj.2017.09.001
Solangi GS, Siyal AA, Babar MM, Siyal P (2019) Groundwater quality evaluation using the water quality index (WQI), the synthetic pollution index (SPI), and geospatial tools: a case study of Sujawal district, Pakistan. Human Ecol. Risk Assess. https://doi.org/10.1080/10807039.2019.1588099
Spandana MP, Suresh KR, Prathima B (2013) Developing an irrigation water quality index for Vrishabavathi command area. Int J Eng Res Technol 2:821–830
Szabolcs I, Darab C (1964) The influence of irrigation water of high sodium carbonate content on soil. In: Szabolics I (ed) Proceedings of the 8th international congress soil science, sodics soils. Res Inst Soil Sci, Agric Chem, Hungarian AcadSci, ISSS Trans II, pp 802–812
Tarabees E, El-Qady G (2016) Sea water intrusion modeling in Rashid area of Nile Delta (Egypt) via the inversion of DC resistivity data. Am J Clim Chang 5:147–156. https://doi.org/10.4236/ajcc.2016.52014
Tiwari N, Manzoor A (1988) River pollution in Kathmandu valley (Nepal), suitability of river water for irrigation. Indian J Environ Prot 8:269–274
Wilcox V (1948) The quality of water for irrigation use. US Department of Agriculture, Technical Bulletin, Washington
Wu Z, Wang Z, Chen Y, Cai Y, Deng J (2018) Assessing river water quality using water quality index in Lake Taihu Basin, China. Sci Total Environ 612:914–922

Sustainability of Groundwater

Agricultural Drainage Strategies in Egypt as a Protection Tool Against Groundwater Contamination by Fertilizers: An Overview

Ahmed Awad, Mustafa El-Rawy, and Aya Hosny Abdelmawgoud

Abstract In Egypt, agricultural lands occupy only about 4% of the total country area, while 85% of the country's freshwater resources are allocated for these lands. This is due to the adoption of perennial irrigation practices in most Egyptian agricultural lands. As these irrigation practices almost provide more water than crops need, groundwater tables started to rise dramatically in agricultural lands of the Nile River valley. As a result, waterlogging and salinity problems started to aggravate in these areas and it was associated with heavy losses in crop yield. In addition and due to the improper application of fertilizers in waterlogged lands, the groundwater quality started to deteriorate as a result of the continuous deep seepage of fertilizers-contaminated excess water to underlain aquifers. This alerted the "poor drainage" problem in agricultural lands in the Nile River valley, thus the need to practice proper drainage strategies to conserve groundwater quality in these lands and ensure better moisture conditions for crops. Therefore, Egypt has adopted a policy for drainage technology represented in the gravity subsurface drainage systems at the farm level and started to implement many drainage projects in the area of the Nile River valley. As a case study; we explored existed drainage strategies in El-Minia governorate that is located in Upper Egypt and comprises an area of more than 157E+3 hectares of agricultural lands served with subsurface drainage systems. In addition, the governorate has a set of open drains with a total length exceeding 1200 km. In Egypt, agricultural drainage strategies became indispensable to conserve the quality of existing

A. Awad (✉) · A. H. Abdelmawgoud
Egyptian Ministry of Water Resources and Irrigation (MWRI), Giza, Egypt
e-mail: eng_ahmed_awad@yahoo.com

A. Awad
College of Water Resources and Hydropower Engineering, Yangzhou University, Yangzhou 225009, China

M. El-Rawy
Civil Engineering Department, Faculty of Engineering, Minia University, Minia 61111, Egypt
e-mail: mustafa.elrawy@mu.edu.eg; mustafa.elrawy@su.edu.sa

Civil Engineering Department, College of Engineering, Shaqra University, Dawadmi, Ar Riyadh 11911, Saudi Arabia

A. M. Negm and M. El-Rawy (eds.), *Sustainability of Groundwater in the Nile Valley, Egypt*, Earth and Environmental Sciences Library,
https://doi.org/10.1007/978-3-031-12676-5_9

aquifers that underlain agricultural lands alongside ensuring the sustainability of the agricultural process.

Keywords Agricultural drainage · Subsurface · Groundwater · Nile valley · El-Minia governorate · Egypt

1 Introduction

In Egypt, surface water resources are the River Nile, rainfall, and flash floods. Each year, the River Nile provides Egypt with a water share of 55.5 billion m^3, representing about 97% of Egypt's renewable water resources (El-Rawy et al. 2020a). Rainfall and flash floods provide about 1 billion m^3 of water annually that can be utilized for agricultural purposes (Abdel-Shafy et al. 2010). The other renewable water resources in Egypt are groundwater, domestic wastewater, and desalinated water. Among these resources, groundwater is considered the most preferred water source in various user sectors based on its near-universal availability and the low capital cost (Elnashar 2014).

However the annual population growth rate in Egypt exceeded 2.01%, which absolutely will increase significantly the water demand for domestic and agricultural consumption; the country's annual share of water is the same since more than fifty years ago without any increase (Bakr and Bahnassy 2019). This, in turn, will aggravate the water scarcity problem in Egypt. Not with standing this projected scarcity of water, the flooding irrigation is the adopted practice in many agricultural areas in Egypt, the thing that places agriculture at the top of the country's water-consuming activities.

Agricultural lands, which occupy an area of no more than 4% of the total land in Egypt, consume about 85% of the country's total freshwater resources (El-Rawy et al. 2020a). In most cases, the amount of water provided by the flooding irrigation technique exceeds crop water requirements; as a result, there is great possibility for excess water to exist and accumulate in agricultural lands. This excess water will infiltrate into the soil, causing groundwater tables to rise dramatically to unacceptable levels. When groundwater tables rise near to the soil surface, some fertilizers and soil sediments may mix with the groundwater and dissolve in it, causing remarkable deterioration in its quality (Dubrovsky et al. 2010; Abd-Elaty et al. 2020). Many studies reported that the possibility of groundwater contamination might increase in the shallow water table farmlands (El-Rawy et al. 2019a, 2020b; Awad et al. 2020). In addition, such a rise in groundwater tables can cause waterlogging and salinity problems to aggravate in agricultural lands resulting in severe losses in crop yield (Singh 2017). Since the flooding irrigation technique is being practiced in many agricultural lands in the Nile River valley and result in waterlogging conditions, the thing that might increase the possibility of groundwater contamination by fertilizers and harm crop yield; it is necessary to enhance the control of groundwater tables to conserve its quality alongside ensuring the desired moisture conditions in crops' root zone.

Many strategies were introduced to enhance groundwater aquifers' management, especially in arid regions like Egypt. The aquifer management strategies include, but not limited to, the conjunctive use of groundwater with adjoining streams, canals, drains, and natural reservoirs (El-Rawy et al. 2016; Al-Maktoumi et al. 2016, 2020; Salem et al. 2020), managed aquifer recharge via injection/recovery (El-Rawy et al. 2019b), and bank infiltration (Hiemstra et al. 2003; Abdelrady et al. 2020). Such practices aim to ensure the abundance of sufficient groundwater quantities, as it is an essential water source that is being relied upon to provide irrigation water for many agricultural lands in Egypt. But in agricultural lands, recharge of aquifers is not the only target of groundwater management and, in most cases, the main aim is to strict any recharge of aquifers to limit unacceptable rises of groundwater tables that can harm both agricultural production and the groundwater quality. Therefore, in such a case, the tool that can control groundwater tables' fluctuation is considered the first desirable strategy for ensuring better groundwater management.

Agricultural drainage is considered a vital tool that allows better management of groundwater tables in both agricultural and urban lands (Abegaz 1995; Blann et al. 2009; Abdelaty et al. 2010; Hansen et al. 2019; Singh 2019). Many studies, like (Ritzema et al. 2008), reported that proper drainage strategies can overcome several poorly-drained farmlands' problems (such as waterlogging and deep seepage of fertilizers-contaminated excess water) by lowering groundwater tables to the desired levels. Thus, these practices can secure the target of groundwater management in agricultural lands and lessen the possibility of aquifers contamination by fertilizers (Dubrovsky et al. 2010). In addition. agricultural drainage strategies can allow the control of aquifers' recharge potential in drained lands (Smith and Berg 2020). Agricultural drainage has been practiced for thousands of years all over the world and it was reported as an instrument to grow agricultural production and an effective tool for the conservation of land resources (Ritzema et al. 2006). It was also reported that drainage systems can be employed to help in coping with adverse impacts of climate change on agriculture and the surrounding environment (Awad et al. 2021). Therefore, Egypt started to adopt drainage strategies in agricultural lands in the Nile River valley to ensure better groundwater management, for both quantity and quality, alongside high agricultural productivity.

Accordingly, this chapter presents the following:

- An overview of the agricultural drainage practices.
- The "poor drainage" conditions and their adverse impacts on groundwater quality and the surrounding environment.
- Effect of practicing proper agricultural drainage strategies on groundwater quality, agricultural productivity, and the surrounding environment.
- Agricultural drainage strategies in Egypt.

2 Agricultural Drainage Practices

2.1 What Is Agricultural Drainage?

Excess water in agricultural lands represents one of the major threats to the yield of several crops (Osakabe et al. 2014). "Agricultural drainage strategies" are such practices that allow a natural/artificial flow of excess water out of agricultural lands (Madramootoo et al. 1997).

Following the rainfall or irrigation events, water starts to infiltrate into the soil causing groundwater tables to rise and create saturation conditions in the soil layers. The more precipitation or irrigation, the more soil depth that becomes saturated. Under poor drainage conditions, groundwater tables will continue to rise causing the saturated depth to exceed the field capacity of the soil, which creates waterlogging conditions (Fig. 1). It was reported that many crops can't grow well or even survive under such waterlogging conditions (Singh 2015). Therefore, agricultural drainage practices aim to:

- Lower groundwater table and bring soil moisture down from saturation to field capacity, which is the optimal soil moisture content for most crops to grow well.
- Enhance soil structure, as many soils' structures may collapse under very wet conditions (Bronick and Lal 2005).

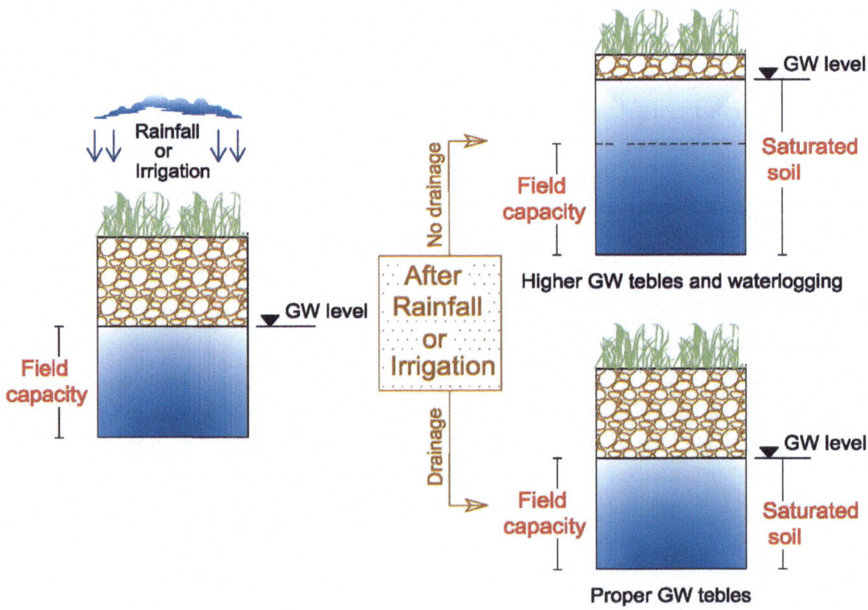

Fig. 1 Flocculation in groundwater tables after a rainfall or irrigation events, under proper and poor drainage conditions

- Enhance soil workability during planting and harvesting seasons.
- Shorten the waiting time for soil dryness between cropping seasons.

2.2 Agricultural Drainage Systems

The system by which excess water is being drained can exist above or under the ground underground, and this leads to the distinction between what so-called "surface drainage systems" and "subsurface drainage systems".

2.2.1 Surface Drainage Systems

Surface drainage systems were the first way to practice agricultural drainage strategies in Egypt. Surface drainage practices aim to remove the low spots where water can accumulate on the land surface. This can be achieved through land forming and smoothing and as a result, excess water on the soil surface will be allowed to runoff to the nearest surface pathways of water (Osman 2012). In these systems, there is a system of ditches that are dug in the land to collect such excess water runoff from the soil surface and carry it to the adjacent branch/main water bodies (Fig. 2). Therefore, a subsurface drainage system comprises a set of excavated ditches that collect excess water that exists on the land surface through runoff and then get rid of it in the nearest drains or other water bodies. Surface drainage systems are a vital tool for draining excess water in humid flatlands characterized by limited hydraulic gradients to nearby rivers or other disposals. Also, such systems are useful in agricultural lands that have low infiltration rates and/or low permeable soils.

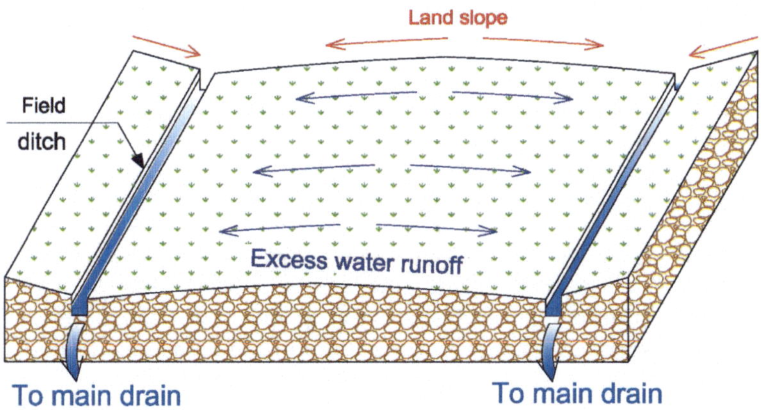

Fig. 2 Surface drainage practices, modified after Mejía (2007)

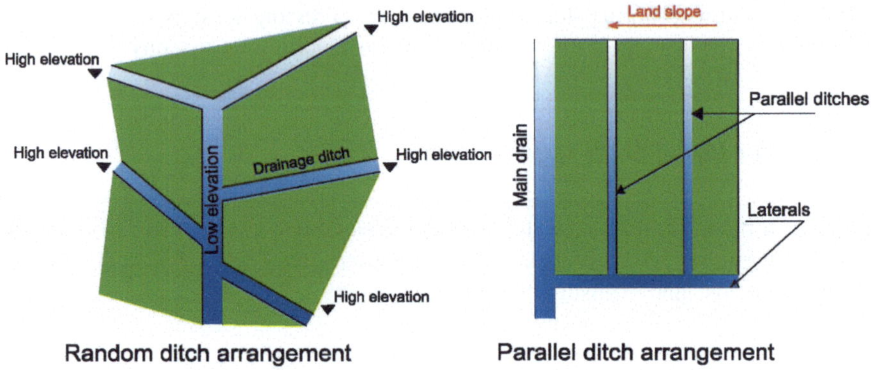

Fig. 3 Arrangement of ditches in surface drainage systems

Arrangement of Ditches

Ditches (drains) can be arranged whether in a random or parallel way. The random arrangement of drainage ditches is only used in scattered wetlands that require drainage, so that each ditch runs to any lower areas in the land. The parallel ditch arrangement is the most practiced surface drainage system (Ghane 2018) and this arrangement is being used in agricultural lands characterized by the semi-flat topology in where ditches are perpendicular to the land slope. These parallel ditches deliver excess water to the laterals which then discharge into the main ditch (Fig. 3).

Advantages of Surface Drainage Systems

Since the main task of surface drainage systems is to remove the excess water that exists on soil surface through runoff, the advantages of these systems are:

- Minimizing the duration of ponded water that immerse crops.
- Minimizing the prolonged saturation conditions of soil, which may restrict oxygen and carbon dioxide exchange between the soil and crops' roots.
- Surface drainage systems are easy to be installed. Also, the cost of it is lower compared to subsurface drainage systems.

Disadvantages of Surface Drainage Systems

In surface drainage systems, ditches are the only element in which excess water can be collected and then discharged out of the land. So, ditches installation represents the backbone of surface drainage systems. This points to the main disadvantage of these systems that is the area that is being wasted from the farmlands to dig these ditches.

Surface drainage systems are effective in collecting excess water that exists on the soil surface, while the efficiency of these systems decreases in high-permeable-soil lands as excess water will infiltrate from the land surface down to the various deep soil layers.

2.2.2 Subsurface Drainage Systems

Subsurface drainage systems are such practices that allow excess water in the soil surface to be drained. As these systems operate underground; it is, therefore, a vital tool for controlling and adjusting groundwater tables as desired. Subsurface drainage can be practiced through many techniques such as:

- The buried corrugated and perforated plastic or clay pipes (Tile system).
- Mole drain.
- Tube wells.

In Egypt, the typical gravity drainage system consists of open and tile drains, and therefore this chapter will focus on this type. Tile drainage practices were first introduced in America in 1838 when a farmer, dubbed "the father of tile drainage", installed clay pipes on his farm to overcome poor drainage conditions. The tile drainage system comprises a set of buried pipes (with a certain diameter) that collect excess water and discharge it into larger-diameter pipes and then to the main ditches out of the land (Fig. 4). Pipes are dug at a certain depth and spacing based on land and soil characteristics, field hydrology, adopted management practices, etc.

Drain-pipe Shape and Materials

Drain pipes are usually round/cylindrical, and sometimes their cross-section is square or rectangular. The diameter of the drain pipes ranges from 2 to 18 inches. To allow a point of entry for excess water into these pipes, they are perforated on their surface. Several materials were used in manufacturing the drain pipes such as; clay, concrete, and plastic (Fig. 5). To prevent the clog of pipes' surface-holes and mitigate the possibility of soil deposits to be accumulated inside these pipes, they must be surrounded by a proper filtering material, which can be:

- Mineral: such as gravel, sand, crushed stone, and clinker.
- Organic: such as moss, scobs, heather, straw mats, poorly decomposed peat, and divot.

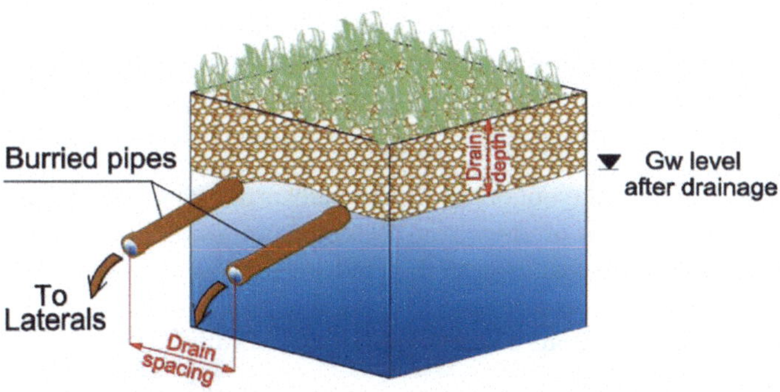

Fig. 4 Subsurface drainage systems (tile drainage), modified after Blann et al. (2009)

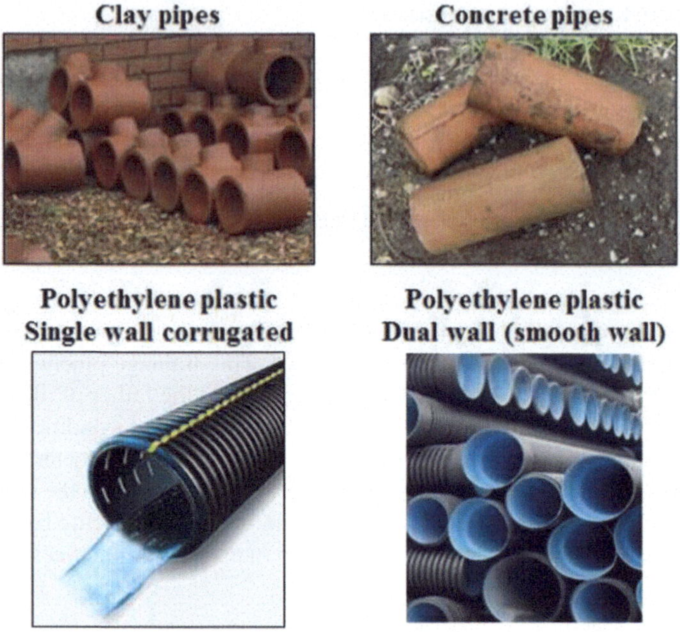

Fig. 5 Some types of drain-pipe materials

- Artificial mineral: such as glass cloth, polyethylene cloth, nitrone and chlorine cloth.

General Layout of Subsurface Drainage Systems

In subsurface drainage system, drain pipes are the first receiver of excess water that exists in the soil profile. This water enters drain pipes through their surface-holes, and then these pipes deliver excess water to other larger-diameter pipes or sometimes to the main collectors or ditches. Finally, excess water is discharged into surface water bodies through what is called "the main outlet" to take its turn in the water cycle (Fig. 6). Since the layout of subsurface drainage systems affect much the water balance in agricultural lands, there are many important considerations for the proper layout selection of subsurface drainage systems and can be concluded as follows:

- It's preferable to fit the layout of a particular subsurface drainage system (in terms of lengths and direction of slope, for both drain pipes and open ditches) to the layout of infrastructure that exist in the land.
- The outlet level of a collector pipe (the yellow-circled points in Fig. 6) has to be at or above the highest water level in the collector drain.

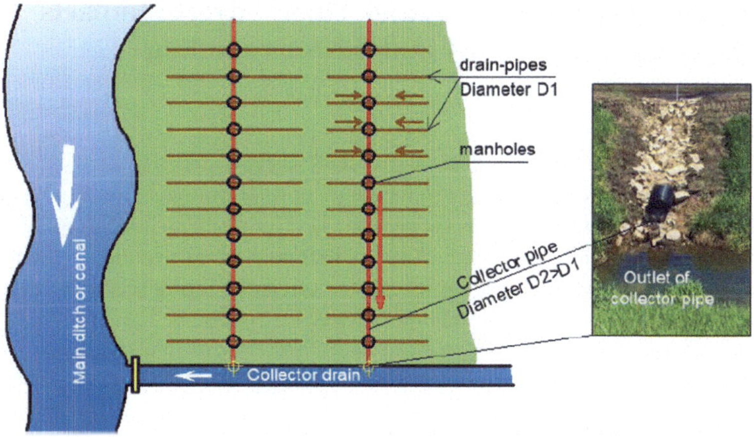

Fig. 6 A small scale typical layout of a subsurface drainage system; brown and white arrows indicate the drainage water pathway

Advantages of Subsurface Drainage Systems

- Since the main task of subsurface drainage systems is to remove excess water that exists beneath the land surface, the significant advantage of these systems is their efficiency in lowering groundwater tables in agricultural lands, and thus creating better conditions for crops' root zone alongside mitigating the deep percolation of fertilizers-contaminated excess water to aquifers.
- As long as most components of subsurface drainage systems are existed and installed underground, practicing these systems will save large areas in agricultural lands to be planted, rather than being lost in digging too much ditches like under surface drainage systems.

Disadvantages of Subsurface Drainage Systems

- Expensive to be implemented and maintained.
- Subsurface drainage systems are not effective in lands characterized by shallow compacted soils where infiltration rates are low enough to hinder excess water on the land surface from percolating down to drain pipes.
- Drainage flux from subsurface drainage systems may contain agriculture nutrients, sediments, and unacceptable contaminants, which threats water quality in waterways that receive such flux.

3 Poor Drainage Conditions and Their Adverse Impacts on Groundwater Quality and the Surrounding Environment

Irrigated lands that lack a good drainage system always suffer from what so-called "poor drainage conditions". The "twin menace" of poor drainage conditions is salinization and waterlogging problems (Singh 2015). Threats from poor drainage conditions are increasing and more than 30% of irrigated lands around the world was reported to be affected by waterlogging and/or soil salinization problems due to the lack of proper drainage practices (Heuperman et al. 2002).

3.1 Waterlogging

When there is no proper drainage system in a particular farmland, the continuous irrigation and/or the successive precipitation events will cause a remarkable rise in groundwater tables. In this case, water starts to replace air voids in the crop root zone until the stage in which there will be no sufficient Oxygen available for plants in the root zone; this is known as "Waterlogging" (Parent et al. 2008). Therefore, waterlogging can be defined as the conditions under which roots cannot respire due to excess water that exists in the crops' root zone (Valipour 2014). Waterlogging is a major abiotic factor that affects the development and growth of crops and many studies reported the adverse impacts of waterlogging on crop growth and agricultural productivity (Barman et al. 2011; Watson et al. 1976).

Besides the aforesaid adverse impacts of waterlogging on agricultural production, waterlogging conditions represent a major threat for groundwater quality and the surrounding environment as following:

- Concerning the groundwater quality.
- In waterlogged lands where groundwater tables are high enough to reach the upper soil layers, there is a great possibility for groundwater to be contaminated by fertilizers, under the improper application rates of these fertilizers (Ashraf et al. 2019; Srivastav 2020). This may cause considerable degradation in the quality of aquifers that underlain waterlogged farmlands, which in turn will result in the loss of an important water source in Egypt.
- Concerning the environment.
- Waterlogged lands were reported to produce high runoff rates compared to drained lands (Skaggs et al. 1982). Such higher runoff rates may carry significant amount of fertilizers and sediments (Khaleel et al. 1980). When this runoff enters water channels, such fertilizers and contaminants cause remarkable deterioration of the water quality in these channels.

3.2 Salinity

Soil salinity is not a stand-alone problem, rather it is mostly the consequences of waterlogging conditions when existing in agricultural lands. In agricultural lands characterized by salty groundwater, and under poor drainage conditions; groundwater tables will go higher in the soil profile creating waterlogging conditions. When this water evaporates, it leaves its salts behind to accumulate and settle in the soil. The more evaporation, the more salt concentration in the soil profile and on the land surface. Also if the irrigation water is salty enough and there are no proper drainage strategies exist in the land, the same effect will occur. Therefore, soil salinization can be defined as the accumulation of water-soluble salts in the soil profile to a level that adversely impacts agricultural production, environmental health, and economics (Munns et al. 2008). It was reported that salinity could affect the germination, vegetative growth, and reproductive development of plants (Parihar et al. 2015).

Based on the aforesaid, groundwater—in terms of quantity and quality—can be adversely impacted by poor drainage conditions in agricultural lands. Therefore, proper drainage strategies are indispensable in agricultural lands for ensuring better conditions for groundwater resources alongside high agricultural productivity.

4 Effect of Practicing Agricultural Drainage on Groundwater Quality, Agricultural Productivity, and the Surrounding Environment

As above-mentioned, practicing subsurface drainage allows much control on groundwater behavior (in terms of depth and quality) in agricultural lands. But to understand well how groundwater can be affected and benefit under subsurface drainage systems, it is necessary to look at the different ways with which farmers can practice subsurface drainage systems, as the behavior of groundwater differs under each way. Subsurface drainage systems can be practiced in two ways:

- Conventional drainage practices.
- Controlled drainage and sub-irrigation practices.

4.1 Conventional Drainage Practices

Conventional drainage practices are designed to allow excess water that enters drain pipes to move freely towards the nearest drainage outlet, through the supposed path for the drainage flux (e.g., from drain pipes to collector pipesm and then to the outlet).

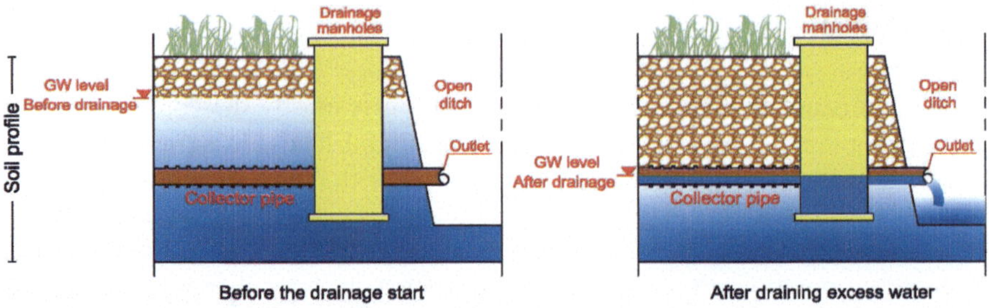

Fig. 7 The conventional drainage practices

4.1.1 Effect of Conventional Drainage Practices on Groundwater

As shown in Figs. 6 and 7, excess water enters the drain pipes through their surface holes. This water then takes the drainage path to the nearest outlet, based on the supposed layout of the system. After some time, groundwater tables start to drop, as excess water is being drained.

Under the conventional practices, excess water will continue to discharge from the outlet until groundwater tables drop to the same level as the collector pipe. Therefore, conventional drainage practices are a useful tool in agricultural lands where it is desirable to lower groundwater tables to a certain level.

4.1.2 Factors Affecting Groundwater Tables Under Conventional Drainage Practices (Relating to the Layout of the Drainage System)

As shown in Fig. 8, the rate of groundwater tables' drop under conventional drainage practices depends on the drainage system design (Skaggs et al. 1999). There are some effective design parameters that can determine and control the rate of groundwater tables' drop, these parameters are:

- The installation depth of drain pipes (Referred to as "d" in Fig. 8)

The depth at which drain pipes is buried affects much the drainage flux and thus the drop rate of groundwater tables (Skaggs et al. 2012). Under the same spacing between drain pipes, it was reported that increasing the depth of drain pipes would increase the subsurface drainage flux and thus causing rapid drops in groundwater tables (Sands et al. 2008). Therefore in Fig. 8, the deeper the drain-pipe's installation depth, the faster to get smaller "m" values and thus a rapid drop in groundwater tables.

- The spacing between drain pipes (Referred to as "L" in Fig. 8)

The spacing between drain pipes refers to the designed distance that must be existed in the land between the centerlines of each two adjacent drain pipes. It was reported

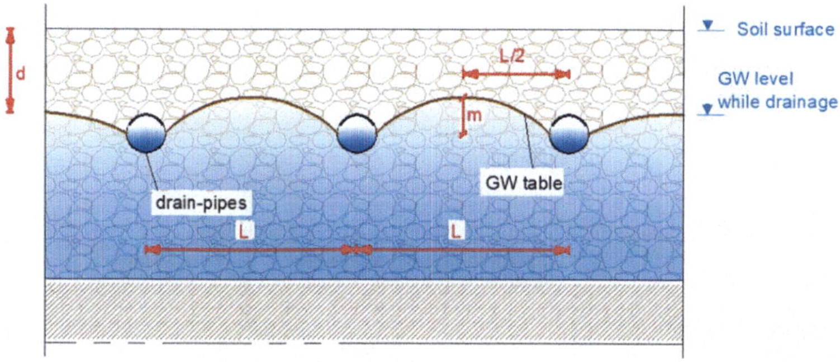

Fig. 8 Design parameters and layout of the subsurface drainage system, modified after Skaggs et al. (2012)

that under the same depth of drain pipes, drainage flux is inversely proportional to the spacing between drain pipes. There would be higher drainage flux under narrow drain-spacings than under large drain-spacings (Kladivko et al. 2004). Therefore under the narrow drain-spacing, groundwater tables will drop faster than under larger drain-spacing values.

4.1.3 Disadvantages of Conventional Drainage Practices

- Despite that conventional drainage practices are a vital tool in lowering groundwater tables to a certain desirable depth (the depth of the collector pipe); these practices do not allow full control on groundwater tables to be adjusted above or under this desired depth (which is the depth of the collector pipe).
- As groundwater tables rise, it mixes with some fertilizers that exist in the upper soil layers. The low control of drainage flux may result in intensive drainage for water that exists in the soil profile. This, in turn, causing massive losses in fertilizers that exist in the root zone as such fertilizers will dissolve in soil–water and then be drained before crops can make use of it. In addition to depriving crops of these fertilizers, the presence of such fertilizers in the drainage flux will increase the non-point source pollution probabilities for surface water bodies that receive this flux.

4.2 Controlled Drainage Practices

Rather than having a constant depth of the drainage outlet (like under the conventional drainage systems), controlled drainage practices allow raising the depth of the drainage outlet to control (decrease) drainage fluxes when desirable (Skaggs et al.

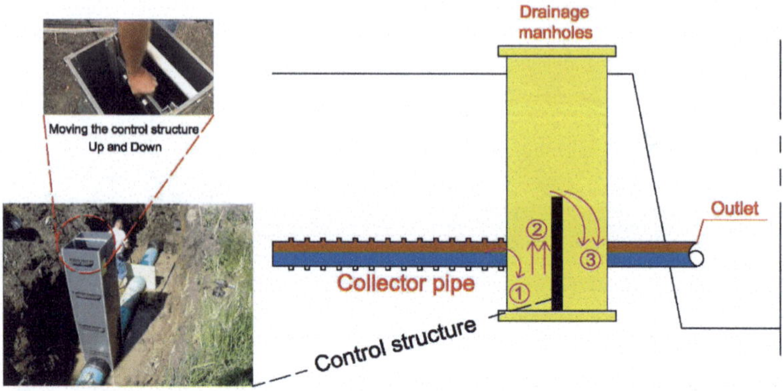

Fig. 9 Controlled drainage practices; pink arrows indicate the path of drainage water

1999). This is done using a water control structure to adjust the desired depth of the drainage outlet (Fig. 9).

As shown in Fig. 9; when drainage begins, excess water in the collector pipe starts to flow into the manhole. Under conventional drainage practices, excess water will then flow directly into the outlet. But under controlled drainage practices; excess water starts to accumulate in manholes (Fig. 10b) until the level of water goes up higher than the level of the control structure. At this stage, excess water starts to flow above the control structure directly to the outlet to be discharged (Fig. 10c).

4.2.1 Effect of Controlled Drainage Practices on Groundwater

Considering Fig. 10; under the controlled drainage practices, groundwater tables go through four stages:

- Phase (a): This phase represents the time before the drainage starts. In this case, groundwater tables are still at high levels in the soil profile and would cause waterlogging conditions in the root zone (Fig. 10a).
- Phase (b): This phase represents the time when excess water starts to enter drain pipes and flow to collectors. In this case, excess water flows into manholes and accumulates in them (Fig. 10b), and groundwater tables start to drop, but still, there is no flux from the drainage outlet.
- Phase (c): This phase represents the time when excess water accumulates in manholes up to the control structure level and starts to flow from the drainage outlet (Fig. 10c). In this case, groundwater tables start to drop rapidly more than phase (b).
- Phase (d): This phase represents the time when the drainage flux ends, as groundwater tables in the soil profile were lowered to the same level as the control structure (Fig. 10d). There will be no additional flux from the drainage outlet until the soil profile receives more water (e.g., from irrigation, rainfall, and so

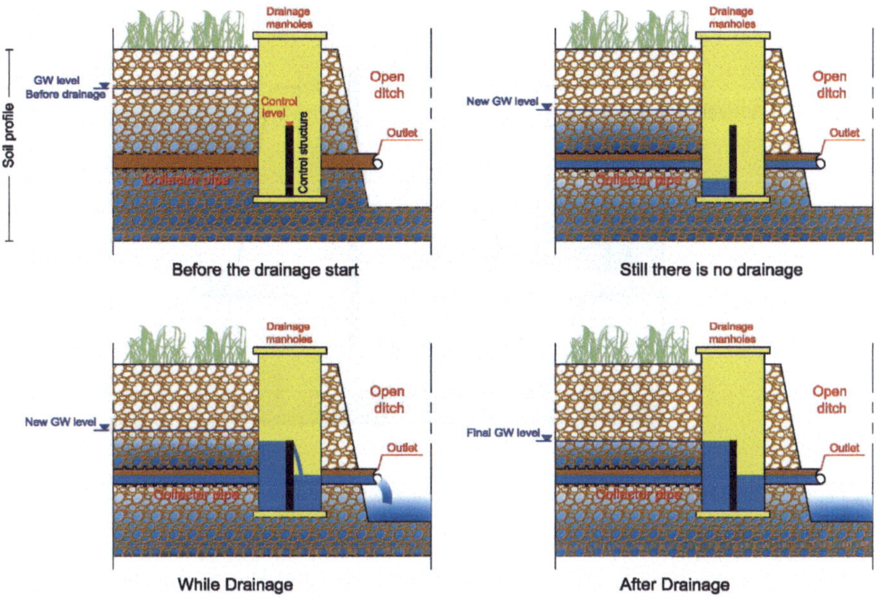

Fig. 10 Flow of excess water from collectors to outlets under controlled drainage practices alongside how that impacts groundwater tables

on) and cause the accumulated excess water in manholes to rise higher than the control structure, or until the level of the control structure will be lowered than the current level.

Under certain circumstances, the control structure could be closed so as not to allow any drainage from the soil profile (Fig. 11). This can occur under the following conditions:

- In arid lands that lack sufficient water resources: In these lands and following a precipitation or irrigation event; it is a not-recommended planning to quickly lower groundwater tables and drain the water from the soil profile, as it is a big challenge to have such soil–water availability to the plant in most of growing season periods based on the climate pattern in such lands. Therefore, the control structure is closed for a certain time after precipitation or irrigation events to keep groundwater tables higher to allow the plant's roots making the best use of water before being lost in drainage (Ayars et al. 2006).
- Mitigating nutrients losses: When fertilizers are applied in agricultural lands, it starts to dissolve in the water that exist in the upper soil layers to be then uptaken by crops' roots. Following the application of these fertilizers; if there is a drainage flux, soil–water that contains such fertilizers could be lost in drainage. Under these conditions, there will be significant losses in these fertilizers before crops can make use of them. Furthermore, the drainage water that is contaminated by these fertilizers will cause remarkable deterioration in the quality of surface water

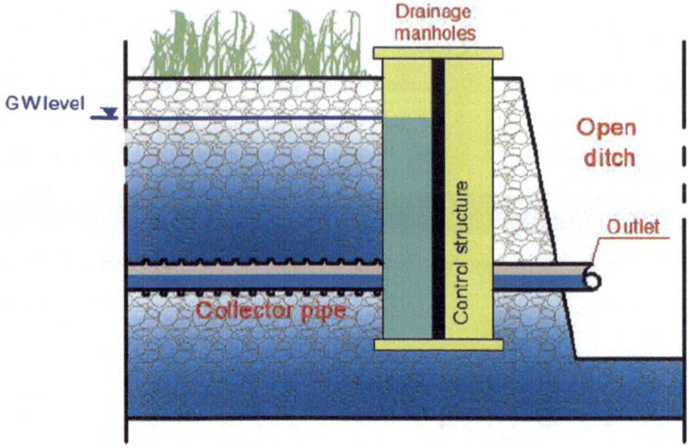

Fig. 11 Closing of the control structure to stop drainage flux

bodies that receive the drainage flux. Therefore, practicing controlled drainage systems in agricultural lands after the application of fertilizers could mitigate nutrients' losses in drainage flux, thus ensuring higher agricultural productivity alongside better quality for adjacent water bodies (Carstensen et al. 2019).

Based on the aforesaid, controlled drainage systems play a vital role in adjusting groundwater tables in agricultural lands at the desirable levels, which benefits both agricultural production and the surrounding environment.

Nowadays, agricultural drainage practices mutated from merely a way to drain excess water from the soil profile to become an intelligent tool that can be used to supply the root zone with water when needed through what so-called "sub-irrigation practices". In sub-irrigation practices (Fig. 12), water is pumped back to the drainage network to rise groundwater tables and increase the moisture in the root zone (Wesström et al. 2014).

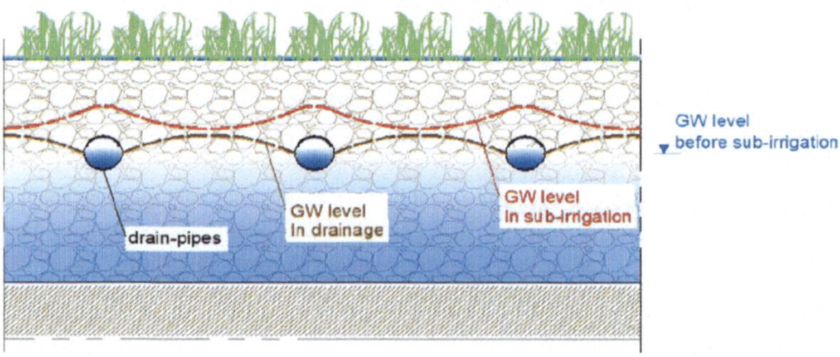

Fig. 12 Effect of sub-irrigation practices on groundwater tables, modified after Skaggs et al. (2012)

4.3 *Summary of Subsurface Drainage Practices' impacts on Groundwater*

The impact of subsurface drainage practices on groundwater management can be listed as in Table 1.

Table 1 Impacts of subsurface drainage practices on groundwater management and the surrounding environment

	Conventional drainage practices	Controlled drainage practices	Sub-irrigation practices
Degree of controlling groundwater tables	Low control: Groundwater tables will automatically drop until it reaches the same level as the drainage outlet	High control: Groundwater tables will continue to drop until it reaches the same level as the control structure (which can be easily adjusted to the desired level)	High control: Groundwater tables will continue to rise until the supply of water stops. Thus, the continuous monitoring of both groundwater tables and the water supply can help in adjusting and controlling the rise of groundwater tables under these practices
Mitigation of nutrients losses through drainage flux (Fig. 13)	Can't mitigate nutrients losses through drainage flux	Can mitigate nutrients losses by raising the control structure level after the application of nutrients, or by a complete block for the drainage outlet to stop any drainage from the soil	
Effect on the quality of surface water in adjacent water bodies that receive the drainage water (Fig. 13)	High probability to contaminate the adjacent water bodies due the contaminated drainage flux	Low probability to contaminate the adjacent water bodies as the drainage flux can be stopped when it's contaminated by nutrients	
Effect on groundwater quality in the areas where groundwater is recharged by the surrounding surface water bodies (Fig. 13)	High probability to have a low-quality groundwater due to the deterioration of surrounding water bodies by agricultural nutrients	Mostly result in high-quality groundwater, as the probability to have a contamination in surface water bodies by agricultural nutrients is very low	

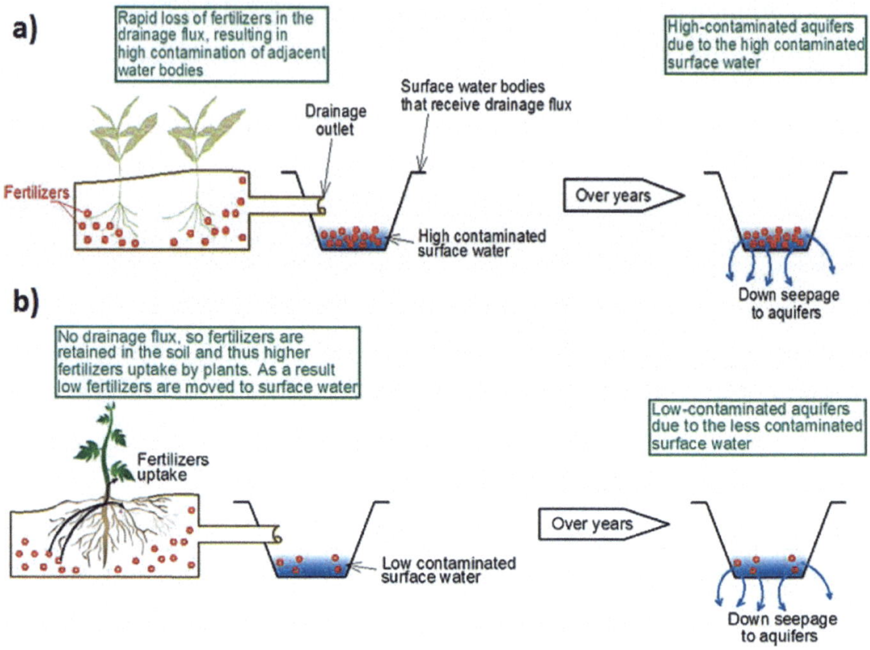

Fig. 13 Impacts of agricultural practices (in terms of ferilizers application) on surface water and groundwater contamination under **a** conventional drainage strategies and **b** controlled drainage strategies

In addition to the aforesaid, one of the main advantages of practicing agricultural drainage strategies is ensuring better groundwater quality, as follows:

- In coastal areas where soils are highly saline; soil salts will be leached down to the existed aquifers when irrigation is applied in these lands or as a result of precipitation events. But if there are proper drainage strategies being practiced in these lands, then the soil salts will be leached in the drained water instead of being leached down to the groundwater (Kamra 2015).
- In waterlogged lands, unmanaged application rates of fertilizers can contaminate the groundwater exist in the upper soil layers, while proper drainage strategies can lower groundwater tables and lessen the possibility of such contamination (Burri et al. 2019).

5 Agricultural Drainage in Egypt

5.1 *Brief History and Authorities*

Until the mid-twentieth century, farmers were responsible for getting rid of the excess water that exists in their farmlands without any obligation from the state, while the responsibility of the state was limited to the main ditches. The importance of agricultural drainage strategies, as a basis for better groundwater management and high agricultural productivity, has then imposed to appoint the establishment, operation, maintenance, and replacement of drainage systems to the state. Therefore the Republican Decree No. (2434) of 1969, the first Republican Decree for the work of an organizational structure for drainage practices in Egypt, was issued to establish "The Public Authority for Covered Drainage Projects in the Nile River Delta". Then in 1971, "The Egyptian General Authority for Drainage Projects in Upper Egypt" was established. In 1973, the two authorities were merged in one body, named "The Egyptian Public Authority for Drainage Projects (EPADP)". In Egypt, the typical gravity drainage system consists of open and tile drains. Drainage water is collected from agricultural lands by the subsurface drain pipes and then is discharged into open ditches, which finally discharge this excess water into the River Nile or its branches, or to the Northern Lakes and the Mediterranean Sea (Abdel-Dayem 1987).

5.2 *Responsibilities and Targets*

The EPADP is responsible for the following activities:

5.2.1 Concerning the Surface Drainage

EPADP is responsible for the implementation of the agricultural drainage policy in Egypt, which aims to widen the use of drainage networks in all farmlands in both the northern and southern areas. EPADP is also responsible for the construction, expansion, and deepening of open public drains alongside the establishment of necessary industrial works on these drains, such as bridges, culverts, etc. There are about 8 million acres farmlands in which the Authority aims to establish new public drains alongside expanding and deepening the existed ones.

5.2.2 Concerning the Subsurface Drainage

- Implementation of new subsurface drainage systems

It includes the installion of the new buried-pipes' networks at the desired depths to get rid of the excess water in the root zone. These pipes are given different slopes (depending on the pipe's diameter) to collect and carry the excess water until it discharges into the exposed drains. The EPADP's strategy for subsurface drainage projects aims to implement subsurface drainage systems in 6.4 million acres in Upper and Lower Egypt, of which 4.6 million acres in Lower Egypt and 1.8 million acres in Upper Egypt. Until June 2018, the implemented subsurface drainage projects covered an area about 4.260 million acres in Lower Egypt and 1.722 million acres in Upper Egypt.

- Rehabilitation and renewal of the existing subsurface drainage networks

EPADP rehabilitates and replaces the subsurface drainage networks when their efficiency start to decrease according to the adopted criteria of replacement and renewal as follows:

- If groundwater tables continue to rise despite of the existence of subsurface drainage systems in the land.
- Increase in soil salinity.
- Repeating the maintenance work for the network within short periods.
- Increasing the number of farmers' complaints from the region regarding crop yield despite adopting the proper agricultural practices.
- The end of the supposed network life.

The current average annual rehabilitation and renewal of subsurface drainage networks in Egypt is about 90,000 acres; till June 2018, an area of about 1960 thousand acres has been rehabilitated and renewed (MWRI 2020). In addition; the authority has the plan to carry out periodical maintenance works that are necessary to maintain the desired efficiency of both surface and subsurface drainage networks; this includes:

- Clearing the surface drainage network to maintain the design-sections of drains by bulldozing approximately 11 million cubic meters annually (MWRI 2020).
- Cleansing and removal of all weeds' kinds from drains, with annual target of about 20 thousand kilometers of drains' lengths.
- Washing all the subsurface drainage networks that have been implemented to ensure high efficiency in their performance by removing any blockages in the network.

5.2.3 Concerning the Reuse of Drainage Water for Irrigation

EPADP is also responsible for the cooperation and liaison with the funding bodies to provide grants to improve the quality of water in open drains and reuse it for irrigation purposes. The annual amount of agricultural drainage water in Egypt is about 18.5 billion m^3; officially, about 9.5 billion m^3 of this water is being reused for irrigation purposes (Barnes 2014).

5.3 Exploring Ways of Practicing Agriculture Drainage Strategies in the Nile Valley: A Case Study of El-Minia Governorate

Many studies reported a remarkable degradation in the water quality in many provinces in Egypt due to the improper and excessive application of fertilizers by farmers. As an example, Zaki et al. (2015) reported that the concentration of heavy metals in surface water bodies and groundwater in El-Minia Governorate is most significantly affected by the leachate of many pollutants from agricultural lands. Therefore, El-Minia governorate pays great attention to the implementation and rehabilitation of agricultural drainage projects to ensure high quality for its water resources alongside high agricultural productivity. In the following, a brief discussion will be given about how agricultural drainage strategies are being practiced in El-Minia Governorate. The executive and administrative structure of different authorities which concern the agricultural drainage system in the governorate will also discussed.

El-Minia is considered an agrarian Upper Egyptian governorate. It starts from about 230 km south of Cairo (Fig. 14). The north–south length of the governorate is about 128.5 km (Abdelmawgoud et al. 2020) and it is bounded on the north by Beni Suef Governorate, on the south by the Governorate of Assiut and New Valley, and it extends to the eastern governorate of the Red Sea and the west Giza Governorate. Administratively, El-Minia governorate comprises nine municipal divisions; five at the northern (Samalut, Matay, Bani Mazar, Magagha, and Idwah), and four at the southern (El-Minia, Abu Qurqas, Mallawi, and Dayr Mawas). Agricultural drainage in El-Minia governorate is mainly managed by the "General Administration for the drainage of North El-Minia and General Administration for the drainage of South El-Minia". The General Administration for the drainage of North Minia serves an area of about 170,000 ha, while the General Administration for the drainage of South El-Minia serves an area of about 150,000 (MWRI 2020). According to ELDeeb et al. (2015); in 2010, agriculture practices in El-Minia governorate consumed about 2582.3 million m^3/year (representing about 85% of the total water volume discharged from the the governorate).

5.3.1 Surface and Subsurface Drainage Layout in Southern El-Minia

The General Administration for the drainage of South Minia has nine main drains with a total length exceeding 170 km. Some of these main drains discharge into the

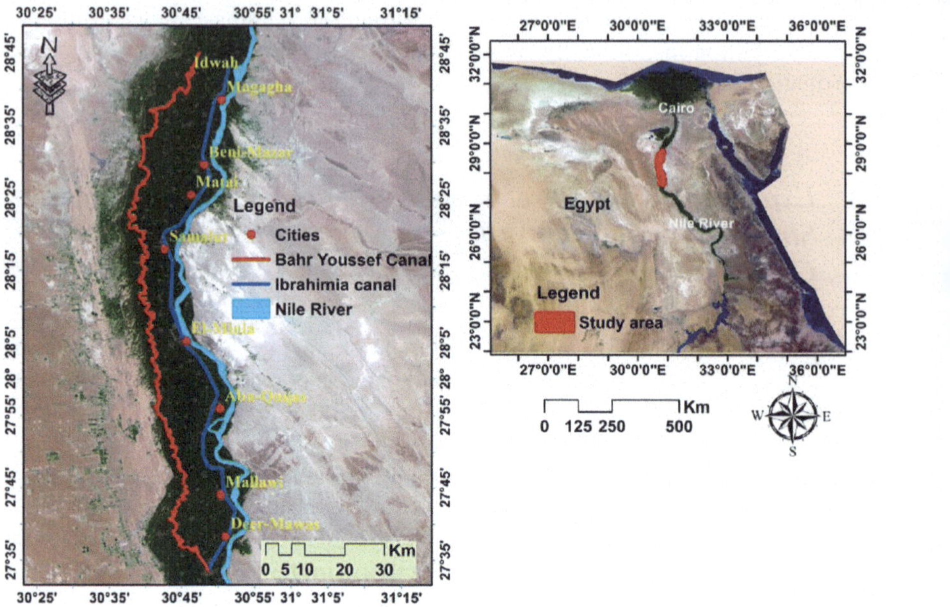

Fig. 14 Location of the study area in Egypt

Nile River, while the others discharge into the Bahr Youssef Canal (Fig. 15). In case that the level of the drain is lower than the Nile River or Bahr Youssef Canal (at the point of intersection with the end of the drain), a pump station is used to pump water from the drain whether to the Nile River or the Bahr Youssef Canal. There are four pump stations under the control of the General Administration for the drainage of south Minia. In addition to the main drains, there is a large set of branch drains that discharge into these main drains (Fig. 15). The total length of drains in southern El-Minia is about 557 km (MWRI 2020).

The subsurface drainage practices in the south of El-Minia governorate exist in the whole municipal divisions (Fig. 16). The General Administration for the drainage of South Minia has about 45 implemented drainage projects; distributed as follows: 12 in El-Minia, 10 in Abu Qurqas, 15 in Mallawi, and 8 in Dayr Mawas (MWRI 2020).

The total served area in each municipal division is shown in Fig. 17. The figure also shows the portion of the served area that had rehabilitation works for the existed subsurface drainage systems.

5.3.2 Surface and Subsurface Drainage Layout in Northern El-Minia

The General Administration for the drainage of North Minia has 14 main drains with a total length of about 166 km. As in Southern El-Minia, some of these main drains

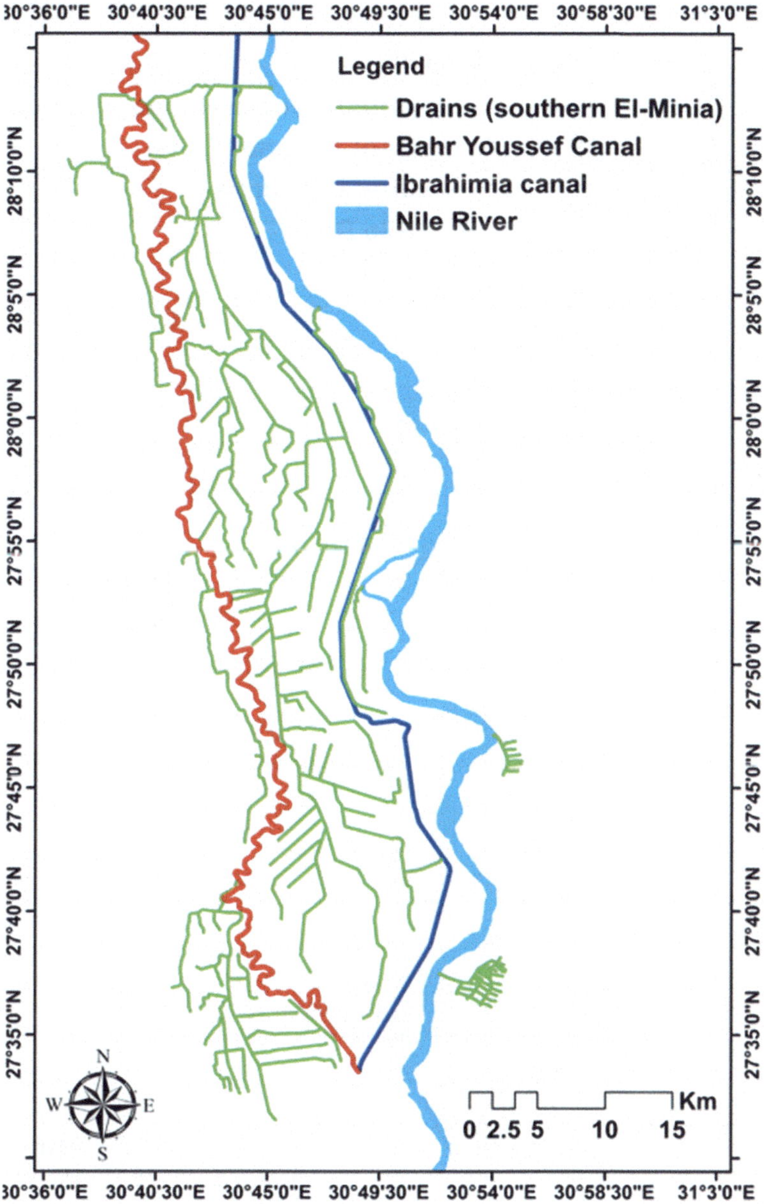

Fig. 15 Drains distribution in southern El-Minia governorate

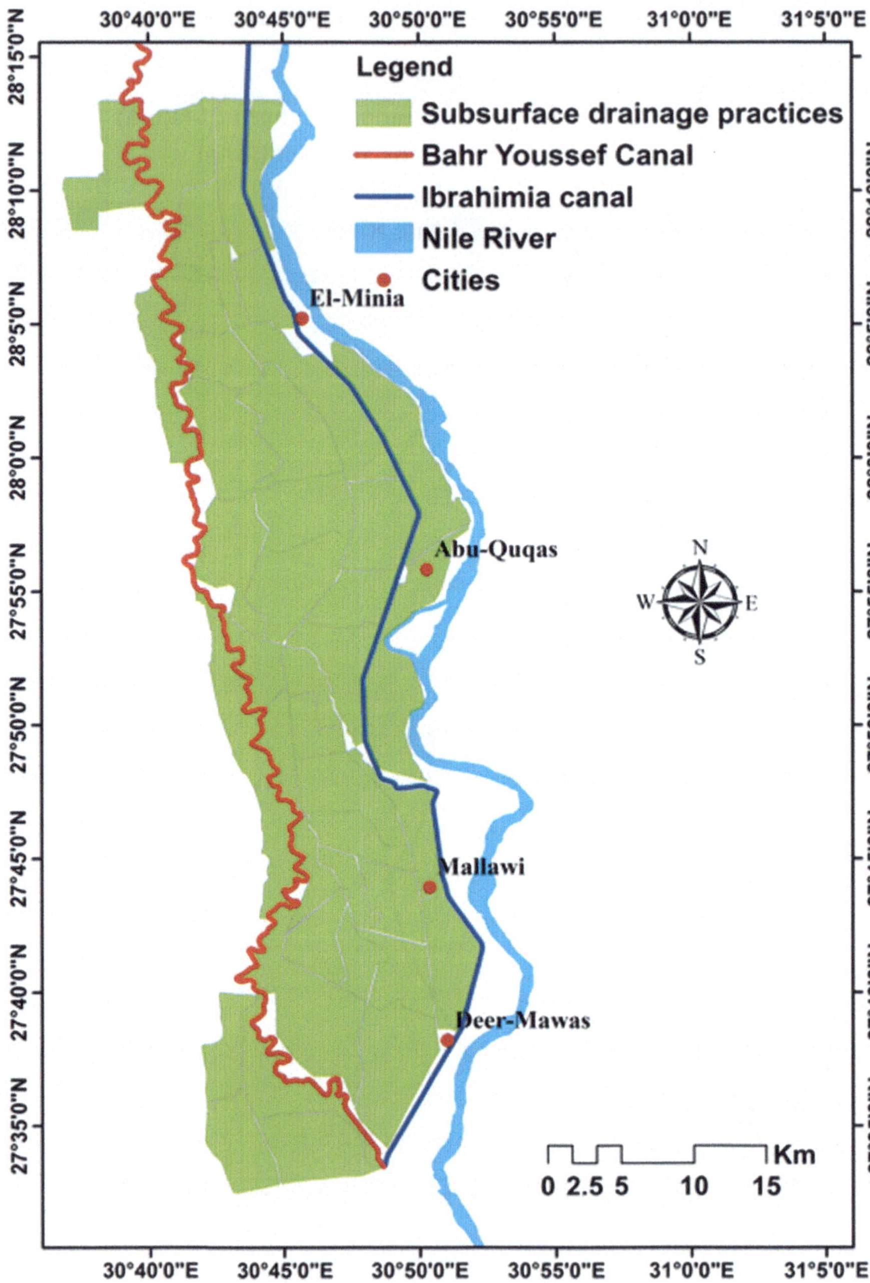

Fig. 16 Implemented drainage projects in southern El-Minia governorate

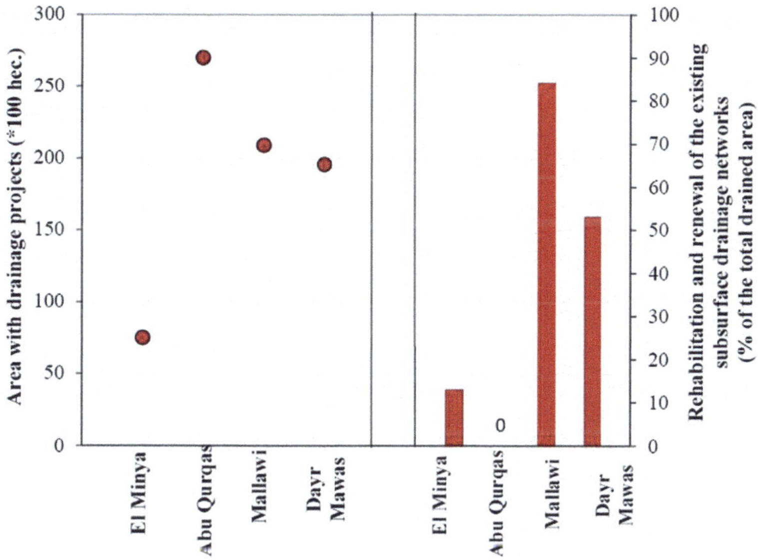

Fig. 17 The total area of subsurface drained farmlands in the south of El-Minia governorate along with the development of rehabilitation works

discharge into the Nile River, while the others discharge into the Bahr Youssef Canal (Fig. 17). Also, the General Administration for the drainage of North Minia has three pump stations that exist at the end of certain drains in which water levels are lower than those in the Nile River or Bahr Youssef Canal (at the points of intersection, which locate at the end of these drains) to pump the water from these drain whether to the Nile River or the Bahr Youssef Canal. In addition to the main drains; there is a large set of branch drains that discharge into these main drains (Fig. 18). The total length of drains in northern El-Minia is about 663 km (MWRI 2020).

The General Administration for the drainage of north Minia has 35 implemented drainage projects serving an area of about 82,000 ha. Till 2020, the administration carried out 26 rehabilitation projects to renew the drainage networks that serve an area of about 36,000 ha (MWRI 2020). The progress per year (since 2000) in the implementation and rehabilitation works in the General Administration for the drainage of north Minia is shown in Fig. 19.

6 Conclusions

In Egypt, the rise of groundwater tables alongside the excessive application of fertilizers increase the possibility of groundwater contamination by these fertilizers. In addition, such a rise in groundwater tables can also hinder or stop the respiration of

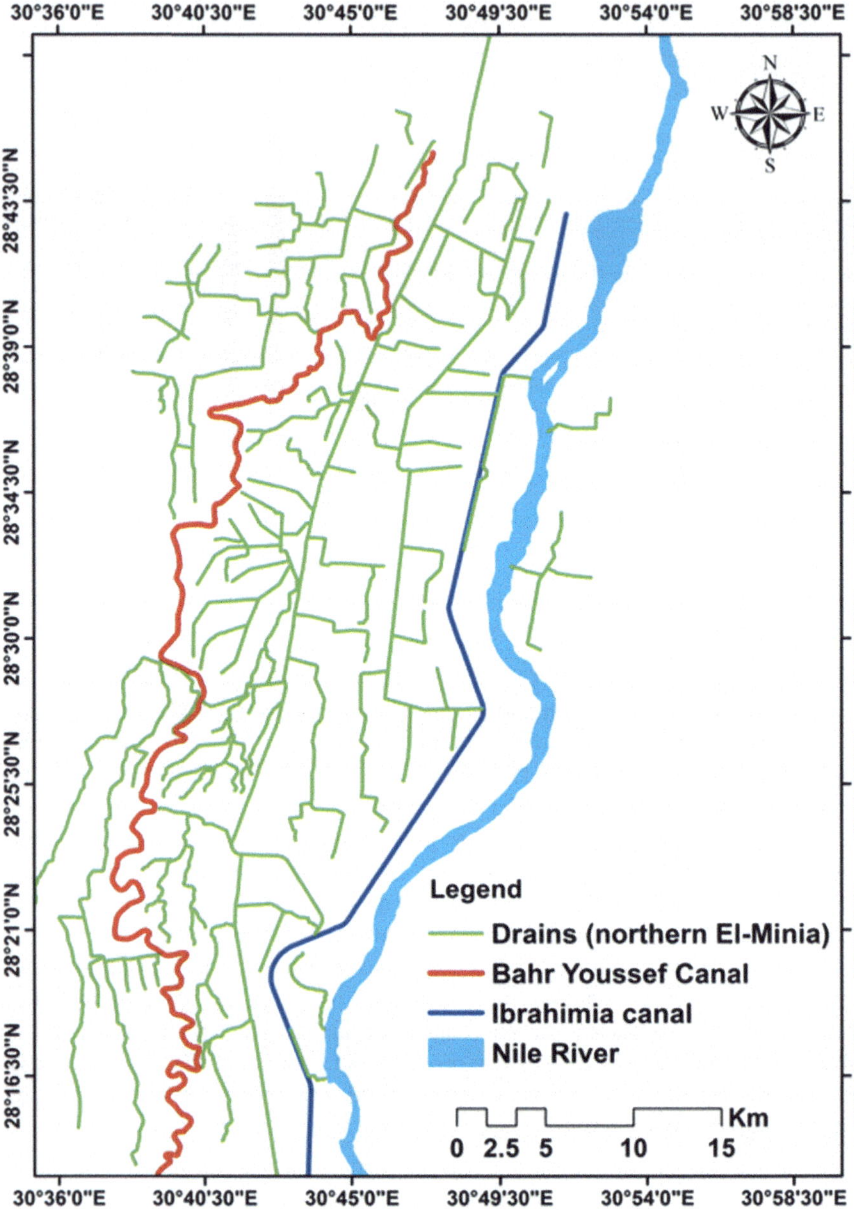

Fig. 18 Distribution of drains' networks in northern El-Minia governorate

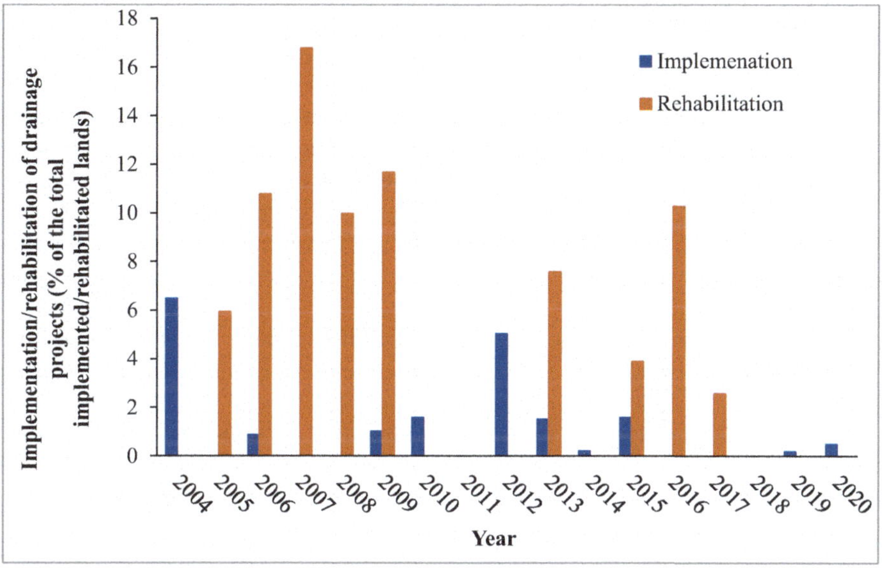

Fig. 19 Total work carried/year, from 2004 to 2000, in the implementation and rehabilitation works of subsurface drainage networks in northern El-Minia governorate

crops' roots, which threats the agricultural production. Therefore, drainage strategies are necessary in Egyptian farmlands to lower and adjust groundwater tables and moisture conditions at the desired levels. In the Nile valley in Egypt, many agricultural lands have been suffering from poor drainage conditions, and as a result, there were heavy losses in crop yield alongside a remarkable degradation in groundwater quality. Therefore, the country decided to adopt drainage strategies in these lands to ensure better management of groundwater resources, and high agricultural productivity. As a case of study of El-Minia governorate, there is a massive subsurface drainage network that serves an area of more than 157,000 ha. The drainage flux from these subsurface networks flows into a set of open drains with a total length of about 1220 km, and then dicharges into the Bahr Youssef canal or the Nile River. Practicing agricultural drainage in farmlands of the Nile River valley in Egypt led to remarkable increases in crop yield and enhanced the quality of aquifers that underlain these lands.

7 Recommendations

- In Egypt, most of the drainage projects are of high costs and are mostly carried out through grants and external loans; therefore, the possibility of renewing such projects that begian to suffer from shortages in their performance is not easy. So, we recommend using more efficient irrigation techniques that conserve the available water, thus result in less excess water in farmlands. This will help the low-efficiency drainage networks to continue operating until a proper budget is provided for their replacement.

- The inadequate use of fertilizers and pesticides in Egypt may cause drainage water to contain high amounts of nitrates, which are the most severe persistent pollutants for surface water and groundwater (Taha et al. 2004). Therefore, we recommend policymakers in Egypt to widen the use of controlled drainage strategies, especially after fertilizers' application, as this will mitigate the loss of nutrients in drainage water and thus ensuring high quality for both surface water and groundwater.
- It is recommended to ncrease the frequency of the washing schedules of subsurface drainage networks, as this will increase the lifetime and efficiency of these networks.
- It is recommended to have continuous monitoring for groundwater tables and soil salinity levels in farmlands to judge the efficiency of the existed subsurface drainage networks and to early discover any shortage in their elements.
- Finally concerning the pump stations that pump water from main drains whether to the Nile River or main canals; the presence of weeds, garbage, or animal carcasses in these drains may cause the pumps in these stations to malfunction or stop, which may lead to unacceptable rises in the water level in these drains. If these water levels rise higher than the outlets of drainage networks that flow into such drains, these outlets will be submerged and cannot discharge excess water into such drains, which will lead to the return of excess water to agricultural lands instead of getting rid of it. Therefore, we recommend setting up deterrent laws to prevent any waste from being dumped into the open drains' networks to preserve the efficiency of the pumping stations and the reasonable water levels in these drains.

References

Abdelaty E, Sallam GA, Abdelaal G, Waheed-Eldin O (2010) Environmental impact assessment of subsurface drainage projects in Egypt. Egypt Int J Eng Sci Technol 13(2):411–418

Abd-Elaty I, Pugliese L, Zelenakova M, Mesaros P, Shinawi AE (2020) Simulation-based solutions reducing soil and groundwater contamination from fertilizers in arid and semi-arid regions: case study the Eastern Nile Delta. Egypt Int J Environ Res Public Health 17(24):9373

Abdel-Dayem M (1987) Development of land drainage in Egypt. In: Vos J (ed) Proceeding, symposium 25th international course on land drainage. Twenty-five years of drainage experience, pp 195–204

Abdelmawgoud AH, El-Rawy M, Moussa Abu Bakr A (2020) Evaluating the suitability of groundwater quality for drinking and irrigation purposes in El-Minia governorate, Egypt. J Adv Eng Trends. https://doi.org/10.21608/JAET.2020.25561.1003

Abdelrady A, Sharma S, Sefelnasr A, El-Rawy M, Kennedy M (2020) Analysis of the performance of bank filtration for water supply in arid climates: case study in Egypt. Water 12(6):1816

Abdel-Shafy H, El-Saharty A, Regelsberger M, Platzer C (2010) Rainwater in Egypt: quantity, distribution and harvesting. Mediterr Mar Sci 11:245–258

Abegaz F (1995) Effect of sub-surface drainage system on groundwater table, soil salinity and crop yield in Melka Sadi pilot drainage scheme. In: On proceedings of the 1st and inaugural conference of the agronomy and crop physiology society of Ethiopia, pp 139–148

Al-Maktoumi A, El-Rawy M, Zekri S (2016) Management options for a multipurpose coastal aquifer in Oman. Arab J Geosci 9(14):636

Al-Maktoumi A, Zekri S, El-Rawy M, Abdalla O, Al-Abri R, Triki C, Bazargan-Lari MR (2020) Aquifer storage and recovery, and managed aquifer recharge of reclaimed water for management of coastal aquifers. Desalin Water Treat 176:67–77

Ashraf I, Khan T, Rashid N, Ramzan S, Khurshid N (2019) Role of some emerging agro-chemicals in groundwater contamination. Contaminants in agriculture and environment: health risk and remediation. Agro Environ Media 1:9–20

Awad A, Eldeeb H, El-Rawy M (2020) Assessment of surface and groundwater interaction using field measurements: a case study of Dairut City, Assuit, Egypt. J Eng Sci Technol 15:406–425

Awad A, Luo W, Zou J (2021) DRAINMOD simulation of paddy field drainage strategies and adaptation to future climate change in lower reaches of the Yangtze river basin. Irrig Drain 2021:1–13. https://doi.org/10.1002/ird.2564

Ayars JE, Christen EW, Hornbuckle J (2006) Controlled drainage for improved water management in arid regions irrigated agriculture. Agric Water Manage 86:128–139

Bakr N, Bahnassy MH (2019) Egyptian natural resources. In: The soils of Egypt. Springer, pp 33–49

Barman TS, Baruah U, Deka D, Borah D, Lahon T, Saikia J (2011) Selection and evaluation of waterlogging tolerant tea genotypes for plantation in marginal land

Barnes J (2014) Mixing waters: the reuse of agricultural drainage water in Egypt. Geoforum 57:181–191

Blann KL, Anderson JL, Sands GR, Vondracek B (2009) Effects of agricultural drainage on aquatic ecosystems: a review. Crit Rev Environ Sci Technol 39:909–1001

Bronick CJ, Lal R (2005) Soil structure and management: a review. Geoderma 124:3–22

Burri NM, Weatherl R, Moeck C, Schirmer M (2019) A review of threats to groundwater quality in the anthropocene. Sci Total Environ 684:136–154

Carstensen MV, Børgesen CD, Ovesen NB, Poulsen JR, Hvid SK, Kronvang B (2019) Controlled drainage as a targeted mitigation measure for nitrogen and phosphorus. J Environ Qual 48:677–685

Dubrovsky NM, Burow KR, Clark GM, Gronberg J, Hamilton PA, Hitt KJ, Mueller DK, Munn MD, Nolan BT, Puckett LJ (2010) The quality of our nation's waters—nutrients in the nation's streams and groundwater, 1992–2004. US Geol Surv Circ 1350(2):174

ELDeeb H, El-Rawy M, Habib E (2015) Water resources management: case study of El-Minya governorate. Egypt Int J Sci Eng Res 6(6):48–55

Elnashar WY (2014) Groundwater management in Egypt. J Mech Civil Eng 11:69–78

El-Rawy M, Zlotnik VA, Al-Raggad M, Al-Maktoumi A, Kacimov A, Abdalla O (2016) Conjunctive use of groundwater and surface water resources with aquifer recharge by treated wastewater: evaluation of management scenarios in the Zarqa River Basin, Jordan. Environ Earth Sci 75:1146

El-Rawy M, Al-Maktoumi A, Zekri S, Abdalla O, Al-Abri R (2019a) Hydrological and economic feasibility of mitigating a stressed coastal aquifer using managed aquifer recharge: a case study of Jamma aquifer, Oman. J Arid Land 11:148–159

El-Rawy M, Ismail E, Abdalla O (2019b) Assessment of groundwater quality using GIS, hydrogeochemistry, and factor statistical analysis in Qena Governorate, Egypt. Desalin Water Treat 162:14–29

El-Rawy M, Abdalla F, El Alfy M (2020a) Water resources in Egypt. In: Hamimi Z, El-Barkooky A, MartínezFrías J, Fritz H, Abd El-Rahman Y (eds) The geology of Egypt. Regional geology reviews. Springer Nature, Cham, Switzerland, pp 687–711

El-Rawy M, Fathi H, Abdalla F (2020b) Integration of remote sensing data and in situ measurements to monitor the water quality of the Ismailia Canal, Nile Delta, Egypt. Environ Geochem Health 42:2101–2120. https://doi.org/10.1007/s10653-019-00466-5

Ghane E (2018) Agricultural drainage extension bulletin E3370

Hansen A, Jakobsen R, Refsgaard J, Højberg A, Iversen B, Kjaergaard C (2019) Groundwater dynamics and effect of tile drainage on water flow across the redox interface in a Danish Weichsel till area. Adv Water Res 123:23–39

Heuperman AF, Kapoor AS, Denecke HW (2002) Biodrainage: principles, experiences and applications, vol 6. Food & Agriculture Org

Hiemstra P, Kolpa RJ, van Eekhout JM, van Kessel TA, Adamse ED, van Paassen JA (2003) Natural recharge of groundwater: bank infiltration in the Netherlands. J Water Supply Res Technol—AQUA 52:37–47

Kamra S (2015) An overview of subsurface drainage for management of waterlogged saline soils of India. Water Energy Int 58(6):46–53

Khaleel R, Reddy K, Overcash M (1980) Transport of potential pollutants in runoff water from land areas receiving animal wastes: a review. Water Res 14:421–436

Kladivko E, Frankenberger J, Jaynes D, Meek D, Jenkinson B, Fausey N (2004) Nitrate leaching to subsurface drains as affected by drain spacing and changes in crop production system. J Environ Qual 33:1803–1813

Madramootoo CA, Johnston WR, Willardson LS (1997) Management of agricultural drainage water quality, vol 13. Food & Agriculture Org

Mejía D (2007) FAO (Food and Agriculture Organization of the United Nations). Acta Physiol Plantarum 35:931–939

Munns R, Goyal S, Passioura J (2008) The impact of salinity stress. PlantStress. http://www.plantstress.com/Articles/index.asp

Osakabe Y, Osakabe K, Shinozaki K, Tran LSP (2014) Response of plants to water stress. Front Plant Sci 5:86

Osman KT (2012) Soils: principles, properties and management. Springer Science and Business Media

Parent C, Capelli N, Berger A, Crèvecoeur M, Dat JF (2008) An overview of plant responses to soil waterlogging. Plant Stress 2:20–27

Parihar P, Singh S, Singh R, Singh VP, Prasad SM (2015) Effect of salinity stress on plants and its tolerance strategies: a review. Environ Sci Pollut Res 22:4056–4075

Ritzema H, Nijland H, Croon F (2006) Subsurface drainage practices: from manual installation to large-scale implementation. Agric Water Manage 86:60–71

Ritzema H, Satyanarayana T, Raman S, Boonstra J (2008) Subsurface drainage to combat water-logging and salinity in irrigated lands in India: lessons learned in farmers' fields. Agric Water Manage 95:179–189

Salem A, Dezső J, El-Rawy M, Lóczy D (2020) Hydrological modeling to assess the efficiency of groundwater replenishment through natural reservoirs in the Hungarian drava river floodplain. Water 12(1):250

Sands GR, Song I, Busman LM, Hansen BJ (2008) The effects of subsurface drainage depth and intensity on nitrate loads in the northern cornbelt. Trans ASABE 51:937–946

Singh A (2015) Soil salinization and waterlogging: a threat to environment and agricultural sustainability. Ecol Ind 57:128–130

Singh A (2017) Waterlogging and salinity management for sustainable irrigated agriculture. I: Overview, implication, and plant response. J Irrig Drain Eng 143:04017035

Singh A (2019) Poor-drainage-induced salinization of agricultural lands: management through structural measures. Land Use Policy 82:457–463

Skaggs R, Nassehzadeh-Tabrizi A, Foster G (1982) Subsurface drainage effects on erosion. J Soil Water Conserv 37:167–172

Skaggs RW, Van Schilfgaarde J, Bartels J, Hatfield JL, Volenec JJ, Bigham J (1999) Agricultural drainage. American Society of Agronomy, Madison, WI, USA

Skaggs RW, Youssef M, Chescheir G (2012) DRAINMOD: model use, calibration, and validation. Trans ASABE 55:1509–1522

Smith EA, Berg AM (2020) Potential groundwater recharge rates for two subsurface-drained agricultural fields, southeastern Minnesota. US Geological Survey, pp 2016–2018

Srivastav AL (2020) Chemical fertilizers and pesticides: role in groundwater contamination. In: Agrochemicals detection, treatment and remediation. Elsevier, pp 143–159

Taha A, El-Mahmoudi A, El-Haddad I (2004) Pollution sources and related environmental impacts in the new communities southeast Nile Delta, Egypt. Emirates J Eng Res 9:35–49

Valipour M (2014) Drainage, waterlogging, and salinity. Arch Agron Soil Sci 60:1625–1640

Watson E, Lapins P, Barron R (1976) Effect of waterlogging on the growth, grain and straw yield of wheat, barley and oats. Aust J Exp Agric 16:114–122

Wesström I, Joel A, Messing I (2014) Controlled drainage and subirrigation—a water management option to reduce non-point source pollution from agricultural land. Agric Ecosyst Environ 198:74–82

Zaki R, Ismail EA, Mohamed WS, Ali AK (2015) Impact of surface water and groundwater pollutions on irrigated soil, El Minia Province, Northern upper Egypt. J Water Resour Prot 7(17):1467

Pliocene Aquifer Characterization Using TEM and VES Geophysical Techniques: Case Study at the Area to the East of Wadi El-Natrun City, West Nile Delta, Egypt

Ismael M. Ibraheem, Amal Othman, and Hosni Ghazala

Abstract The present work aims to characterize the Pliocene groundwater bearing formation at the region to the east of Wadi El-Natrun city and to the northwest of El-Sadat city, West Nile Delta, Egypt by using a combined interpretation of DC vertical electrical sounding (VES) and transient electromagnetic (TEM) data. The VES survey has been carried out during several campaigns, where a sum of 57 VES stations was established along eight profiles. In order to collect the geoelectric data, Schlumberger array was used with AB/2 values varying from 1 to 500 m. Very close to the VES stations, a single-loop configuration of $50 \times 50\ m^2$ was used to measure 47 TEM stations. The results revealed the aquifer geometries and its spatial extents and determined the most promising sites for drilling new productive boreholes. They also indicate that the Pliocene groundwater bearing layer, which is divided into two sub-aquifers, is the main aquifer in the survey region. The upper Pliocene aquifer is found at a depth varying from 30 to 70 m, has a thickness of 20–60 m, and resistivity values of 28–70 Ω.m. The lower Pliocene aquifer is encountered at depths ranging from 95 to 135 m, and has resistivity values ranging from 33 to 57 Ω.m. The findings of this research can serve as a valuable groundwater information base to help establishing new cities and land reclamation projects supported by the Egyptian government plan of the sustainable development strategy for the year 2030.

Keywords Wadi El-Natrun · El-Sadat city · West Nile Delta · TEM · VES · Groundwater

I. M. Ibraheem (✉)
Institute of Geophysics and Meteorology, University of Cologne, Pohligstraße 3, 50969 Cologne, Germany
e-mail: ismael.ibraheem@geo.uni-koeln.de

A. Othman · H. Ghazala
Department of Geology, Faculty of Science, Mansoura University, Mansoura 35116, Egypt
e-mail: osman83@mans.edu.eg

A. M. Negm and M. El-Rawy (eds.), *Sustainability of Groundwater in the Nile Valley, Egypt*, Earth and Environmental Sciences Library,
https://doi.org/10.1007/978-3-031-12676-5_10

1 Introduction

The main freshwater resources in Egypt are the River Nile, groundwater, and precipitation (El-Rawy et al. 2020). To overcome the problem of increasing population growth and shortness of water in the main cities of Egypt and also to reclaim desert regions in the area of the Nile valley and its surroundings, the Egyptian government has carried out strenuous efforts to create new urban communities and large agricultural projects. Egypt depends mainly on water from the Nile River due to the restricted amounts of rainfall. Therefore, there is an intensive demand for searching for groundwater needed for reclamation and reconstruction new cities and also for agricultural purposes. Consequently, intensive geophysical studies should be executed to detect the groundwater aquifers, delineate the most suitable localities for drilling groundwater wells, and also to evaluate the quality of the groundwater at the most promising potential areas. This satisfies the Egyptian government's plan of the sustainable development strategy for the year 2030.

The study area is located to the northwest of El-Sadat city and the east of Wadi El-Natrun city, West Nile Delta, Egypt (Fig. 1). It occupies the region located between latitudes 30° 16′ 48″ N and 30° 30′ 40″ N and longitudes 30° 21′ 36″ E and 30° 31′ 12″ E, covering an area of about 325 km^2. El-Sadat is one of Egypt's most important new industrial cities. Due to the presence of groundwater supplies, mild relief, and soil potentialities, great efforts have been made to develop such prospective places.

The area along the Nile River and its delta include the best fertilizer lands in Egypt. A desert plateau is found on the western side of the Nile Delta, where the main source of the water there for drinking, agricultural, domestic, and industrial purposes is the groundwater (Dawoud et al. 2005). Nowadays, the surface water of the Nile becomes insufficient to the Egyptian demands due to the overpopulation problem. So, searching for more groundwater reservoirs can help to solve the current problem of shortage in surface water. To achieve this goal, the current study focuses on defining groundwater aquifers' characterizations in the investigated area using a combination of vertical electric sounding and transient electromagnetic data.

2 Geomorphological and Topographic Settings and Climatic Conditions

The survey area is characterized by gentle reliefs and smooth topography. The topographic map (Fig. 2) derived from the Shuttle Radar Topography Mission (SRTM) digital elevation model (DEM) with 90 m resolution reveals that the ground elevation in the study area alters from 20 to 53 m above the mean sea level. The southeastern sector represents the highest elevation, while southwestern one towards Wadi El-Natrun depression shows the lowest elevation all over the study area. Due to this

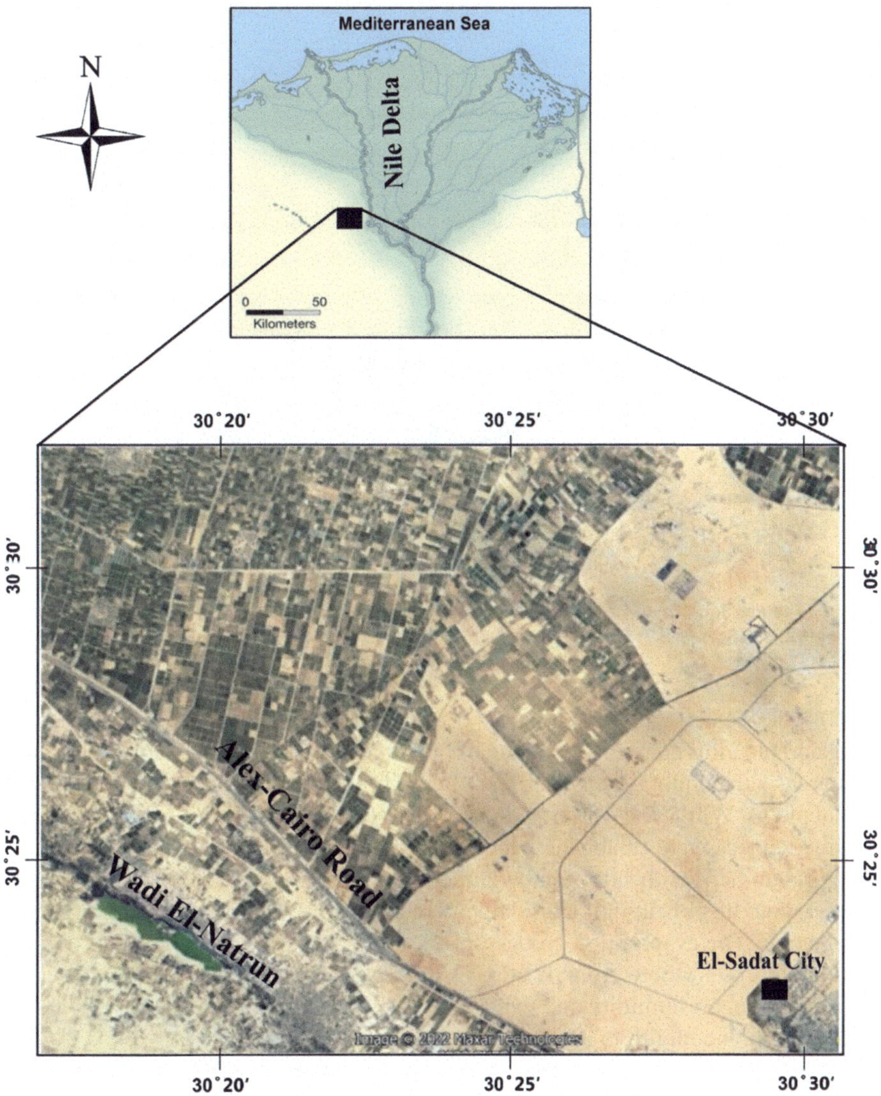

Fig. 1 A Map showing the location of the study area. Image © 2022 Maxar Technologies and CNES/Airbus. Used in accordance with the Google Maps/Earth terms of service for research purposes

moderate state of morphology, new reclamation projects and agricultural improvements have been promoted in this area. Many studies have been executed on the geomorphological settings of area of the western Nile Delta (e.g., Philip 1955; Shata and El Fayoumi 1967; Abu Al-Izz 1971; Sanad 1973; Attia 1975; El Shazly et al. 1975; El-Ghazawi 1982; Abdel Baki 1983; Embaby 1995; Ibraheem 2009, among

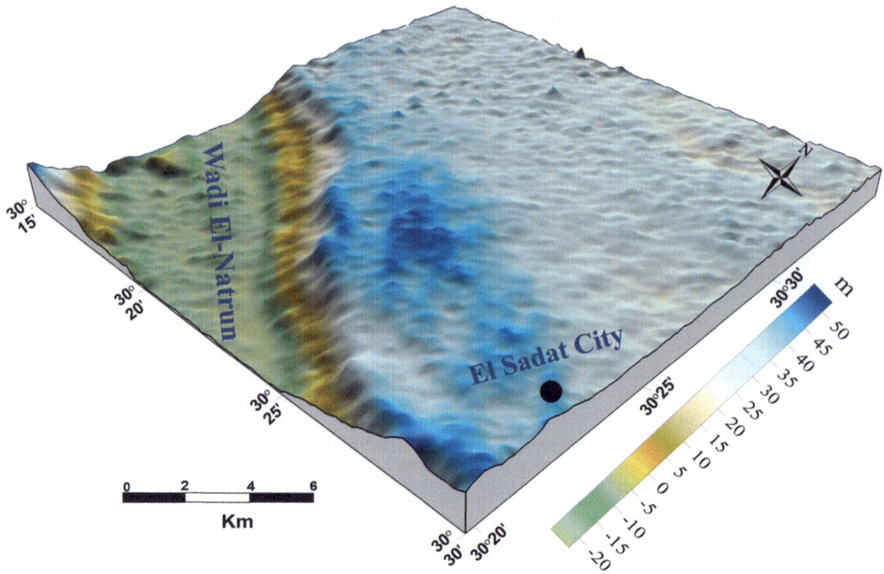

Fig. 2 3D topographic map of the survey area extracted from the SRTM DEM with a resolution of about 90 m (USGS 2017)

others). The Western Nile Delta area is divided geomorphologically into four units: young and old alluvial plains, sand dunes, and fanglomerates (Said 1962). The cultivated areas surrounding the Nile River and its branches are occupied by the young alluvial plains. While, the western fringes of the Nile Delta, where most of the latterly reclaimed places were established, are covered by the old alluvial plains.

Long summers with high temperatures and humidity, and short warm winters with low amounts of rainfall and a high percentage of evaporation characterize the study region. The evaporation process has a significant impact on the water quality, especially in cases where groundwater is located at very shallow depths. It is affected by temperature, air humidity and wind speed. Evaporation rate (4.68–13.4 mm\day) changes from one month to another (Sharaky et al. 2017).

3 Geological and Tectonic Settings

The geological and tectonic settings of the West Nile Delta including the surveyed area have been studied by several researchers such as: Shata (1953, 1955, 1961), El Fayoumy (1964), Attia (1975), Zaghloul (1976), Shata and El Fayoumy (1967), El-Ghazawi (1982), Abdel Baki (1983), Shedid (1989), Harms and Wray 1990, Abdel Aal et al. (1994, 2000), Sarhan and Hemdan (1994), Sarhan et al. (1996), Garfunkel (1998), Guiraud and Bosworth (1999), Dolson et al. (2001), Dawoud et al. (2005),

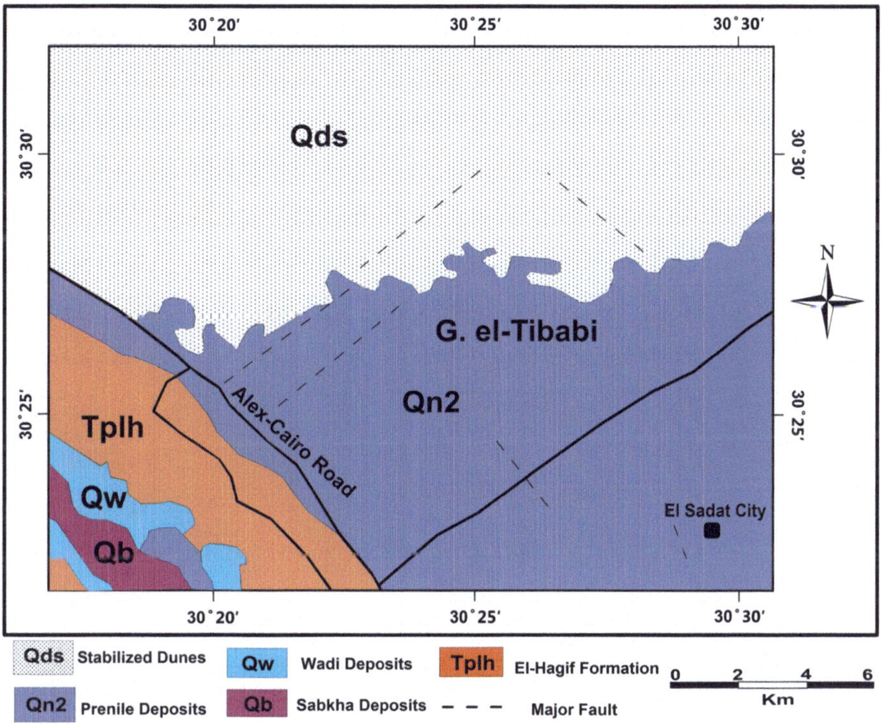

Fig. 3 Geology map of the investigated area (modified from Conoco 1987)

and Abd El-Kawy et al. (2011). In general, the Quaternary Prenile deposits and stabilized dunes dominate the survey area (Fig. 3).

Towards the Delta, the Quaternary deposits range in thickness from 25 to 350 m. It constitutes of sandy gravel and gravely sand. Nile Delta basin contains the thickest Quaternary deposits recorded in Egypt. It is divided into several units overlying unconformably on each other (Said 1990). The Nile Delta sediments had been classified according to Said (1981) into Neonile, Prenile/Neonile, Prenile, and Protonile (Bilqas, Abbassia, Mit Ghamr, El-Wastani formations, respectively). The subsurface stratigraphy includes sediments representing the period from Triassic to Quaternary. A thick Late Cretaceous to the Quaternary sedimentary sequence was determined in the West Nile Delta with a thickness of roughly 4000 m overlying the crystalline basement (Sharaky et al., 2017). Some available boreholes located at the study area (W2, W3, W5, W6, W6 new, W7, W25, W61, and El Shaheed well) are well presenting different lithology of Pleistocene and Pliocene deposits which mainly made up of gravels, sand, clay, and sandy clay. Most of the boreholes reach a depth of about 100 m (Fig. 4).

Tectonically, Egypt is subdivided into two provinces: the unstable shelf which is distinguished by the existence of several faults and folds, and the stable shelf which is characterized by the presence of the relatively less deformed horizontally bedded

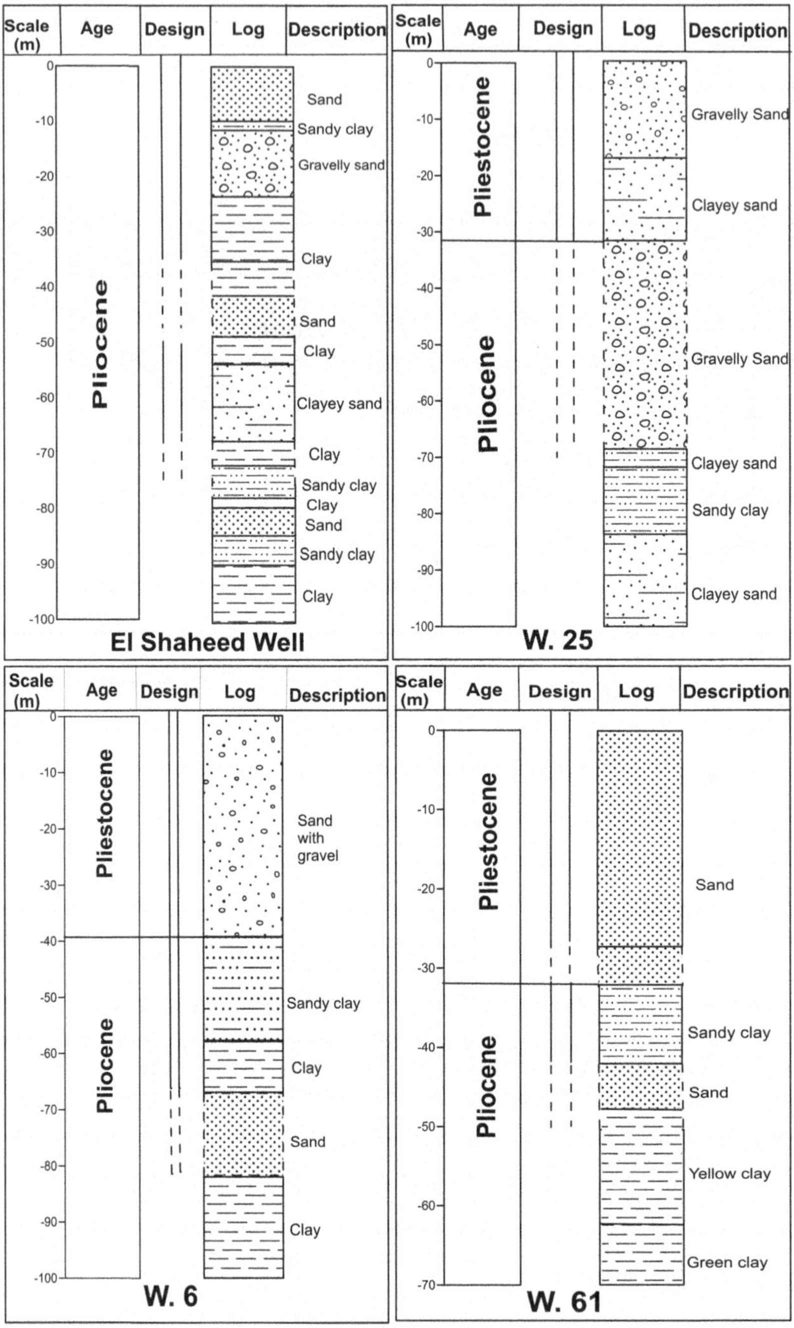

Fig. 4 Subsurface lithology of selected boreholes (W6, W25, W61, and El Shaheed wells) in the study area (Othman 2018). Refer to Fig. 9 for the locations of the boreholes

sediments of Mesozoic-Tertiary times (Said 1990). According to Dolson et al. (2001), the Nile Delta and its fan had been developed as a result of the extensive Missinian Salinity Crisis event occurred in Late Miocene. El Shazly et al. (1975) has classified the trends of faults and fractures in the West Nile Delta based on the remote sensing analysis into three main sets: N85° W–S85° E, N55° W–S55° E, and N75° E–S75° W. According to Said (1981), the Nile River had its origin as a result of a major tectonic event during the Late Miocene time which affected the Mediterranean Sea. Moreover, Nile Delta tectonics is highly related to the Oligocene emergence and Gulf of Suez rift (Abu El-Ella 1990). Lower Miocene tectonic activity (by its resultant NW–SE faults) may probably be the reason for the construction of the horst and graben of Wadi El-Farigh and Wadi El-Natrun structures (Geirnaert and Laeven 1992). Many authors have investigated the faulting system of the West Nile Delta (Shata et al. 1962; El Shazly et al. 1975; ElGalladi et al. 2009; Ibraheem et al. 2018 and others). Structurally, the Nile Delta Plio-Pleistocene sediments are affected by the following striking faults in directions of E–W, N–S, NE–SW, and SW–NE (Harms and Wray 1990; Sarhan and Hemdan 1994; Sarhan et al. 1996).

4 Hydrogeological Background

Several researchers have concerned with the hydrogeological settings of the West Nile Delta (e.g., El Fayoumy 1964; Abdel Baki 1983; Al-Kilany 2001; Embaby 2003; Massoud et al. 2014; Ibraheem et al. 2016; Salem and El-Bayumy 2016; Ibraheem and El-Qady 2019; Othman et al. 2019). The area of the Western Nile Delta has four main water-bearing formations (Fig. 5) as following: Oligocene, Miocene, Pliocene, and Quaternary. The Quaternary aquifer is made up of sand and gravel layers. Clay layers subdivide it into smaller confined sub-aquifers (Mohamed and Hua 2010). This aquifer includes the Recent and the Pleistocene aquifers. The Recent aquifer is well known at Wadi El-Natrun depression and consists of sand with some intercalations of aeolian deposits (Ibraheem 2009). The Pleistocene aquifer is the most significant aquifer in El-Sadat region (Al-Kilany 2001). The Pliocene aquifer appears at Wadi El-Natrun area, where this aquifer terminates. The Pliocene aquifer is made up of several layers of sand and clay (Omara and Sanad 1975). Moghra Formation, the second main aquifer in the western part of the Nile Delta, represents the Miocene aquifer. It has thickness of 50–250 m at Wadi El-Farigh (Dawoud et al. 2005) and consists of sand, sandstone, interbeds of clay with silicified wood, and remains of vertebrate (Said 1962). It overlies the shales or basalt sheet rocks of Oligocene. Sand, gravel, clay interbeds, and thin bands of limestone make up the Oligocene aquifer (Salem and El-Bayumy 2016) which is found in the western parts of Cairo. The main recharge sources for this aquifer are the rainwater (Abdel Baki 1983) and the paleo-water (El Abd 2005). A hydrogeological cross-section along the SW–NE direction, including the investigated area is represented in Fig. 6. It shows that the Quaternary and the Pliocene are the main aquifers in the survey area.

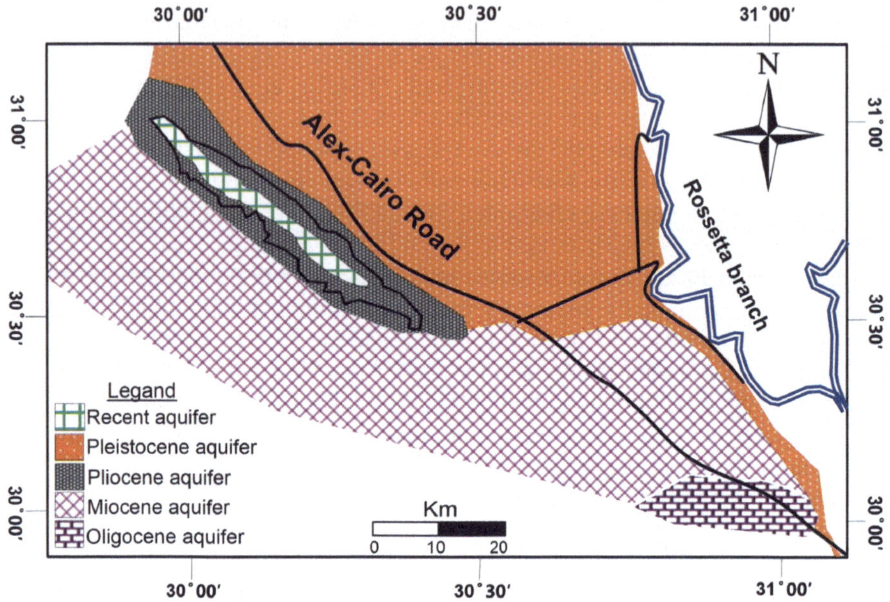

Fig. 5 A map showing the main groundwater-bearing formations at the West Nile Delta (modified from Abdel Baki 1983)

5 Literature Review

Several geophysical and hydrogeochemical studies have been carried out in the region of the West Nile Delta. Ahmed et al. (2011) concluded that the groundwater in El-Sadat industrial city is not consistent and varies in salinity and ionic composition. The groundwater salinity increases gradually from northeast to southwest in the direction of the groundwater movement. They claimed that shallow groundwater has high nitrate concentration caused by agriculture procedures and the groundwater at southwestern and northeastern parts of the industrial region has high heavy metals concentrations because of the industrial activities. Massoud et al. (2014) conducted a joint inversion of TEM and VES data at El-Sadat city to explore the characterization of the groundwater aquifers. The study revealed two aquifers; thick shallow Pleistocene aquifer exceeding 200 m where the main aquifer was encountered at a depth of 6–36 m and has resistivity values of 27–115 Ω.m and groundwater salinity varies from 217 to 925 ppm. The deep Pliocene aquifer has low resistivity values represent the second aquifer. Both aquifers are continuous without any structural features affecting the surfaces of this continuity. The groundwater hydrochemistry of the area east of Wadi El-Natrun was examined by Salem and El Bayumy in (2016). Except for few samples with high TDS values, they reported that most groundwater samples are suitable for irrigation. The samples show low to medium salinity and low sodium

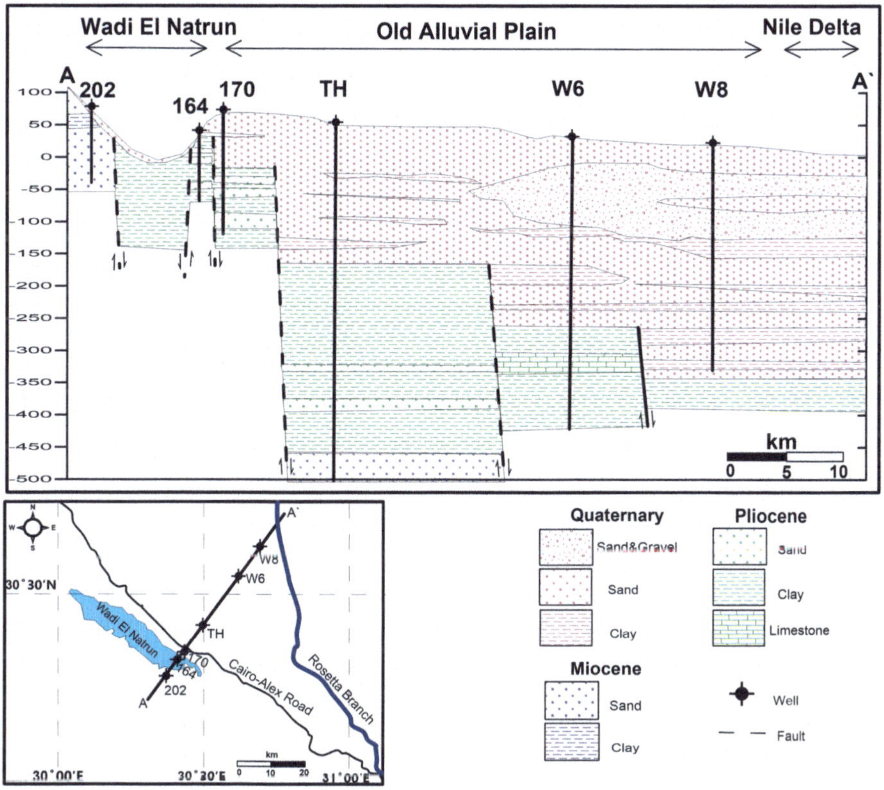

Fig. 6 Hydrogeological cross-section crossing the study area and extends in SW–NE direction (after Sharaky et al. 2017)

absorption ratio values. The main groundwater types in the area are: Na-Cl, Na-HCO_3, Mg-Cl, Ca-HCO_3 and Mg-HCO_3, where Sodium-Chloride-Carbonate water type is dominant.

Ibraheem and El-Qady (2019) carried out a hydrogeophysical survey on the area of El-Nubariya–Wadi El-Natrun to the west of the survey area. They conducted VES and TEM measurements and reported two main groundwater aquifers: Pleistocene aquifer, which is consisting of gavels and clayey sand deposits, and Pliocene aquifer, which constitutes of water-bearing sandy layers intercalated by clay facies. The depth to the main aquifer was calculated, where it ranges from 6 m at the northern part of the study area near to El-Nubariya city to 90 m at the south part close to the Wadi El-Natrun city. They described groundwater of the shallower Pliocene aquifer at the northern parts as brackish water, while groundwater of the deeper Pliocene aquifer at the southern part is classified as freshwater. Ibraheem et al. (2018) used aeromagnetic data to study the structural setting of Wadi El-Natrun area. They mentioned that the thickness of the sedimentary cover varies from 2.25 to 5.43 km and the NW, NNW,

NE, and ENE are the predominant trends of the tectonics for deep structures, while NW, ENE, and NNE are the main tectonic trends for the shallow levels.

6 Geophysical Surveys

Electrical and electromagnetic techniques are the most widely and frequently used geophysical methods in investigating and characterization the relatively shallow subsurface structures particularly in groundwater exploration (Meju 2005; Tezkan 1999, 2009). The present study was mainly based on conducting VES and TEM geophysical surveys through several field trips during the period from Dec. 2014 to Feb. 2015.

6.1 Electrical Resistivity of Minerals and Rocks

Rocks have an extremely wide range of resistivities varying from about 10^{-6} Ω.m for minerals (Graphite as an example) to more than 10^{12} Ω.m for dry quartzite rocks. Basement rocks generally have very high resistivities. On the contrary, sedimentary rocks usually have resistivities much lower than the other rock types. Many factors affecting the electrical resistivity of rocks such as: mineral composition, porosity, the existence of clay and shale, water content, and the water salinity in the pores (Yungul 1996; Parasnis 1986). An inverse relationship is noticed between resistivity and the water content, clay content, and water salinity (Burger 1992).

Generally, clay has low resistivity, whereas dry sand is more resistive for the electric current flow. Also, bedrocks have a larger resistivity compared to that of overlaid sediments. Figure 7 shows the resistivity values for different rock types and materials, where the overlap between the resistivity values can be seen. Most minerals consisting of different rocks are insulators. Therefore, the electric current would be carried out by ion movement through the pores. Hence, the electrolytic process is the most common for most of the rocks (Keary et al. 2002).

6.2 Resistivity Survey

Geoelectrical methods are widely used to map the resistivities of the geological formations encountered in the subsurface and play a major role in hydrogeological studies (Ernstson and Kirsch 2009). The electrolytic property of water within the formation represents an important factor which affects highly the resistivity of the formation. The resistivity technique is considered as one of the most widely used in geoelectric surveys. It is based on injection a direct current through current electrodes (A, B) into the earth and measuring the potential difference at voltage

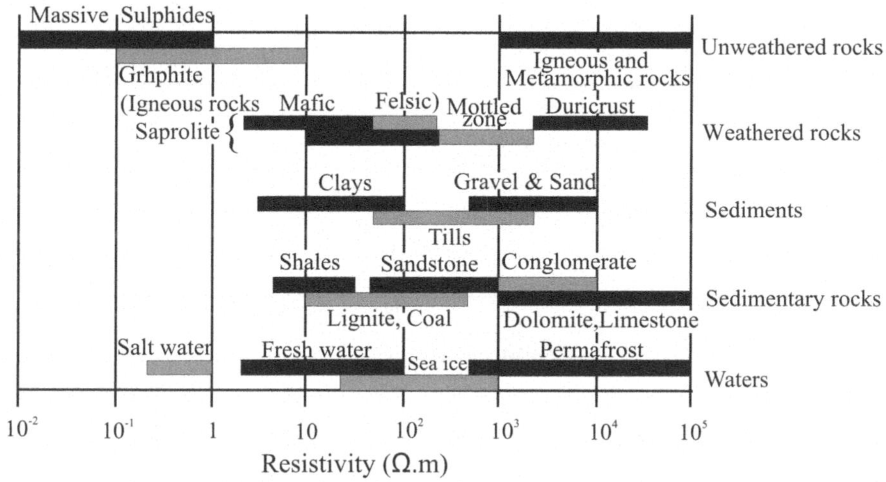

Fig. 7 Resistivities of typical earth material (after Palacky 1987)

electrodes (M, N) on the surface. Subsequently, apparent resistivity will be obtained. By inverting these apparent resistivities, true resistivities, thicknesses, and depths of subsurface layers can be identified. Generally, resistivity techniques are classified into two main methods based on the field procedures; electrical resistivity sounding and electrical profiling. In vertical electrical sounding method, the locations of electrodes are changed with respect to the sounding point whereas in profiling method all the electrodes are moved together along the traverse line. VES method is very powerful in detecting subsurface vertical changes while profiling method is used for revealing lateral variations. More details about the electrical resistivity methods can be obtained in the geophysical literature (e.g., Parasnis 1986; Telford et al. 1990; Burger 1992; Keary et al. 2002; Ernstson and Kirsch 2009; Reynolds 2011).

In hydrogeophysical studies, the VES technique is applied to define the properties of the groundwater aquifers. Generally, the Schlumberger, Wenner, and dipole–dipole electrode configurations (Fig. 8) are the most well-known configurations used for resistivity measurements. Schlumberger array is essentially formed by keeping the voltage probes M and N fixed at the center of the array and moving the current electrodes A and B gradually outwards systematically in steps. By increasing the AB spacing, the penetration depth accordingly increases, leading to the detection of the deeper-seated layers. To avoid the drop in voltage, at some stages, the distance MN is increased. Repeating some measurements with the same AB values before and after changing the MN distance is recommended. Subsequently, the different segments on the sounding curve resulted due to the change of MN distance should be suitably shifted to get a smooth VES curve. In the Wenner array, all four electrodes have to be moved where electrode spacing remains equal. In the dipole–dipole array, the electrodes A, B, M, and N form a fixed current dipole and a moving potential dipole. The Schlumberger array was elected for carrying out the present research

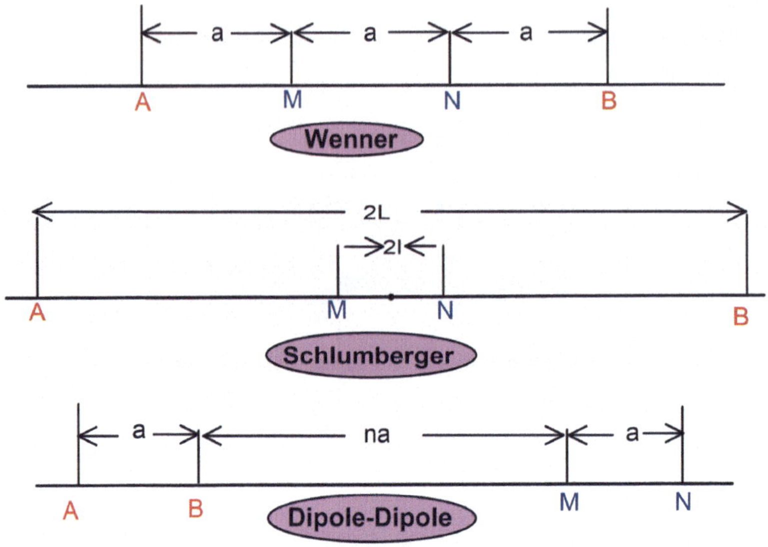

Fig. 8 Common configurations used for resistivity measurements

because it has the advantages that only the current electrodes need to be moved for most of the reading which significantly reduces the required time for measuring the station, and besides, it is less sensitive to lateral resistivity variations in the top soil. Moreover, soundings using Schlumberger configuration generally have better resolution. The apparent resistivity is calculated using Eq. (1):

$$\rho_a = k\frac{\Delta V}{I} \tag{1}$$

where ΔV is potential difference in volts, I is electric current in Amperes, and k is configuration geometric factor. In case of Schlumberger array it is given by Eq. (2):

$$k = \pi\frac{\left(\frac{AB}{2}\right)^2 - \left(\frac{MN}{2}\right)^2}{MN} \tag{2}$$

where AB and MN are current and potential electrode spacing in meters, respectively.

The equivalence principle states that a model which fits the obtained data is not the only possible model. Therefore, the measured resistivity data could have several solutions. Equivalence analysis shows how the model can be interpreted in different ways from the best fit to a close model that fits the data. Thus, equivalence analysis allows for generating a set of equivalent models. Occam's inversion (Constable et al. 1987) is used to minimize the different model solutions or roughness by applying constraints on the resistivity model. It aims to reach the best model fitting the obtained data. Occam's inversion requires only the resistivity of the first layer and the number of layers and does not need any other constraints or any previous knowledge of the subsurface.

6.2.1 VES Measurements

Several campaigns were executed in order to carry out the VES survey covering the study area. As a result, a set of 57 VES stations were conducted along eight profiles trending in NW–SE and SW–NE directions (Fig. 9). The VES survey was performed using the Syscal R2 unit designed for DC electrical surveys applied to groundwater exploration, mineral exploration and environmental studies. The Schlumberger configuration was used during the measurements by applying AB/2 values of 1–500 m. To overcome any drop in potential difference observed at M and N, the MN/2 values were increased a few times from 0.2 to 20 m for every few readings. Because of the change in MN spacing, the resulting sounding curve for each VES station has several segments. As a result, the VES curves were smoothed using the IPI2Win software before being inverted further. Some VES stations were measured near drilled boreholes to acquire information on lithology and depth to water-bearing formations. This information is essential for calibrating the geoelectrical measurements. Some selected VES curves for the area are shown in Fig. 10.

6.2.2 Geoelectric VES Interpretation

In order to depict the hydrogeological characterizes of the investigated area, qualitative and quantitative interpretation of the obtained VES data was performed.

Qualitative Interpretation

The Qualitative interpretation of resistivity data is an important step for giving a preliminary picture of the horizontal and vertical variations of the electrical resistivities of the subsurface rocks and to highlight the geological and hydrogeological settings of the survey region (Ibraheem and El-Qady 2019). Twelve iso-apparent resistivity maps have been established at different values of AB/2 (1, 4, 8, 10, 30, 60, 80, 100, 140, 200, 300, and 400 m), as shown in Fig. 11. These maps were created to describe the subsurface lithological and geological variations laterally and vertically and provide an overview of the distribution of water-bearing layers in the investigated area.

The iso-apparent resistivity maps demonstrate that the surface layers (up to AB/2 = 10 m depth) have a wide variety of resistivity values. This can be explained by heterogeneities of surface and near-surface lithology and the moisture conditions of the area. Moreover, the inspection of the twelve maps implies that the measured resistivities generally have high values at shallow depths and they decrease gradually with depth. Furthermore, remarkable change of apparent resistivity values happen

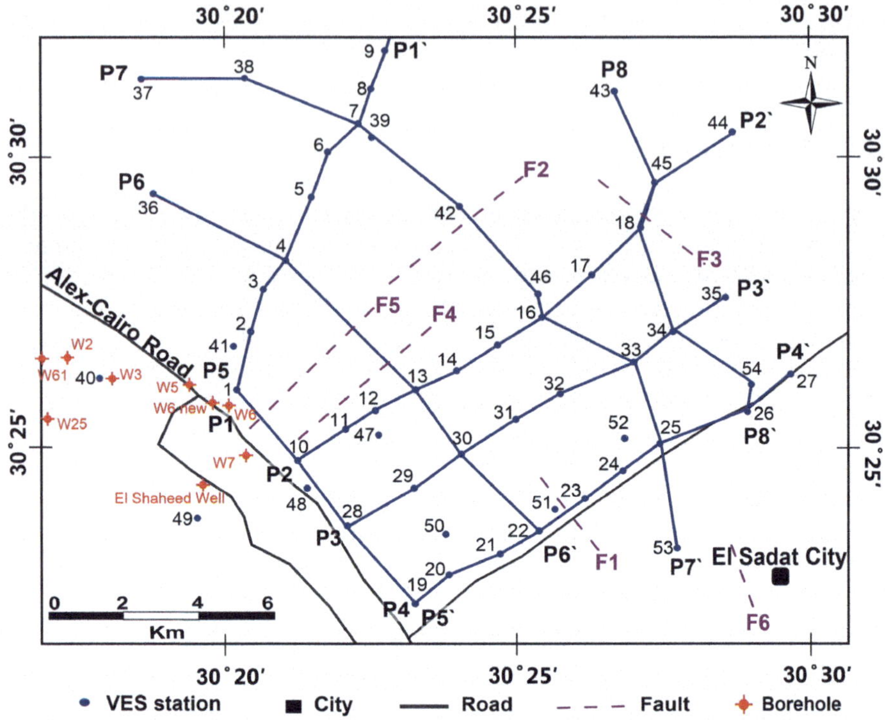

Fig. 9 Distribution of VES stations over the survey area in addition to the nearby boreholes

starting from depths corresponding with AB/2 value more than 80 m, which can refers to the top of the Pliocene aquifer (see Fig. 4). Low resistivity values at shallow depths may reflect clay and/or sandy clay nature of the sediments or local accumulations of Pleistocene water-bearing lenses. Besides, high resistivities at the southeastern part of the survey area indicate dry gravelly sediments.

Quantitative Interpretation

The aim of the quantitative interpretation of VES curves is to define the thicknesses and true resistivities of subsurface layers below each VES station. The available lithological, hydrogeological, and previous geophysical information can be used as constraints during the processing and interpretation of geoelectrical data. Hence, to overcome ambiguity in the geoelectrical sounding data, geophysicists try to apply some constraints to the inversion process in order to get a logical real final solution (Constable et al. 1987). Geoelectric station VES 1 was measured very close to well W6 new. Therefore, the lithological log of this borehole was used as an initial model to calibrate the 1D inversion process (Fig. 12). A good matching between the obtained model and the borehole information can be noticed, considering that the boundaries

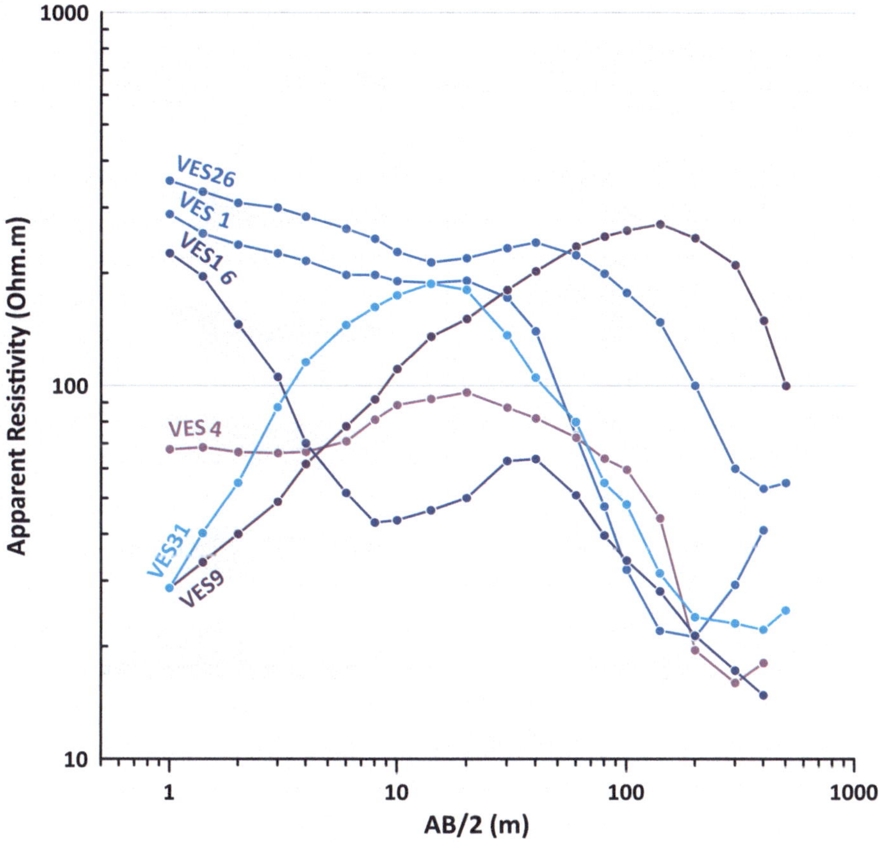

Fig. 10 Selected VES curves for the investigated area

between the layers in the lithological log may differ from that of the geoelectric log. The trial-and-error approach was utilized to get a best-fitting between the calculated and measured resistivity data with a good RMS value of 3.33%. The inversion was executed with Interpex Ltd.'s IX1D software, which uses a ridge regression method to solve the inversion (Inman 1975). The Occam and Equivalence models (assuming 30 layers) were calculated and considered in the initial model before the 1D inversion process. The results of the 1D inversion of chosen VES stations are shown in Fig. 13.

The results of the VES survey separate the subsurface into various geoelectric layers based on their resistivities. Eight geoelectric cross-sections were constructed using the geoelectric models derived from the 1D inversion of VES data (Fig. 9). The geoelectric profiles trending nearly in NE–SW and NW–SE directions were selected to cover most of the investigated area. Three geoelectric cross-sections were elected and presented in Figs. 14, 15, and 16.

The geoelectric cross-section P1–P1′ (Fig. 14) was constructed based on the interpretation of 9 VES stations (VES 1 to VES 9). It extends along SW-NE direction and

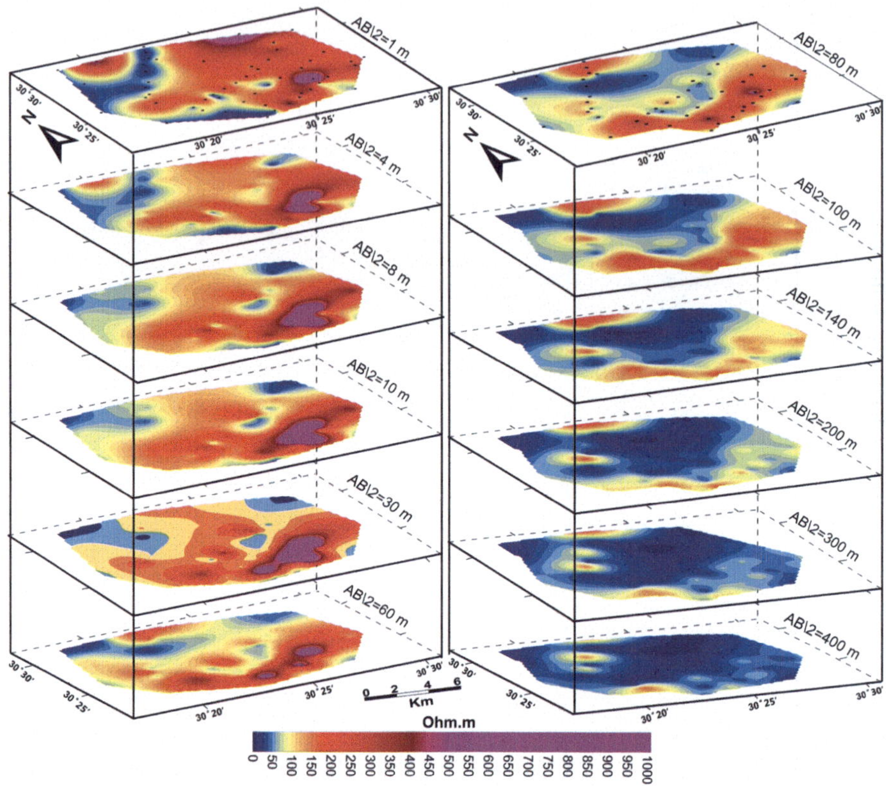

Fig. 11 3D perspective of iso-apparent resistivity maps for AB/2 values of 1, 4, 8, 10, 30, 60, 80, 100, 140, 200, 300, and 400 m

has a length of 13.5 km. Six geoelectric units were detected below this profile. The surface layer is made of intercalation of gravels, clay, and sand. It has a wide range of resistivity values changing from 1 to 450 Ω.m. The second layer is mainly sandy gravel with a thickness of about 40 m and resistivities ranging from 81 to 1610 Ω.m. The resistivities of the third layer range from 2 to 20 Ω.m, indicating clay and sandy clay sediments with a thickness value of about 30 m. The fourth geoelectric layer represents the upper Pliocene aquifer, which consists of sand with a thickness range of 35–45 m. The resistivity values of this water-bearing layer range from 33 to 64 Ω.m. This layer can be encountered at a depth of about 55 m. A geoelectric unit made up of clay and sandy clay forms the bottom of this aquifer. It has low resistivities range from 1 to 21 Ω.m. The last layer consists of sand with resistivity value varies from 33 to 66 Ω.m. This layer has been detected only beneath VES stations VES 1 to VES 6. It represents the lower the Pliocene aquifer.

The geoelectric cross-section P4–P4′ (Fig. 15) was created based on the interpretation of the following set of VES stations: 19, 20, 21, 22, 23, 24, 25, 26, and 27. It extends parallel to the profile P1–P1′ with a length of 12.25 km. The inspection

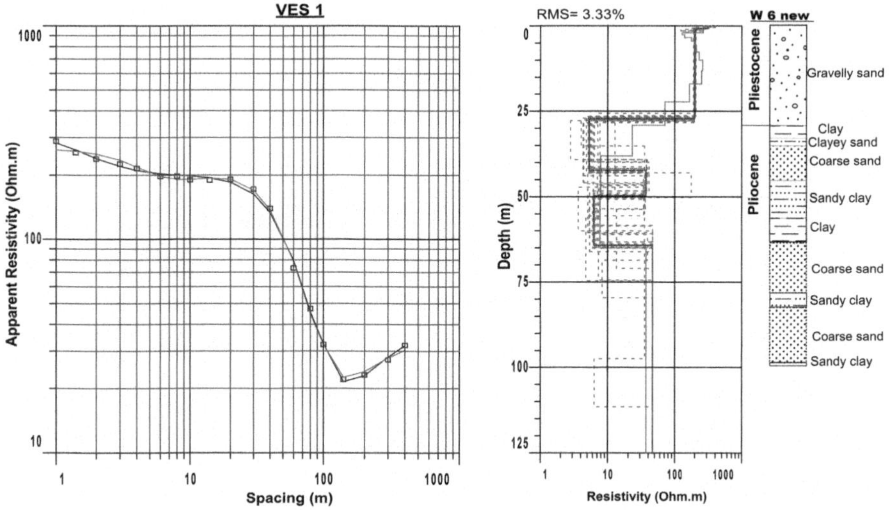

Fig. 12 Interpreted 1D model beneath geoelectric station VES 1 and corresponding lithological log of the nearby borehole W6 new

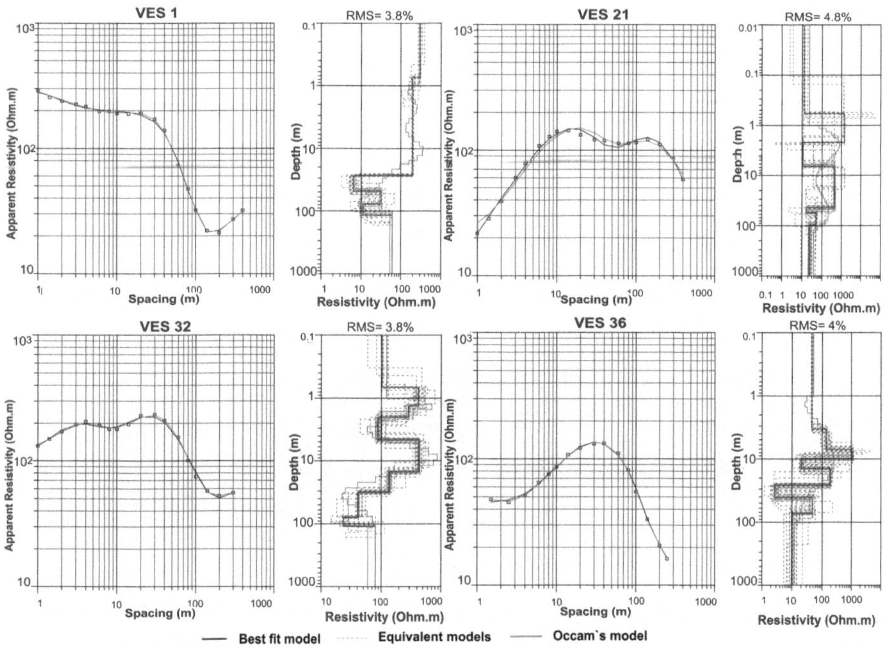

Fig. 13 1D inversion results of chosen VES data. Equivalence models, Occam's inversion, and RMS values are presented for each VES station

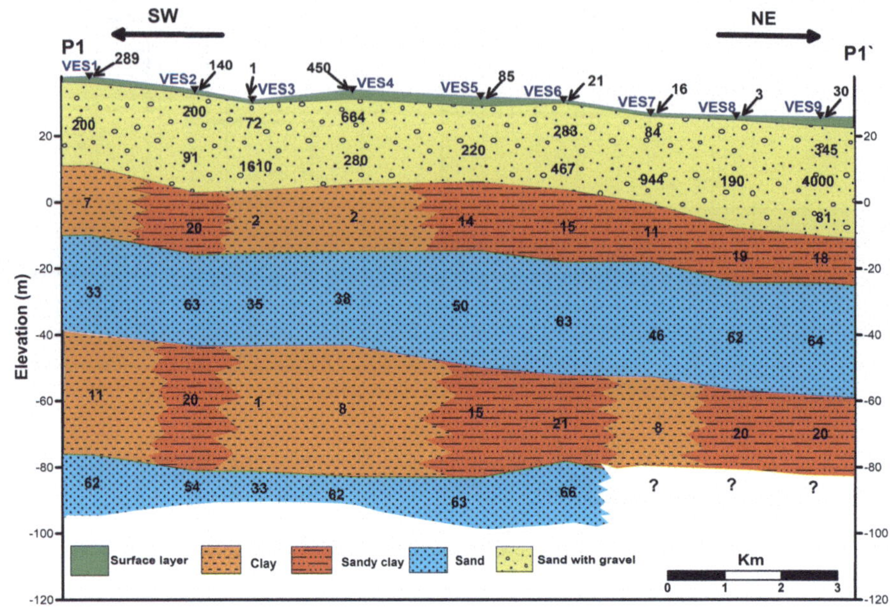

Fig. 14 Geoelectric cross-section along profile P1-P1′

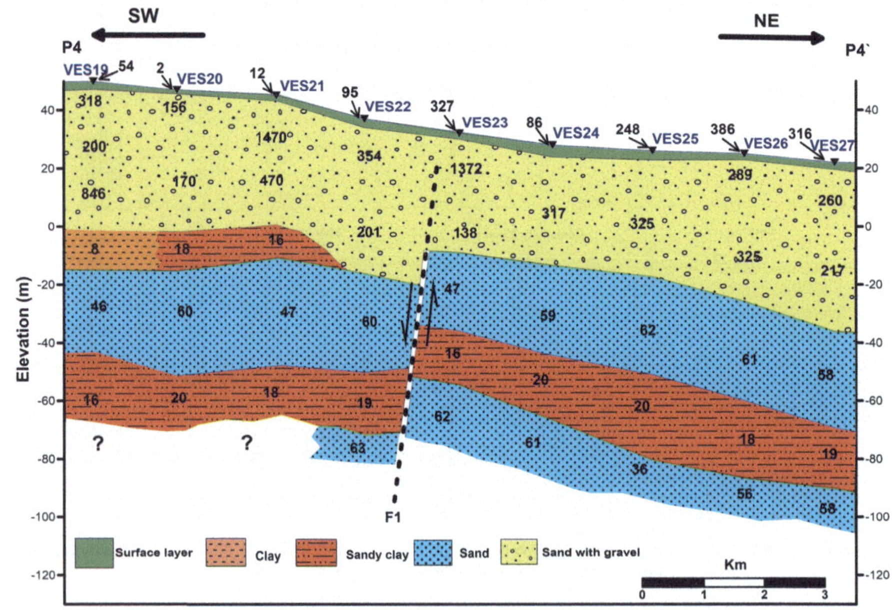

Fig. 15 Geoelectric cross-section along profile P4-P4′

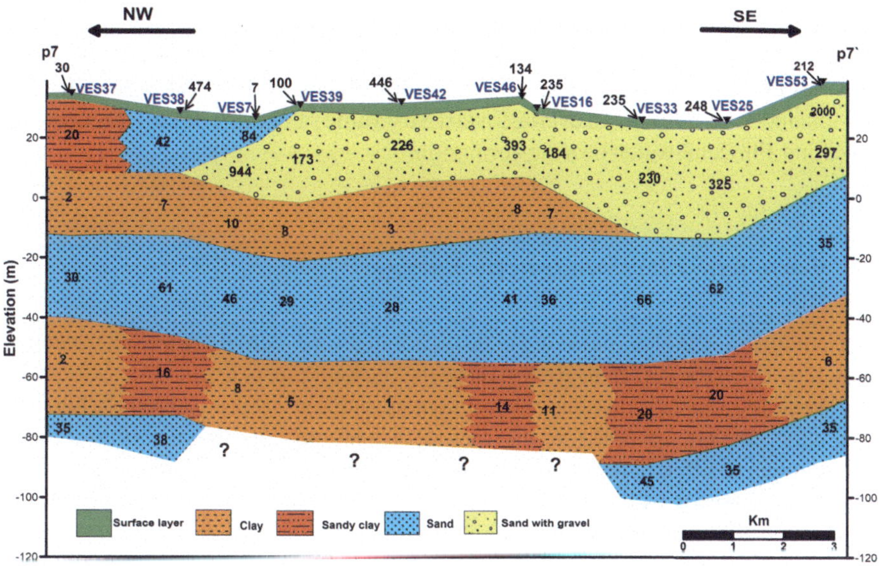

Fig. 16 Geoelectric cross-section along profile P7-P7′

of this cross-section shows the presence of six different geoelectric units. A thin surface layer with variable resistivity range of 2 and 386 Ω.m represents the top of the succession. The second layer has very high resistivity values up to 1470 Ω.m. It is made up of 50-m-thick gravels. The third layer appears only beneath VES 19, VES 20, and VES 21. It has low resistivity values (8–18 Ω.m) and made up of clay and sandy clay. The fourth geoelectric layer constitutes the first aquifer with resistivities range between 46 and 60 Ω.m and an average thickness of 35 m. This groundwater aquifer is underlain by a 20 m layer of clay and sandy clay of low resistivity values. The second Pliocene aquifer constitutes the bottom of the succession and consists of sand. Its depth varies between 85 m below VES 23 and 110 m beneath VES 27. This aquifer has resistivities range from 36 to 63 Ω.m. A normal fault F1 was suggested between VES 22 and VES 23 which can be seen on the geologic map as well (refer to Fig. 3).

Construction of the geoelectric cross-section P7–P7′ (Fig. 16) was based on the interpretation of the VES stations 37, 38, 7, 39, 42, 46, 16, 33, 25, and 53. It spreads along 15 km in NW–SE direction. Six geoelectric formations have been revealed below this profile. A thin surface layer made up of sand and gravel deposits with resistivities vary from 30 to 474 Ω.m. The second geoelectric layer is mainly gravel with sand. It changes laterally to the sand below VES 39 to VES 53 then to sandy clay below VES 37. It thickens towards the southeast and exhibits resistivity values of up to 944 Ω.m. The third geoelectric layer is composed primarily of clay with low resistivity (2–10 Ω.m) and a thickness of 20 m. It disappears below VES 33, VES 25, and VES 53. The fourth layer is 35 m thick on average, with resistivity values ranging from Ω.m beneath VES 42 to 66 Ω.m beneath VES 33. In the research area, this layer is thought to be the upper Pliocene groundwater aquifer. Below this aquifer lies a clay and sandy clay intercalation layer with low resistivity values (2–16 Ω.m). The sixth layer is encountered at 110 m depth beneath VES 37, VES 38, VES 33, VES 25, and VES 55 with an average resistivity value of about 37 Ω.m. It represents the second Pliocene water-bearing formation.

6.3 Transient Electromagnetic Survey

Electromagnetic (EM) methods are classified into a time-domain (transient) electromagnetics (Christiansen et al. 2009) where the measurements are made as a function of time and a frequency-domain electromagnetics (FEM) (Siemon 2009), where the instruments use one or more frequencies. EM methods can also be either passive such as magnetotellurics (MT) or active such as TEM and very low frequency (VLF) methods. In contrast with electrical methods, the EM methods do not require direct contact with the ground, which gives them much superiority regarding the speed of the survey. The range of applications for EM surveying is very wide as they are used in mineral exploration, hydrocarbon exploration, groundwater surveys, landfill surveys, geological mapping, unexploded ordnance (UXO) detection, and archaeological investigations (Reynolds 2011). Since it has been developed in the mid-1980, the TEM method has been applied in environmental and hydrogeological explorations (Meju 2005; Goldman et al. 1994; Yogeshwar and Tezkan 2017) due to its ability for defining good conductors (Auken et al. 1994; Pullan et al. 1994). EM method, especially TEM has various advantages: it is a fast method because it does not require a huge spread, nor does it require direct contact with the ground or electrode penetration into the earth's surface. Moreover, the TEM method has a greater penetration depth than the DC resistivity method (Sharma 1997). The fundamental physical concepts of the method can be found in Telford et al. (1990), Nabighian and Macanae (1991), Christiansen et al. (2009), Reynolds (2011), among others.

6.3.1 TEM Measurements

A SIROTEM MK3 tool was used to collect 47 TEM stations (Fig. 17). Locations of the TEM stations were chosen to be close to the VES station locations. A single-loop (Nabighian and Macanae 1991) with size of 50 m × 50 m was used during the TEM survey. After the current cut-off, the magnetic field's time rate change (dB/dt) is measured by the receiver. TEM measurements are thought of as a sequence of output voltage values at each successive time gate, which can range from microseconds to tens or even hundreds of milliseconds. Background noise can impact the signal's latest time, which can be removed using the editing function before the processing phase. The measurement of the present survey at each station was repeated numerous times with different gain factors and electrical current of 10 A. To reduce the noise

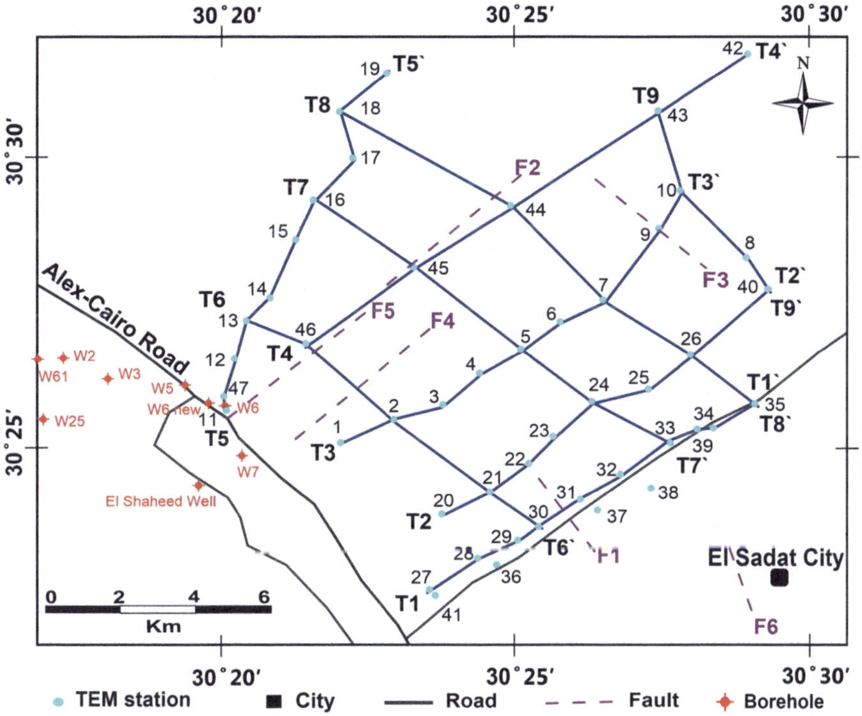

Fig. 17 Location map of TEM stations. The TEM profiles are represented by blue lines

effect of the recorded data, each measurement was stacked 256 times, and the data with the best signal/noise ratio were picked for further processing.

6.3.2 TEM Data Interpretation

IX1D Interpex software and Aarhus SPIA (Auken et al. 2015) were used to invert the TEM data. Firstly, Occam's inversion was applied to the TEM data to create multi-layer models. Then this process was followed by equivalence analysis in order to create models with a small number of layers, reducing the level of uncertainty. For enhancing our results, the obtained VES models were used as initial models to perform the TEM inversion. The lithological and hydrogeological information derived from the log of well W6 was used to calibrate and constrain the geoelectrical model at TEM station 47, as illustrated in Fig. 18. In addition to the equivalency and Occam's models, Fig. 19 illustrates the findings of the 1D inversion of various TEM data and their corresponding models. Depths and true resistivities obtained from each electromagnetic model were gathered and used to create nine geoelectrical cross-sections that covered the area and trended in two directions: NE–SW and NW–SE.

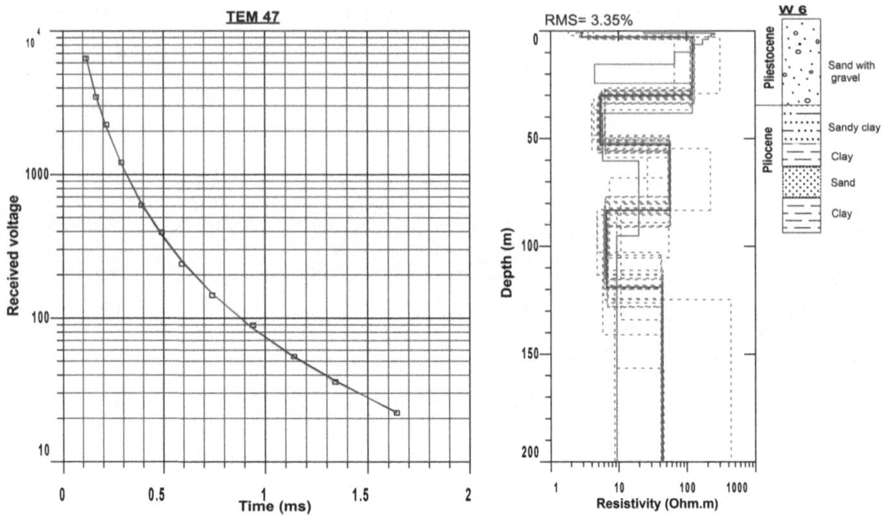

Fig. 18 Interpreted 1D model beneath TEM station no. 47 and corresponding lithology log of the nearby well W6

Three of them were chosen to be presented in this research: T1–T1′, T5–T5′, and T6–T6′.

The geoelectric cross-section T1–T1′ (Fig. 20) was created based on the 1D inverted models of the TEM stations: 27, 28, 29, 30, 31, 32, 33, 34, and 35. It is 12 km long and runs in a SW-NE direction. There are six separate geoelectric layers in this cross-section. The upper one is a thin layer (2–3 m) consisting of clay with low resistivity values. It overlies another layer of high resistivity values change from 110 Ω.m below the station TEM 30–489 Ω.m below TEM station no. 35. The main components of this second layer are gravels and sands. It has a maximum thickness of 62 m below TEM station 31. The third layer is mainly clay with very low resistivities (2–9 Ω.m). It is 20 m thick on average and well recorded beneath TEM stations TEM 27, TEM 28, and TEM 29, and it is disappeared below the rest of TEM stations. The upper Pliocene groundwater aquifer is detected below this layer at a depth of 50–66 m. Its resistivity ranges from 40 to 57 Ω.m and it composes mainly of sand. A clayey layer with low resistivity values (1–12 Ω.m) and an average thickness of around 26 m lies underneath this aquifer. The base of this cross-section is the lower Pliocene aquifer which is encountered at a depth of up to 135 m. It is composed of sand with resistivities range from 34 to 48 Ω.m. Between TEM 30 and TEM 31, the fault structure F1 is also detected.

The geoelectric cross-section T5–T5′ (Fig. 21) extends along SW–NE direction and has a length of 12 km. It has been constructed based on the interpreted electromagnetic models of TEM stations: 11, 12, 13, 14, 15, 16, 17, 18, and 19 as shown in Fig. 17. This cross-section reveals that subsurface is divided into six geoelectric layers. Gravel, sand, and clay make up a thin surface layer with an average thickness

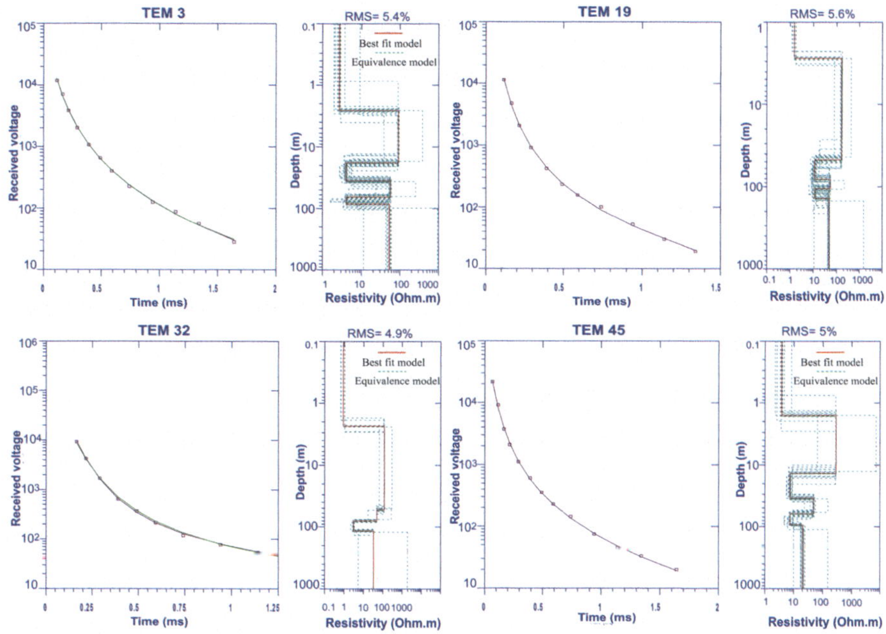

Fig. 19 Interpreted 1D model beneath TEM stations TEM 3, TEM 19, TEM 32, and TEM 45

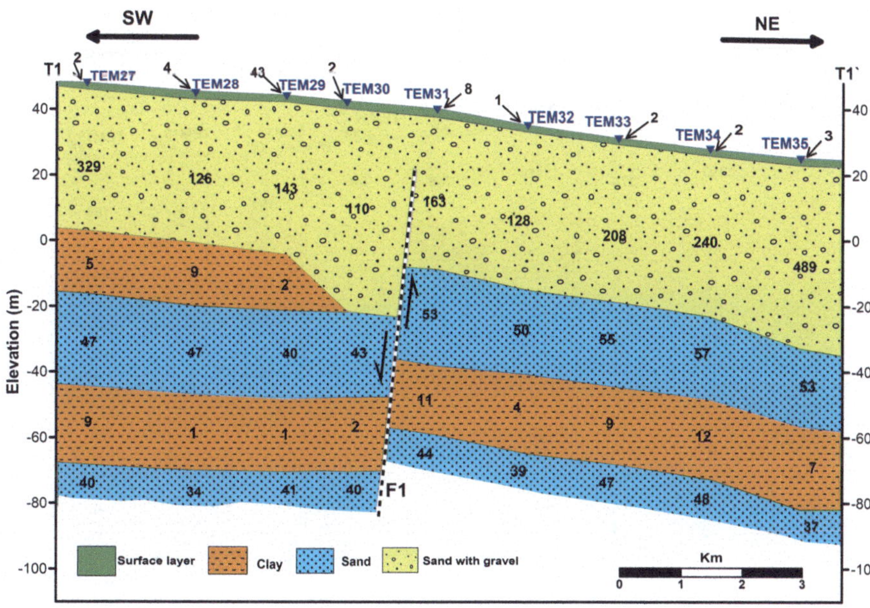

Fig. 20 Geoelectric cross-section along profile T1-T1′

of 2 m and a wide range of resistivity values. Below the surface layer, a gravely layer with high resistivities varying from 98 up to 213 Ω.m is detected. It has a thickness with an average value of 23 m. A formation of clay and sandy clay represents the third geoelectric layer. It has low resistivity values (3–14 Ω.m) and a thickness of 25 m in average. The fourth geoelectric unit was reached at a minimum depth of 46 m beneath TEM 16 and at a maximum depth of 55.5 m below TEM station 19 with an average thickness of 30 m. Its resistivities change from 49 to 59 Ω.m and it consists mainly of sand. This layer is considered as the upper Pliocene groundwater aquifer. Below this sandy layer, a geoelectric formation with low resistivities (2–13 Ω.m) and a thickness up to 37 m was detected. Clay and sandy clay sediments are indicated by these low values. The last layer, with a resistivity range of 42–57 Ω.m, is found at an average depth of 115 m. It is composed of sand and represents the lower Pliocene groundwater aquifer in the research area.

The cross-section T6–T6′ (Fig. 22) extends for 9 km along NW–SE direction. Six different geoelectric units constitute the subsurface image below this profile. A thin stratum (2–3 m) depicts the surface clayey layer of low resistivity values. The second layer consists of gravels and sands and has high resistivity values (110–374 Ω.m). However, this layer starts with relatively low thickness and becomes thicker towards the southeast. Below TEM 30, it reaches a maximum value of 60 m. The third is a clayey layer having low resistivity values (6–10 Ω.m). It has been detected below TEM stations 13, 46, and 2 while it is disappeared beneath TEM 21 and TEM 30.

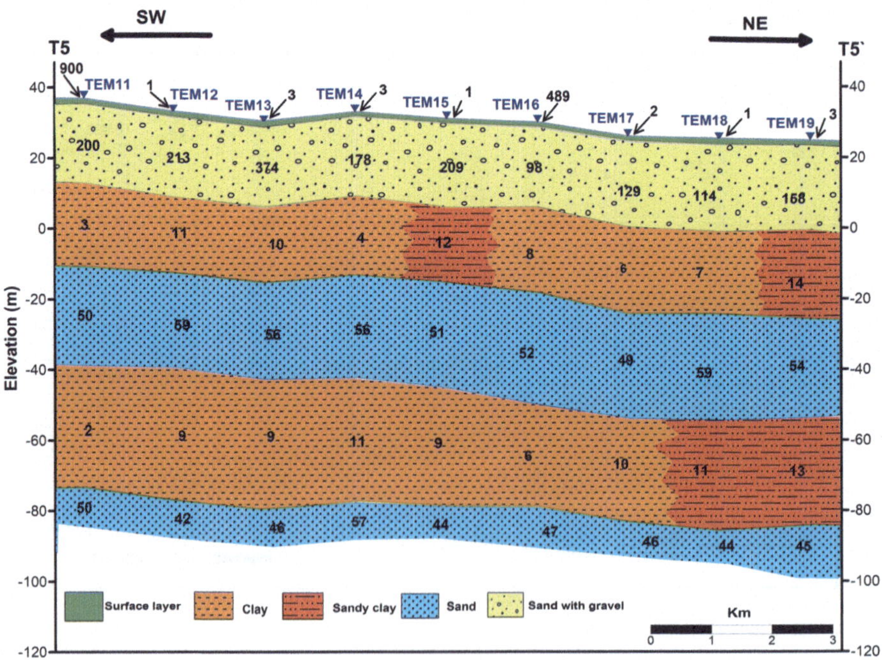

Fig. 21 Geoelectric cross-section along profile T5-T5′

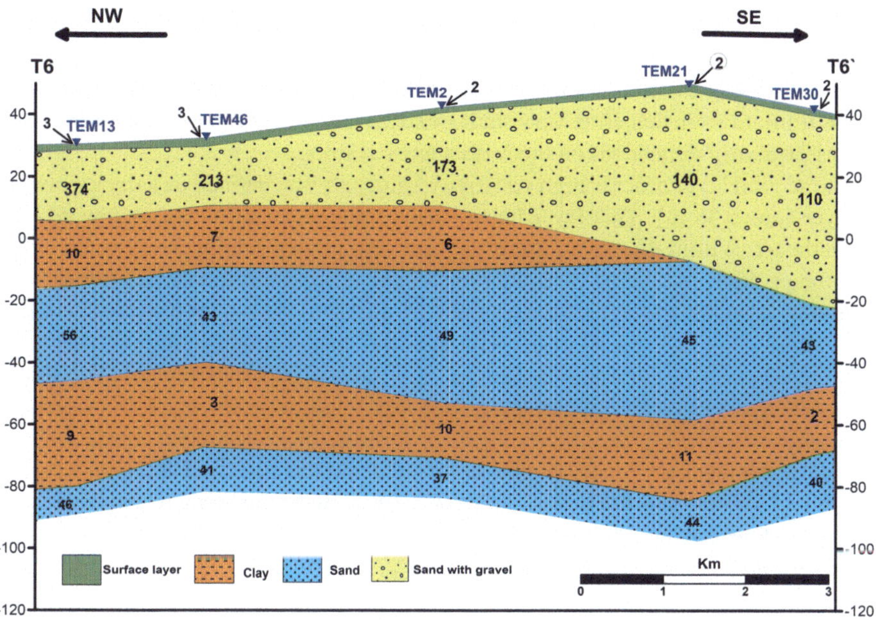

Fig. 22 Geoelectric cross-section along profile T6-T6′

The fourth layer has resistivity varies between 43 and 56 Ω.m and consists of sand. It can be detected at a minimum depth of 43 m at TEM station 46 and up to 64 m below TEM 30. This layer represents the first Pliocene water-bearing formation. The fifth layer is mainly clay with low resistivity values (2–11 Ω.m). It has a maximum thickness of 35 m below TEM station 13. The final layer has moderate resistivity values (37–46 Ω.m). This layer is made up of sand and has an average depth of 115 m. It represents the second Pliocene aquifer.

7 Results and Discussion

Eight geoelectric cross-sections from the VES survey and nine geoelectric cross-sections from the TEM survey were created using the VES and TEM inverted models. A considerable similarity between the results of VES and TEM surveys concerning the depths of the geoelectric layers and their true resistivities was found. The VES technique improved its power in detecting shallow subsurface targets, while the TEM method was more convenient in depicting the deeper layers with a higher resolution. Therefore, a description of the subsurface layers was performed based on the results of both VES and TEM data. The results of the VES and TEM surveys were combined to create six separate geoelectric formations. These layers are as follow (from top to bottom):

- The surface layer is made up of clay, sand, and gravel intercalations with a thin thickness of 1–3 m. Due to the variability of the lithology at the surface and agricultural activities, the resistivities vary greatly. Because the TEM technique has limitations in detecting very shallow structures, the interpretation of this layer was primarily dependent on the VES survey results.
- The second layer reflects high resistivity values. It was defined as a thick gravelly layer (up to 58 m). This layer has a varied thickness that increases towards the southeast. Within this layer, lenses of sand and clayey sand were reported along some profiles, indicating possible Pleistocene groundwater accumulations.
- A clay layer with low resistivity values (1–11 Ω.m) makes up the third geoelectric unit. This layer is 30 m deep on average. In the southern region, this layer decreases until it vanishes. In general, this layer disappears towards the southeast. It constitutes an impermeable base layer that prevents water from seeping downwards, playing an important role in forming the Pleistocene aquifer.
- The fourth layer is primarily made up of sand and can be found at depths ranging from 27 to 64 m. It is the first Pliocene aquifer, with intermediate resistivity values and thicknesses ranging from 24 to 55 m. From the center of the area to the East, the aquifer becomes thicker.
- The fifth layer is composed mainly of clay with low resistivity values (1–11 Ω.m) and clayey sand with a resistivity range of 12–20 Ω.m.
- The second Pliocene groundwater aquifer in the research area is represented by the base layer, which is composed of sand. Its resistivity ranges from 33 to 57 Ω.m. This layer was detected mainly by the TEM survey and partly by the VES survey. The depth of this formation ranges from 98 to 135 m.

Structurally, the impact of the previously known faults in the survey area can be seen along several geoelectric cross-sections (e.g., Faults F1 and F3). These faults can cause a possible connection between the two Pliocene water-bearing horizons. Furthermore, the true resistivities, thicknesses and depths to the water-bearing formations have been deduced from the combined interpretation of 1D inverted TEM and VES models. These data were gridded and mapped in order to provide a comprehensive hydrogeological vision in the research area. The first aquifer's true resistivity (Fig. 23a) manifest values in the range 30–40 Ω.m characterizing the region extended from the southwest to the northeast, whereas the area in the southeast and the northwest exhibits relatively moderate resistivities (50–70 Ω.m). The first aquifer can be detected at depths of 30–70 m. It is found at shallow levels at the eastern portions of the area (Fig. 23b), whereas it is encountered at deep depths in the south and southwest parts of the region. Its thickness ranges from 20 to 60 m (Fig. 23c). True resistivity values of the second aquifer vary from 33 to 57 Ω.m (Fig. 24a). The iso-resistivity map of this aquifer demonstrates relatively low resistivities at the central area along the SW-NE trend comparing with other places in the area. This low resistivity zone means that the groundwater flow direction is probably to the northeast and the southwest. The depth to the second aquifer (Fig. 24b) varies between 98 and 135 m. The relatively low resistivities of the upper aquifer refer to brackish water. These values become greater in the lower aquifer which reflects fresher groundwater.

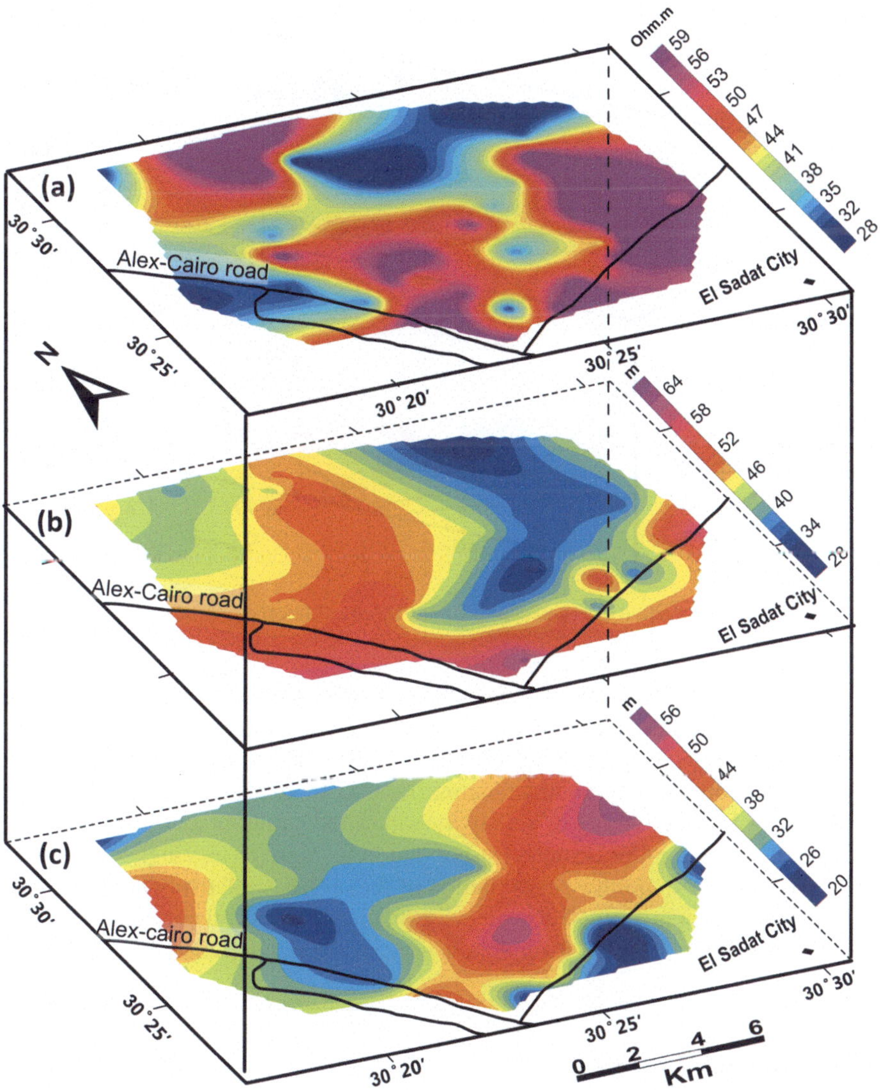

Fig. 23 Iso-resistivity (**a**), depth (**b**), and isopach (**c**) maps of the upper Pliocene groundwater aquifer

8 Conclusions

The geophysical research presented in this work provides essential results of the water-bearing formations in the investigated area in terms of the groundwater potentialities. The Pliocene groundwater aquifer was successfully characterized using a combined interpretation of VES and TEM data, demonstrating that it is the major

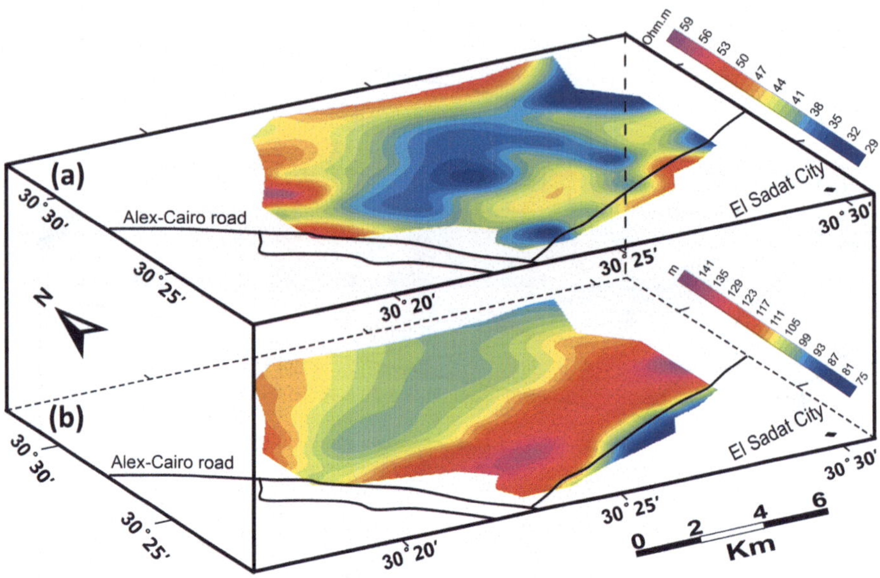

Fig. 24 Maps of Iso-resistivity (**a**) and depth (**b**) of the lower Pliocene groundwater aquifer

aquifer in the survey area and is separated into two water-bearing horizons. The upper one is found at depths ranging from 30 to 70 m, with a thickness range of 20–60 m and a resistivity of 28–70 m. The lower one has resistivity values of 33–57 Ω.m and is observed at depths ranging from 95 to 135 m. The upper Pliocene water-bearing layer exhibits low resistivity values indicating brackish water, while the lower Pliocene aquifer is much fresher due to its more or less higher resistivity values. The aquifer in northwestern portion of the area has relatively high resistivities, suggesting coarse-grained deposits with shallow depths and considerable value of thickness making it the most suitable region for future boreholes.

9 Recommendations

Nevertheless, more studies are necessary to set up a sustainable level of exploitation from this water-bearing formation and assess the vulnerability to avoid any potential pollution associated with excessive groundwater abstraction of the Pliocene aquifer. Furthermore, public awareness programmes concerning the significance of water conservation should be introduced through different media to build up a new water culture based on the conservation principle in a country facing several challenges. The results of current research include a comprehensive description of the hydrogeological characteristics of the Pliocene aquifer at the investigated area and can provide valuable information for professionals, planners, decision-makers, and stakeholders.

Therefore, it can be considered as an essential documentation for further advanced geophysical and hydrogeological investigations.

References

Abd El-Kawy OR, Rød JK, Ismail HA, Suliman AS (2011) Land use and land cover change detection in the western Nile delta of Egypt using remote sensing data. Appl Geogr 31(2):483–494

Abdel Aal A, El Barkooky A, Gerrits M, Meyer H, Schwander M, Zaki H (2000) Tectonic evolution of the eastern Mediterranean Basin and its significance for the hydrocarbon prospectivity in the ultradeepwater of the Nile Delta. Lead Edge 19(10):1086–1102

Abdel Aal A, Price R, Vaitl J, Shrallow J (1994) Tectonic evolution of the Nile Delta, its impact on sedimentation and hydrocarbon potential. In: EGPC 12th exploration and production conference, pp 19–34

Abdel Baki AA (1983) Hydrogeological and hydrogeochemical studies on the area west of Rosetta Branch and south of El Nasr Canal. Ph.D. thesis, Faculty of Science, Ain Shams University, Cairo, Egypt

Abu Al-Izz MS (1971) Land forms of Egypt. The American University in Cairo Press, Dar Al Maaref, Cairo, Egypt

Abu El-Ella R (1990) The Neogene-Quaternary section in the Nile Delta, Egypt: geology and hydrocarbon potential. J Pet Geol 13(3):329–340

Ahmed MA, Abdel Samie SG, El-Maghrabi HM (2011) Recharge and contamination sources of shallow and deep groundwater of Pleistocene aquifer in El-Sadat industrial city: isotope and hydrochemical approaches. Environ Earth Sci 62(4):751–768

Al-Kilany SMH (2001) New geoelectrical studies on groundwater aquifer, El-Sadat area and its vicinities Egypt. M.Sc. thesis, Faculty of Science, Menoufiya University, Egypt

Attia SH (1975) Pedology and soil genesis of the Quaternary deposits in the region west of Nile Delta (north and east of Wadi El-Natrun). Ph.D. thesis, Faculty of Science, Ain Shams University, Cairo, Egypt

Auken E, Christiansen AV, Kirkegaard C, Fiandaca G, Schamper C, Behroozmand AA, Binley A, Nielsen E, Effersø F, Christensen NB, Sørensen KI, Foged N, Vignoli G (2015) An overview of a highly versatile forward and stable inverse algorithm for airborne, ground-based and borehole electromagnetic and electric data. Explor Geophys 46:223–235

Auken E, Christensen N, Sørensen KI, Effersø F (1994) Large scale hydrogeological investigation in the Beder area-a case study. In: 7th EEGS Symposium on the application of geophysics to engineering and environmental problems. European Association of Geoscientists & Engineers. https://doi.org/10.3997/2214-4609-pdb.208.1994_039

Burger HR (1992) Exploration geophysics of the shallow subsurface. Prentice Hall, New Jersey, pp 241–316

Christiansen AV, Auken E, Sørensen K (2009) The transient electromagnetic method. In: Kirsch R (ed) Groundwater geophysics: a tool for hydrogeology, 2nd edn. Springer, Berlin, pp 179–226

Conoco (1987) Geologic Map of Egypt. Egyptian General Authority for Petroleum (UNESCO Joint Map Project), 20 sheets, Scale 1:500,000, Cairo

Constable SC, Parker RL, Constable CG (1987) Occam's inversion: a practical algorithm for generating smooth models from electromagnetic sounding data. Geophysics 52(3):289–300

Dawoud MA, Darwish MM, El-Kady MM (2005) GIS-based groundwater management model for Western Nile Delta. Water Resour Manage 19:585–604

Dolson JC, Shann MV, Matbouly SI, Harwood C, Rashed R, Hammouda H (2001) The petroleum potential of Egypt. In: Downey MW, Threet JC, Morgan WA (eds) Petroleum provinces of the twenty-first century. AAPG Mem 74, pp 453–482

El-Rawy M, Abdalla F, El Alfy M (2020) Water resources in Egypt. In: Hamimi Z, El-Barkooky A, Martínez Frías J, Fritz H, Abd El-Rahman Y (eds) The geology of Egypt. Regional geology reviews. Springer, Cham, pp 687–711

El Abd E (2005) The geological impact on the water bearing formations in the area southwest Nile Delta, Egypt. Ph.D. thesis, Faculty of Science, Menufiya University, Egypt

El Fayoumy IF (1964) Geology of groundwater supplies in Wadi El Natrun area, M.Sc. thesis, Faculty of Science, Cairo University, Cairo, Egypt

El Ghazawi MM (1982) Geological studies of the Quaternary–Neogene aquifers in the area northwest Nile Delta. M.Sc. thesis, Faculty of Science, El Azhar University, Cairo, Egypt

El Shazly EM, Abdel-Hady M, El Ghawaby H, El Kassas K, Khawasik SM, El Shazly MM, Sanad S (1975) Geologic interpretation of Landsat satellite image for west Nile Delta area, Egypt. Remote Sensing Center, Academy of Scientific Research and Technology, Cairo, Egypt

ElGalladi A, Ghazala HH, Ibraheem IM, Ibrahim El-KH (2009) Tectonic interpretation of gravity field data of west of Nile Delta region, Egypt. NRIAG J Astron Geophys, Egypt (Special Issue):487–508

Embaby AAA (1995) Geomorphology and hydrogeology of El Khatatba district, west Nile Delta, Egypt. M.Sc. thesis, Faculty of Science, Mansoura University, Mansoura, Egypt

Embaby AAA (2003) Environmental evaluation for geomorphological situation in relation to the water and soil resources of the region north of the Sadat City, west Nile Delta, Egypt. Ph.D. thesis, Faculty of Science, Mansoura University, Mansoura, Egypt

Ernstson K, Kirsch R (2009) Geoelectrical methods. In: Kirsch R (ed) Groundwater geophysics: a tool for hydrogeology, 2nd edn. Springer, Berlin, pp 85–117

Garfunkel Z (1998) Constrains on the origin and history of the Eastern Mediterranean basin. Tectonophysics 298(1–3):5–35

Geirnaert W, Laeven MP (1992) Composition and history of groundwater in the western Nile Delta. J Hydrol 138(1–2):169–189

Goldman M, Du Plooy A, Eckard M (1994) On reducing ambiguity in the interpretation of transient electromagnetic sounding data. Geophys Prospect 42(1):3–25

Guiraud R, Bosworth W (1999) Phanerozoic geodynamic evolution of northeastern Africa and the northwestern Arabian platform. Tectonophysics 315(1–4):73–104

Harms JC, Wray JL (1990) Nile Delta. In: Said R (ed) The geology of Egypt. A.A. Balkema, Rotterdam & Brookfield, Vermont, pp 329–343

Ibraheem IM, El-Qadi GM, ElGalladi A (2016) Hydrogeophysical and structural investigation using VES and TDEM: a case study at El-Nubariya—Wadi El-Natrun area, west Nile Delta, Egypt. NRIAG J Astron Geophys, Elsevier 5(1):198–215

Ibraheem IM, Elawadi EA, El-Qady GM (2018) Structural interpretation of aeromagnetic data for the Wadi El Natrun area, northwestern Desert, Egypt. J Afr Earth Sci 139:14–25

Ibraheem IM, El-Qady G (2019) Hydrogeophysical Investigations at El-Nubariya–Wadi El-Natrun area, West Nile Delta, Egypt. In: Negm A (ed) Groundwater in the Nile Delta. The Handbook of Environmental Chemistry, Springer, Cham 73, pp 235–271

Ibraheem IM (2009) Geophysical potential field studies for developmental purposes at El-Nubariya—Wadi El-Natrun area, West Nile Delta, Egypt. Ph.D. thesis, Faculty of Science, Mansoura University, Mansoura, Egypt

Inman JR (1975) Resistivity inversion with ridge regression. Geophysics 40:798–817

Keary P, Brooks M, Hill I (2002). An introduction to exploration geophysics, 3rd edn. Blackwell Science, pp 183–207

Massoud U, Kenawy AA, Ragab EA, Abbas AM, El-Kosery HM (2014) Characterization of the groundwater aquifers at El-Sadat City by joint inversion of VES and TEM data. NRIAG J Astron Geophys 3:137–149

Meju MA (2005) Simple relative space-time scaling of electrical and electromagnetic depth sounding arrays: Implications for static shift removal and joint DC-TEM data inversion with the most-squares criterion. Geophys Prospect 53(4):463–479. https://doi.org/10.1111/j.1365-2478.2005.00483.x

Mohamed RF, Hua CZ (2010) Regional groundwater flow modeling in Western Nile Delta, Egypt. World Rural Obser 2(2):37–42
Nabighian MN, Macanae JC (1991) Time-domain electromagnetic prospecting methods. In: Nabighian MN (ed) Electromagnetic methods in applied geophysics. Society of Exploration Geophysicists, Tulsa, OK., USA 2, pp 427–514
Omara SM, Sanad S (1975) Rock stratigraphy and structural features of the area between Wadi El-Natrun and the Moghra depression (Western Desert), Egypt. GEOL. JB., B 16, pp 45–73
Othman A, Ibraheem IM, Ghazala H, Mesbah H, Dahlin T (2019) Hydrogeophysical and hydrochemical characteristics of Pliocene groundwater aquifer at the area northwest El Sadat city, West Nile Delta, Egypt. J Afr Earth Sci 150:1–11
Othman A (2018) Hydrogeophysical study on the area northwest El-sadat City—West Nile Delta—Egypt. Ph.D. thesis, Faculty of Science, Mansoura University, Mansoura, Egypt
Palacky GJ (1987) Resistivity characteristics of geologic targets. In: Nabighian MN (ed) Electromagnetic methods in applied geophysics. Volume 1, Theory, Tulsa, Ok., Society of exploration geophysicists, pp 53–129
Parasnis DS (1986) Principles of applied geophysics, 4th edn. Chapman & Hall, London, pp 104–172
Philip G (1955) Geology of the Pleistocene sediments of the Mediterranean Coast, west of Abu Qir. Ph.D. thesis, Faculty of Science, Cairo University, Cairo
Pullan SE, Pugin A, Dyke LD, Hunter JA, Pilon JA, Todd BJ, Allen VS, Barnett PJ (1994) Shallow geophysics in a hydrogeological investigation of the Oak Ridges Moraine, Ontario. In: Symposium on the application of geophysics to engineering and environmental problems proceedings, pp 143–161. https://doi.org/10.4133/1.2922060
Reynolds JM (2011) An introduction to applied and environmental geophysics. Wiley-Blackwell, Chichester
Said R (1962) The geology of Egypt. Elsevier, Amsterdam
Said R (1981) Geologic evolution of the River Nile. Springer, New York
Said R (1990) The geology of Egypt. In: AA Balkema, Roterdam
Salem ZE, El Bayumy DA (2016) Hydrogeological, petrophysical and hydrogeochemical characteristics of the groundwater aquifers east of Wadi El-Natrun, Egypt. NRIAG J Astron Geophys 5:106–123
Sanad S (1973) Geology of the area between Wadi El-Natrun and the Moghra depression. Ph.D. thesis, Faculty of Science, Assuit University, Assuit, Egypt
Sarhan M, Hemdan K (1994) North Nile Delta structural setting trapping and mechanism, Egypt. In: Proceedings of the12th EGPC exploration and production conference, Cairo, Egypt 1: pp 1–17
Sarhan M, Barsoum F, Bertello F, Talaat M, Nobili M (1996) The Pliocene play in the Mediterranean offshore: structural setting and growth faults-controlled hydrocarbon accumulations in the Nile Delta basin. A comparison with the Niger Delta basin. In: 13th EGPC, Egyptian General Petroleum Corporation conference, Cairo, pp 1–19
Sharaky AM, El Hasanein AS, Atta SA, Khallaf KM (2017) Nile and groundwater interaction in the western Nile Delta, Egypt. In: Negm AM (ed) The Nile Delta. The handbook of environmental chemistry. Springer, Cham 55, pp 33–62
Sharma PV (1997) Environmental and engineering geophysics. Cambridge University Press
Shata AA (1953) New light on the structural development of the Western Desert of Egypt. Bull Inst Desert D' Egypte 3(1):101–106
Shata AA (1955) An introductory note on the geology of northern portion of the Western Desert of Egypt. Bull Inst Desert D' Egypte 5(3):96–106
Shata AA (1961) The geology of groundwater supplies in some arable lands in the desert of Egypt (internal report). Desert Institute, Cairo
Shata AA, El Fayoumi IF (1967) Geomorphological and morphopedological aspects of the region west of the Nile Delta with special reference to Wadi El-Natrun area. Bull Inst Desert D'egypte 13(1):1–38

Shata AA, Pavlov M, Saad FK (1962) Preliminary report on the geology, hydrogeology, and groundwater hydrology of the Wadi El-Natrun and adjacent area (internal report), part 1. Desert Institute, Cairo, Egypt

Shedid AG (1989) Geological and hydrogeological studies of the Sadat area and its vicinities. M.Sc. thesis, Faculty of Science, El Menoufia University, Shibin El Kom, Egypt

Siemon B (2009) Electromagnetic methods—frequency domain. In: Kirsch R (ed) Groundwater geophysics: a tool for hydrogeology, 2nd edn. Springer, Berlin, pp 155–178

Telford WM, Geldart LP, Sheriff RE (1990) Applied geophysics. Cambridge University Press, Cambridge

Tezkan B (1999) A review of environmental applications of quasi-static electromagnetic techniques. Surv Geophys 20:279–308

Tezkan B (2009) Radiomagnetotellurics. In: Kirsch R (ed) Groundwater geophysics: a tool for hydrogeology, 2nd edn. Springer, Berlin, pp 295–317

USGS (2017) Shuttle radar topography mission (SRTM). https://lta.cr.usgs.gov/SRTM (accessed on June 20, 2017)

Yogeshwar P, Tezkan B (2017) Two-dimensional basement modeling of central loop transient electromagnetic data from the central Azraq basin area, Jordan. J Appl Geophys 136:198–210

Yungul S (1996) Electrical methods in geophysical exploration of deep sedimentary basins. Springer

Zaghloul ZM (1976) Stratigraphy of the Nile Delta. In: Proceedings of the seminar on Nile Delta Sedimentology, pp 40–49

Evaluation of Groundwater Potential Zones Using Electrical Resistivity and Hydrogeochemistry in West Tahta Region, Upper Egypt

Esam Ismail, Mustafa El-Rawy, and Hermann Mauritsch

Abstract In this chapter, the geoelectrical and hydrogeochemistry studies were carried out in the western Tahta region to identify the subsurface hydrostratigraphic aquifer units and obtain a proposed potential groundwater development. Twenty-Nine vertical electrical soundings were performed and interpreted to evaluate the geometry of the aquifer. Geoelectrical cross-sections revealed a subsequent pile of layers in the study area from top to bottom: surface layer (dry to wet sand, gravel and rock fragments in the west part and clay in the east one), wet sand, water-bearing sand and gravel, clay, and fractured limestone. From the isopachyte map of the saturated layer, it has been found that the Pleistocene aquifer becoming thicker to the northeast. The resistivity contour map of the Quaternary aquifer showed that average values of about 40 Ω min the region of cultivated land (near to the Nile) increased and reached about 90 Ω m in the west. The depth to water contour map showed that there are decreasing in the values from southwest to northeast. Thirty-Four water samples (6 of surface water and 28 of groundwater) were collected from the study area to detect the source and groundwater type as well as assessment these samples for different uses. The quaternary aquifer is recharged mostly by surface water, particularly from irrigation canals, while the aquifer is released by evaporation. The bulk of the surface water looks to be good for drinking and irrigation, however the majority of the groundwater samples taken are unfit for drinking and half of them are unfit for irrigation due to high salinity.

E. Ismail (✉)
Department of Geology, Faculty of Science, Minia University, Minia 61519, Egypt
e-mail: essam.ismail@mu.edu.eg

M. El-Rawy
Civil Engineering Department, Faculty of Engineering, Minia University, Minia 61111, Egypt
e-mail: mustafa.elrawy@mu.edu.eg; mustafa.elrawy@su.edu.sa

Civil Engineering Department, College of Engineering, Shaqra University, Dawadmi 11911, Ar Riyadh, Saudi Arabia

H. Mauritsch
Applied Geophysics Institute, Montan University Leoben, 8700 Leoben, Austria
e-mail: hermann.mauritsch@unileoben.ac.at

A. M. Negm and M. El-Rawy (eds.), *Sustainability of Groundwater in the Nile Valley, Egypt*, Earth and Environmental Sciences Library,
https://doi.org/10.1007/978-3-031-12676-5_11

Keywords Geoelectrical · West Tahta · Quaternary aquifer · Vertical electrical sounding · Irrigation purposes · Nile Valley

1 Introduction

Egypt is located in North Africa's dry region, which has limited freshwater resources and a rapidly rising population. Water resources in Egypt are mainly from the Nile River which provides 97% of Egyptian's water demands (El-Rawy et al. 2020). One of the most significant natural resources is groundwater to promote human health, economic development, and ecological diversity. It has become an extremely significant and dependable source of water in all climatic regions, including both urban and agricultural areas of developed and developing countries, due to its inherent qualities (e.g. consistent temperature, widespread and continuous availability, excellent natural quality, limited vulnerability, low development costs, drought reliability, etc.) (Todd and Mays 2005). The study area is located in Upper Egypt and has undergone a major reclamation project the last few years. Such agricultural development depends largely on groundwater, as well as a limited supply of surface water from the River Nile and nearby canals. The area under investigation is Sohag Governorate (Fig. 1), which is part of the arid belt of Egypt characterized by long, hot summers, mild winters, and scarce rainfall, with the exception of the occasional storms. Rain is rare and has been randomly distributed throughout the study area. In the summer, the average air temperature is around 36.5 °C, and in the winter, it is around 15.5 °C.

The relative humidity is between 35 and 61%, and it is higher in the winter than in the summer (Ahmed 2009). Monthly evapotranspiration varies from 3.5 mm per day in January to 9.9 mm per day in June. The water distribution system's evaporation loss ranges from 2.1 mm day^{-1} in January to 8.3 mm day^{-1} in June. Many studies have been conducted in the Sohag Governorate on geology, geomorphology, geophysical, hydrogeology, and hydrochemistry. Many authors have examined the area's geology and geomorphology (Said 1981, 1990; Chumakove 1967; Ahmed 1980; Abdel Moneim 1992; Abdel Moneim et al. 1999a, b; Mahran and El-Haddad 1992; Omer 1996; Omer and Abdel Moneim 2001; Hassan 2005; Gomaa 2006).

The hydrogeology of the study area has been studied by a number of authors such as Abdel Moneim (1988, 1992, 1999a, b). He investigated the hydrogeology of the Nile Basin in the province of Sohag, as well as numerical simulation and groundwater management in the Sohag aquifer. Also he described the geoelectrical and hydrogeological investigations of the groundwater resources in the western area of the cultivated land at Sohag. Attia (1985) and Barber and Carr (1981) studied the water resources management in Upper Egypt. Diab et al. (2002) assessed the southern part of Sohag's water resources and soil suitability for development. Korany et al. (2006) established a hydrochemical approach for assessing groundwater recharge routes in the Sohag Nile Valley's Quaternary aquifer. Mousa et al. (1994) used geoelectrical methods, executed a hydrogeological investigation of the Quaternary aquifer in the Nile Valley between Asyut and Sohag Governorate. Zaki (2001) evaluated water

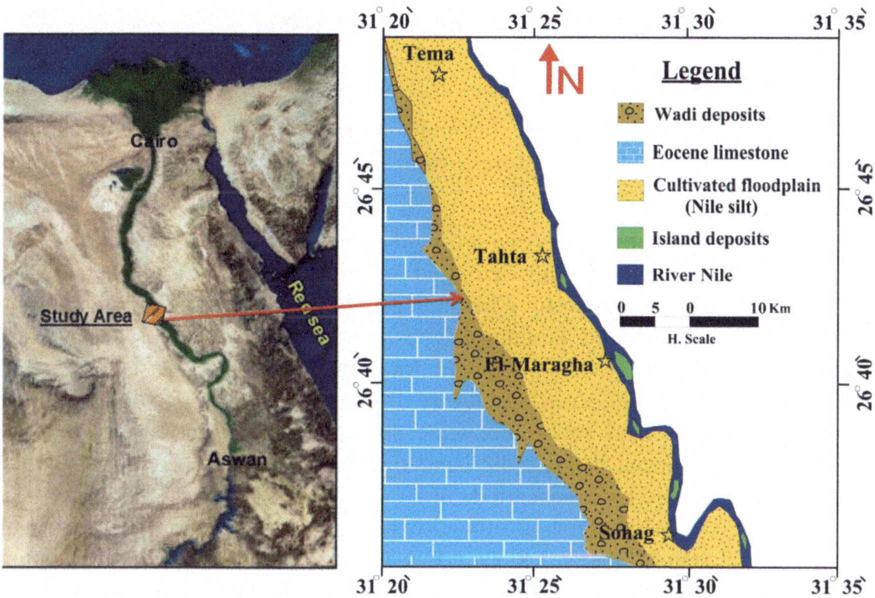

Fig. 1 Location and geologic map of the studied area (after El-Sayed et al. 2010)

resources and land use projects in the southern section of Sohag Governorate using a geographic information system and a hydrogeological research. Omran (2008) developed a method for assessing groundwater potential in the Sohag region that combined remote sensing, geophysics, and geographic information systems. The hydrogeological and geophysical study of the reclaimed regions of Sohag was reported by Gomaa (2006). Awad et al. (1995) carried out chemical and isotopic investigation in the Tahta groundwater aquifer. RIGW and IWACO (1989) published an internal assessment on the Nile Valley's groundwater development for irrigation and drainage (West Tahta). Ahmed (2007) employed lithological modelling tools and groundwater flow modelling to characterise the properties of the Sohag aquifer. Ahmed (2009) used DRASTIC-GIS based generic and pesticide models for the vulnerability assessment of the Sohag Quaternary aquifer. Ahmed and Mohamed (2009) investigated the hydrochemical evaluation and variance of groundwater in the Sohag region, as well as its environmental impact. The flood plain problems in Sohag Governorate were investigated by El-Sayed and El-Shater (2010). Ismail and El-Rawy (2018) investigated the quality of groundwater in Egypt's West Sohag Governorate.

In this chapter, twenty-nine vertical electrical soundings (VES) were performed to figure out the geometry of the aquifer in the research area. Thirty-Four water samples (6 from surface water and 28 from groundwater) were collected and analyzed to study the hydrogeochemical characteristics and to assess the water quality for the drinking and irrigation purposes.

2 Geology and Hydrology of the Study Area

The Eocene plateau in the west and the Nile River in the east outline the study area. According to Said (1981, 1990) the rock units in the study area and its surroundings come from sediments of different ages from Eocene up to Holocene (Fig. 1) as follows:

- **Eocene Deposits**: These deposits are made up of carbonates intercalated with clays, shales, and sands and are delimited by the Nile Valley in the east and west as scarps. The Eocene limestone was named Thebes Formation by Said (1960), who described it as large, laminated limestone with flint bands or nodules and marl rich in Nummulites and planktonic foraminifera.
- **Pliocene Deposits**: Pliocene deposits are divided into two groups: lower marine and upper fluvial. Dark brown clays are interspersed with sandstone lenses to depict them. Pliocene deposits lie beneath Pleistocene sediments, acting as a dense, impermeable layer beneath them.
- **Pleistocene Deposits**, which are found beneath Holocene layers and on top of Pliocene clay and Eocene limestone. They are the primary water-bearing feature in the Nile Valley, made entirely of sand and gravel. Pleistocene layers in the Neonile deposits (Late Pleistocene) are made primarily of silt and clay and lie on the eroded, uneven surface of the flood plains of the Prenile deposits, according to Said (1981, 1990). Prenile deposits (Middle Pleistocene) are made up of a graded sand-gravel unit and are coarse, massive, and thick. They are represented by the Qena Formation and are overlain with Abbassia gravel in a small region, and they are the main aquifer in the research area. Protonile deposits (Early Pleistocene) are fluvial in nature and can be found everywhere on the Nile's western bank.
- **Holocene Deposits**, which contain all types of unconsolidated sediment and have accumulated under a variety of environmental conditions. Nile silt, wadi deposits, and fanglomerates are examples of these deposits.

In a hydrogeological sense, the Nile Valley in the research area is defined by a single aquifer system (Quaternary aquifer) (Fig. 2). Fluvial sands with small conglomerate and clay make up this aquifer. This aquifer consists of fluvial sands with small conglomerate and clay. The permeable thickness of the aquifer system varies from 150 m in the central part of the flood plain to about 59 m in the desert fringes. The aquifer average hydraulic conductivity is about 70 m/day. However, it is only 0.04 m/day for the silty top layer of the aquifer (Omran et al. 2006). The Quaternary aquifer is mainly recharged from the surface water, particularly the irrigations canals (El-Rawy et al. 2019, 2020, 2021a, b; Awad et al. 2020; Ismail et al. 2020, 2021; Abotalib et al. 2021; Snousy et al. 2021a, b; Heleika et al. 2021), which play a key role in the configuration of the water table. During the evaporation process, the aquifer is discharged. In most of its portions, the aquifer-river interaction functions as an effluent stream.

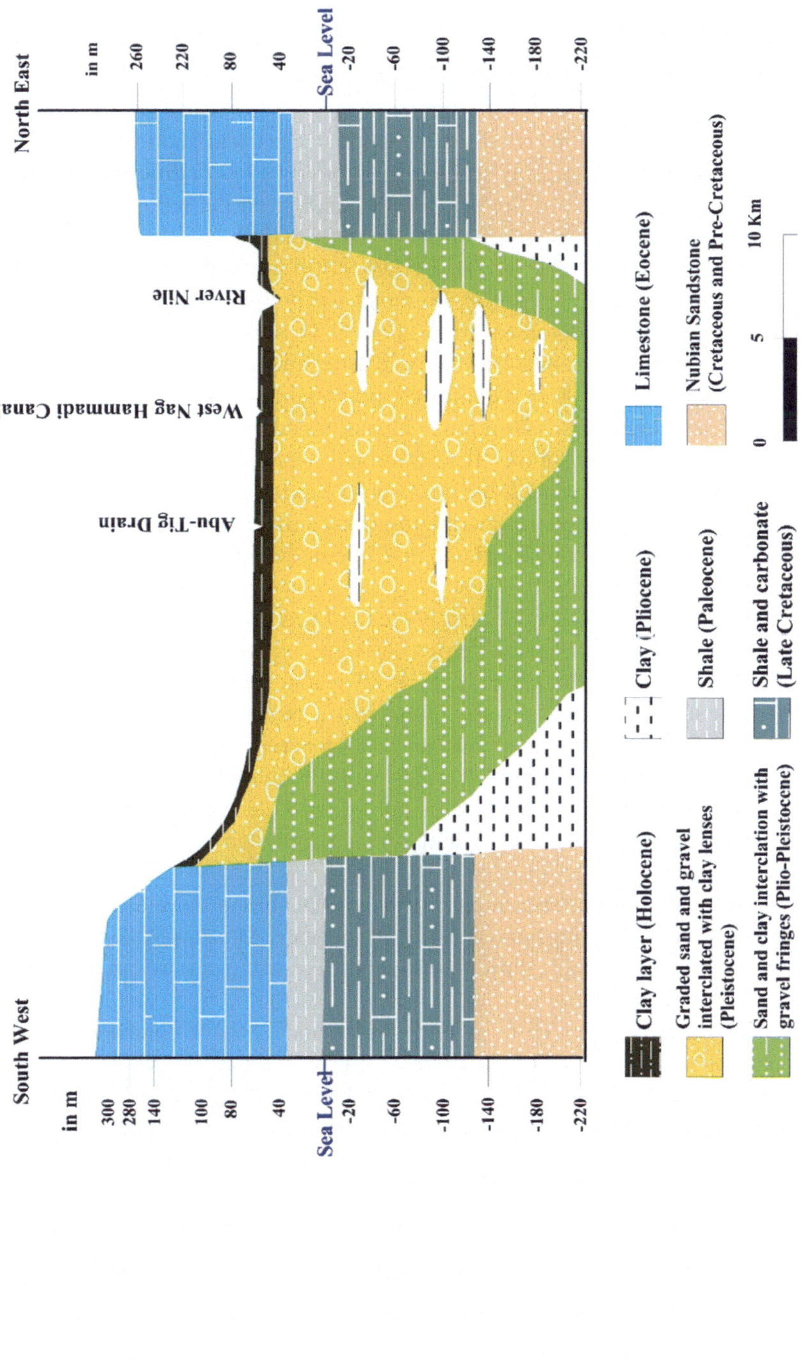

Fig. 2 Schematic hydrogeological cross-section in Tahta area (RIGW and IWACO 1989)

Due to the influence of recharging canals and the depth to the groundwater surface varied from 9 to 56 m and rising to the west, the predominant direction of groundwater flow in the research area is to the northeast, i.e. to the River Nile (Ismail 2013).

3 Sampling, Data, and Methodology

In the area under investigation, twenty-nine vertical electrical soundings (Fig. 3) were made in a Schlumberger array to identify groundwater conditions, including depth, thickness, and location of the aquifer. Depending on the topography of the location, the Schlumberger resistivity sounding was conducted with half-spacing in the range of 300–1000 m. The Terrameter SAS 300 was utilised to conduct the geoelectrical survey in this investigation. Batteries, calibrating resistors for checking the instrument, steel electrodes measuring about 0.85 m in length and 2.5 cm in diameter, field cables with single conductor plastic insulation and low electrical resistance of 0.5 mm^2 (two 500 m reels for the potential lines and two 1000 m reels for the current lines), hammers, a Brunton compass, wood marks, a GPS instrument, and wireless telephone were among the additional accessories. By drawing bold lines through the multi-layer model, the multi-layer model is reduced manually to a simplified one with a number of layers ranging from 3 to 5 layers in this study.

In addition, the available geological and hydrogeological knowledge in the study area influenced this stage.

The simplified model was then iteratively recalculated using a computer software called Velpen's RESIST (1988). The Velpen software uses an iterative technique to estimate the resistivity and thickness of each layer and requires an initial model. It iterates the observed data numerous times until a minor disparity between the observed and predicted resistivity is observed, which is referred to as best fit.

A third computer software called IPI2Win (Moscow State University 2003) was used at the final stage of the analysis (Fig. 4). Vertical electrical sounding and/or induced polarisation data curves 1D reading along a single profile are supported by this software.

It is assumed that an experienced interpreter will be able to solve the geological problem and match the sounding curves. IPI2Win software (Moscow State University 2003) assumed that, the geological problems and the sounding curves can be fit with the experience interpreter.

Thirty-Four water samples (6 of surface water and 28 of groundwater) were collected from the study area (Fig. 3; Table 1). Surface water samples were taken with a sampler from the middle of the stream at a depth of around one metre. The groundwater samples were collected after the water had been pumped out from drilling wells for about an hour. Water samples were contained in one liter polyethylene bottles. To avoid leakage or evaporation, these bottles were carefully sealed with caps, plastic inserts, and cellotape. The pH, total dissolved solids (TDS), and electrical conductivity (EC) of the water were determined in the field using an Ultrameter SM101 instrument shortly after sample. Comprehensive chemical analysis,

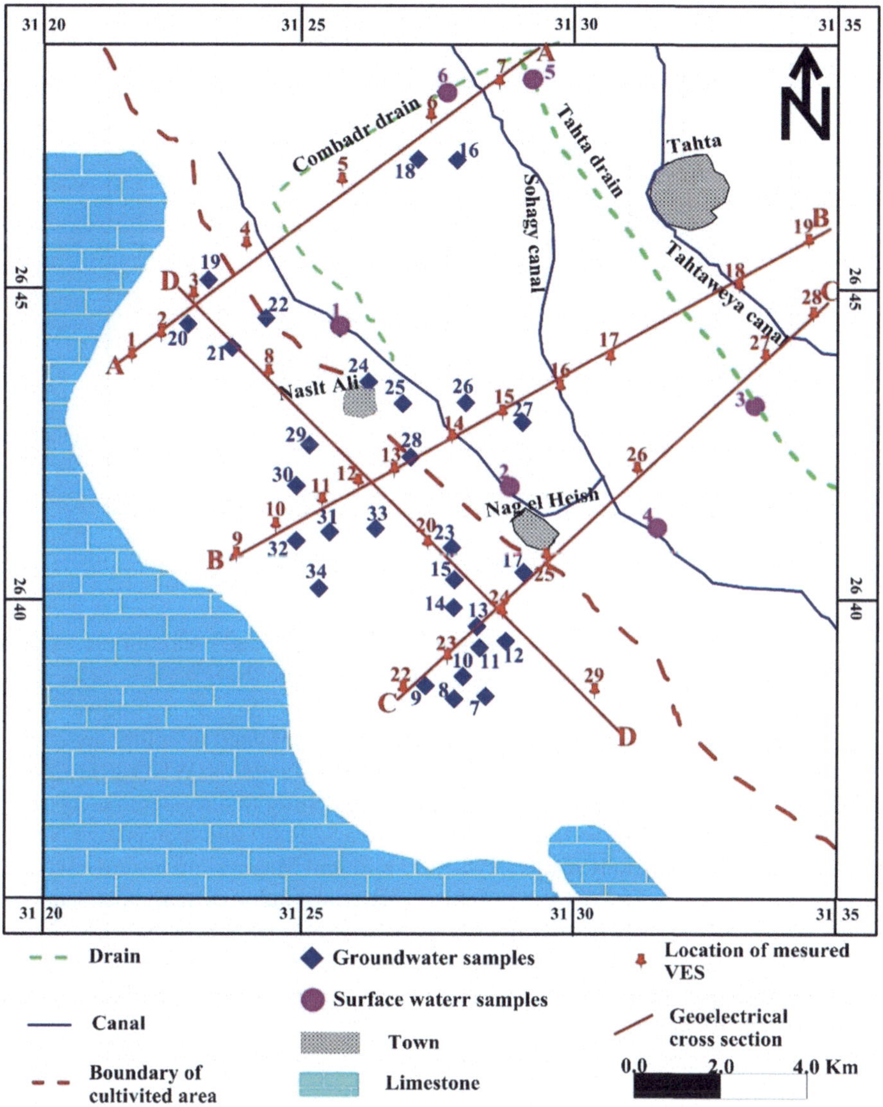

Fig. 3 Location map of the collected water samples and measured VES

including major cations and anions, was performed in the Central Laboratory-Water and Soil Analysis Unit, Desert Research Center, Ministry of Agriculture and Land Reclamation, Egypt, in accordance with the American Public Health Association's standard techniques (APHA 2005). Volumetric titration methods were used to detect calcium (Ca^{2+}), magnesium (Mg^{2+}), bicarbonate (HCO_3^-), and chloride (Cl), while

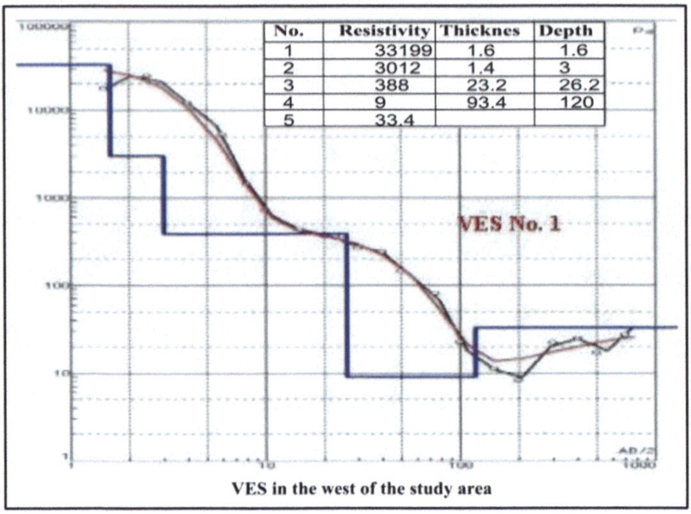

Fig. 4 Example of the qualitative interpretation by using IPI2Win

the flame photometer was used to determine sodium (Na^+) and potassium (K^+), and the UV spectrophotometer was used to determine sulphate (SO_4^{2}).

Various indices have been used to determine the validity of groundwater for domestic consumption and irrigation. The results are evaluated in compliance with the World Health Organization's drinking water quality standards (WHO 2011, 2017) and the Egypt drinking water standards (2007). Irrigation quality parameters (i.e., electrical conductivity EC, sodium absorption ratio SAR, sodium percentage Na %, residual sodium carbonate RSC, Kelley's ratio KR, and Magnesium hazard MH) are calculated using following equations (Table 2). The correlation of analytical data was addressed by the use of several graphical representations such as those of Richards (1954), Gibbs (1970), and Wilcox (1955) to characterize groundwater and its suitability for different purposes by defining different factors based on the chemical characteristics of water.

4 Results and Discussions

4.1 Geoelectrical Analysis

Depending on the geological information and the results of the previous programs a good starting model for each VES station is used during the inverse modeling operation by using IPI2Win program. Thus, all the usable layer thickness obtained from certain wells is added as the ground reality. Reliable geoelectric cross-sections

Table 1 Chemical analysis of the collected surface and groundwater samples

S. No.	pH	E.C	TDS	Unit	Cations				Anions			
					Ca^{++}	Mg^{++}	Na^{+}	K^{+}	CO_3^{-}	HCO_3^{-}	Cl^{-}	SO_4^{--}
1	7.5	239	166	ppm	23.46	7.92	24	6	14.70	119.56	10.77	19.64
2	7.6	241	170	ppm	27.89	9.305	20	5	14.70	116.57	9.57	24.96
3	7.1	930	571	ppm	60.17	29.36	100	7	14.70	295.91	62.23	150
4	7.7	230	163	ppm	21.68	7.916	24	8	11.76	122.54	8.6	20
5	7.5	1535	755	ppm	36.58	24.61	210	9	29.40	266.02	148.4	164.26
6	7.3	2386	1456	ppm	103.7	54.03	300	12	55.86	289.93	153.2	632.07
7	7.0	4360	2655	ppm	224.8	95.33	540	8	5.88	110.59	825.8	900
8	7.3	970	567	ppm	57.25	13.54	130	3	11.76	239.12	122.1	110
9	7.6	1627	1004	ppm	80.1	37.9	210	4	17.64	158.41	225	350
10	7.4	5870	3533	ppm	235.6	55.76	900	7	0.00	146.46	1112	1148.9
11	7.2	7550	4453	ppm	248.8	84.14	1220	11	17.64	170.37	1649	1136.9
12	7.5	1454	818	ppm	54.84	25.3	210	7	14.70	215.2	258.5	140
13	7.3	4550	2808	ppm	218.3	63.27	650	8	11.76	161.4	825.8	950
14	7.6	4110	2493	ppm	243.3	100.2	460	12	0.00	215.2	820.3	750
15	7.5	3300	2379	ppm	150.4	100.2	500	26	20.58	269.01	347.1	1100
16	7.3	2354	1480	ppm	210.7	62.11	190	8	41.16	304.87	220.2	595.2
17	7.3	863	452	ppm	27.01	9.13	132	2	26.46	221.18	28.72	115.9
18	7.66	3490	2124	ppm	175.7	91.22	440	13	29.40	251.07	478.7	770.1
19	7.2	28,544	1858	ppm	100.2	56.1	440	28	29.40	266.02	411.7	660
20	7.7	356	184	ppm	21.1	11.2	30	10	17.64	119.56	19.15	15
21	7.7	1883	1242	ppm	50.2	42.1	330	12	17.64	236.13	172.3	500
22	6.0	6700	4448	ppm	235.8	112.26	1100	6	17.64	134.5	1197	1712
23	7.7	3300	2017	ppm	148.4	20.14	510	3	26.46	218.19	299.2	900.72
24	7.9	3060	2140	ppm	179.7	65.3	420	18	17.64	179.34	299.2	1050
25	7.6	4100	2642	ppm	197.3	58.7	600	7	20.58	239.12	538.5	1100
26	6.6	3860	2196	ppm	176.5	42.87	540	7	32.34	206.24	454.8	839.05
27	7.3	1957	1278	ppm	104.9	25.3	280	11	23.52	230.15	158	560
28	7.6	1154	725	ppm	99.69	18.9	120	7	26.46	260.04	43.08	280
29	7.3	1700	978	ppm	39.83	8.66	280	3	26.46	185.31	71.8	455.55
30	6.8	6820	4063	ppm	338.6	57.76	1020	12	17.64	152.43	1149	1392.1
31	7.0	1187	698	ppm	28.64	19.4	190	3	17.64	188.3	47.87	297.64
32	7.8	4260	2816	ppm	215.5	37.57	660	4	29.40	149.45	394.9	1400
33	7.6	2191	1391	ppm	67.35	39.9	350	2	35.28	304.87	143.6	600
34	7.3	5790	3783	ppm	277	140.1	800	16	29.40	125.53	658.2	1800

Table 2 Equations used for calculating irrigation quality parameters

Items	Equations	References
TH	TH = 2.497 Ca^{2+} + 4.115 Mg^{2+} ions in meq/l	Todd (1980)
SAR	$SAR = Na/\sqrt{(Ca + Mg)/2}$ all ions in meq/l	Richards (1954)
Na (%)	$Na\,\% = \frac{(Na+K)}{(Ca+Mg+Na+K)} \times 100$ all ions in meq/l	Wilcox (1955)
RSC	$RSC = \left(HCO_3^- + CO_3^{--}\right) - \left(Ca^{++} + Mg^{++}\right)$ all ions in meq/l	Richards (1954)
MH	$MH = \frac{Mg}{(Ca+Mg)} \times 100$ all ions in meq/l	Paliwal (1972)
SSP	$((Na^+ + K^+)/(K^+ + Na^+ + Ca^{2+} + Mg^{2+})) \times 100$ all ions in meq/l	Joshi et al. (2009)
PS	$PS = Cl + \sqrt{SO_4}$ all ions in meq/l	Doneen (1964)
Kelly's ratio	$KR = \frac{Na}{(Ca+Mg)}$ all ions in meq/l	Kelly (1940)

are also constructed by re-interpreting some VES according to the results obtained in the vicinity. As well as the information obtained from the drilling wells present in the study area. The subsurface rocks are a pile of sediments consisting, from top to bottom: surface layer (dry sand, gravel and rock fragments in the west and clay in the east), dry sand, water-bearing sand and gravel, clay, and fractured limestone (not defined in the east of the study area). The results obtained from the analysis of the individual sounding curves were used to identify the subsurface of the study area. It was achieved by the development of geoelectrical cross sections and maps. In studying of these cross-sections, one should be aware of that, due to vertical scale limitations, it may be difficult to represent thin layers that are present anywhere along the section, particularly the surface layer. In the same way, more than one geoelectric layers can be represented along the sections as a single layer, so that a particular rock unit can display difference in electrical resistivity for some reason or another. In this case, the layer is characterized by a resistivity spectrum rather than a single resistivity value. Description of these sections is provided as follows:

Geoelectrical cross section A–A; is located in the northern part of the research area and runs from southwest to northeast (Fig. 5). The surface layer in this section is made up of sand, gravel, and rock fragments in the west and clay in the east, with a thickness of 1–5 m. The second layer is made up of wet sand dating from the Holocene. The resistivity ranges from 129 to 325 Ω m. and the thickness varies from 6 m in the east to 34 m in the west. The resistivity of the third layer of the cross section ranges from 29 to 91 Ω m. These resistivity values relate to clay lenses intercalated with water-saturated sand and gravel. This layer's thickness increases to the east, ranging from 18 m in the west under VES No. 1 to 90 m under VES No. 5. Due to instrument limitations, the thickness under VES No. 6 and VES No. 7 cannot be determined. The fourth layer serves as the aquifer's foundation, with resistivities ranging from 0.9 to 5 Ω m. Although it is poorly defined in the east part of the cross section, this clay layer is underlain by limestone (VES No. 4 and VES

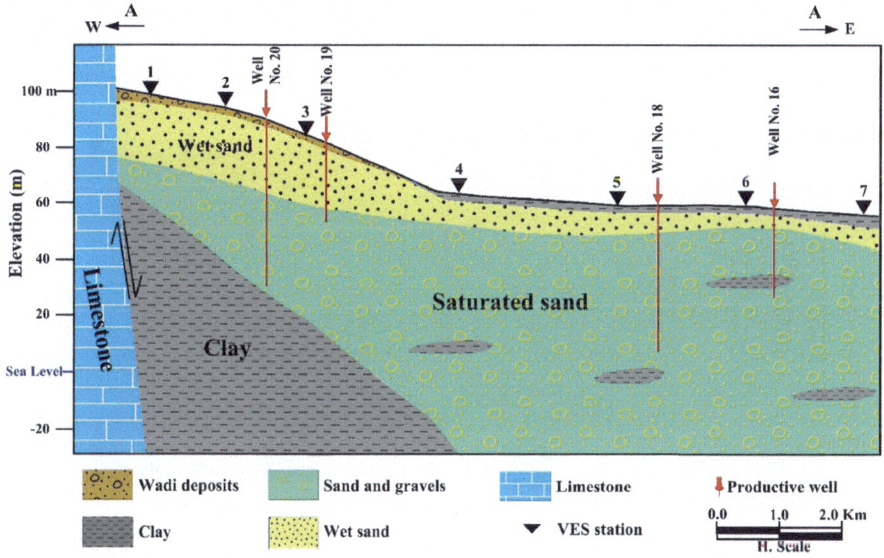

Fig. 5 Geoelectrical cross section A–A

No. 5). The main aquifer in the research area is represented by the third layer of this cross section, which corresponds to the Pleistocene period.

Geoelectric cross section B–B; this section passes through the center of the study area and includes VES stations Nos. 9, 10, 11, 12, 13, 14, 15, 16, 17, 18, and 19 (Fig. 6). The Holocene age is represented by the first layer of this segment, which changes in composition from west to east. Dry sand, gravel, and rock fragments (Wadi deposits) with high resistivities are most abundant in the west, whereas low resistivities reacting to clay are seen in the eastern portion of the cross section. This layer's thickness ranges from 1 to 4 m. High resistivities relating to wet sand and gravel characterise the second layer beneath the surface layer. The thickness of this layer varies between 5 and 18 m. Due to the erosion and uplift of limestone along the fault plain, this layer is not visible under VES No. 9. The third layer is Pleistocene in age and is made up of sand and gravel interspersed with clay lenses. The primary aquifer is represented by the third layer, which has thicknesses varying from 60 m in the west to 143.2 m in the east under VES No. 17, whereas no thickness was detected under VES No. 18. This layer's resistivity ranges from 28.6 to 95.6 Ω m. The fourth layer characterized by low resistivity corresponding to Pliocene impermeable clay acts as the base of the aquifer. This layer's resistivity ranges from 0.8 to 14.8 Ω m, and its thickness is unknown.

Geoelectrical cross section C–C; this section includes VES stations No. 22, 23, 24, 25, 26, 27, and 28 and passes through the southern part of the study area as shown in Fig. 7. Due to the presence of wadi deposits in the west and clay in the east, the first layer (surface layer) exhibits varying resistivity values throughout. This layer is

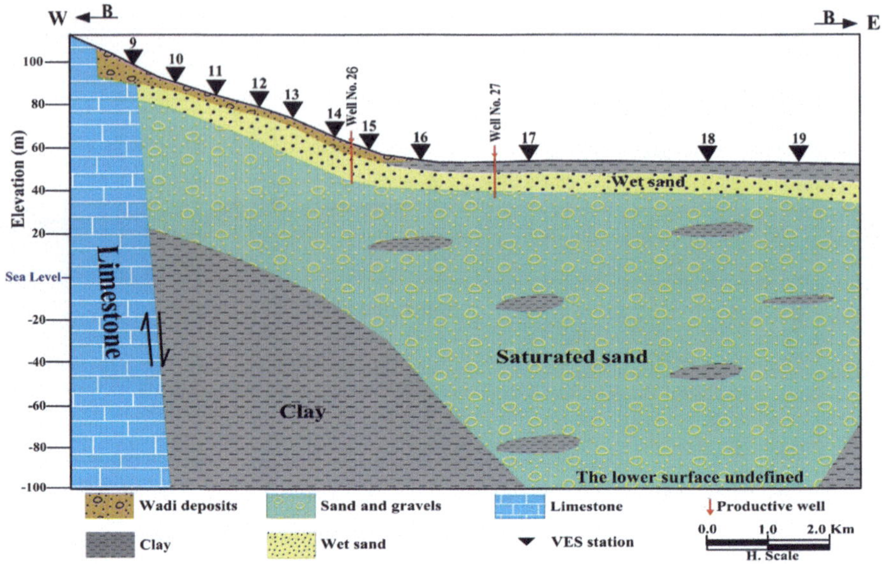

Fig. 6 Geoelectrical cross section B–B

quite thin, ranging from 1 to 2.7 m thick. The second layer has a resistivity range of 111.8–290 Ω m and a thickness range of 10–19 m. This layer is made up of wet sand. The principal water-bearing deposit in this cross-section is the third geoelectrical layer. Sand and gravel with resistivity values ranging from 30 to 99 Ω m make up this mixture. The thickness of this layer reduces in the west of the cross section and increases in the east. However, it is not described in VES Nos. 26 and 27. VES No. 22 has a thickness of 30 m, while VES No. 25 has a thickness of 110 m. The base of the water-bearing layer is represented by the fourth geoelectrical layer, which dates from the Pliocene age and its resistivity values range from 1.7 to 12.9 Ω m.

Geoelectrical cross section D–D; this profile crosses the other profiles and is oriented NW–SE (Fig. 8). The surface layer is formed of clay in the centre and wadi deposits (sand, gravel, and rock fragments) toward the north and south of the section, where the section approaches the escarpment, in this cross section. This layer's thickness ranges from 1 to 6 m. Wet sand makes up the second layer, which has resistivities ranging from 129 to 325 Ω m and thicknesses ranging from 5 m at VES No. 29 to 34 m at VES No. 3. The water-bearing deposit (Quaternary aquifer) is represented by the third layer, which is made up of sand and gravel. The resistivities of this layer range from 28 to 91 Ω m, and its thickness extends from 60 to 90 m. The last layer in the cross section is impermeable clay, which serves as the aquifer's base. The typical resistivity is in the range of 1–12 Ω m, with an undefined bottom.

Isopachyte map of the saturated layer; the thickness of the Quaternary aquifer in the research area is displayed on this map. The Pleistocene aquifer thickens towards

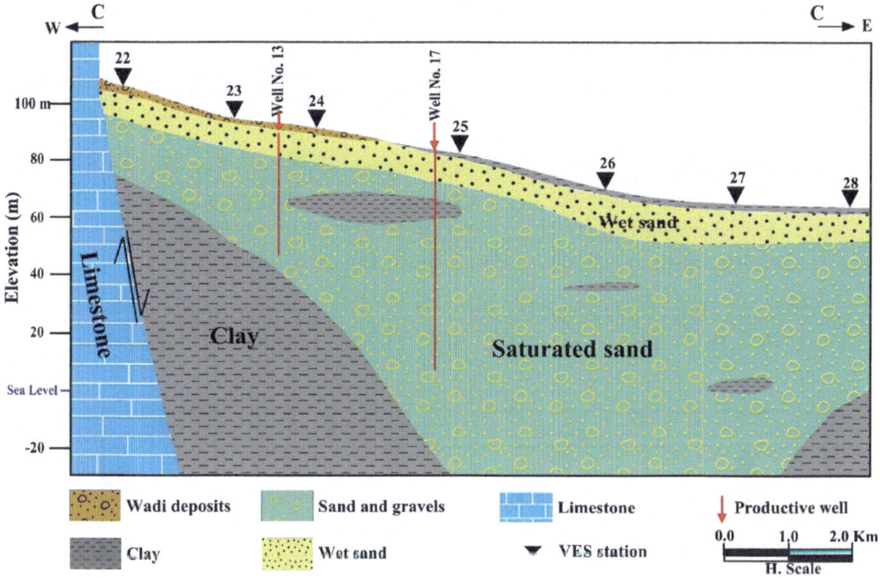

Fig. 7 Geoelectrical cross section C–C

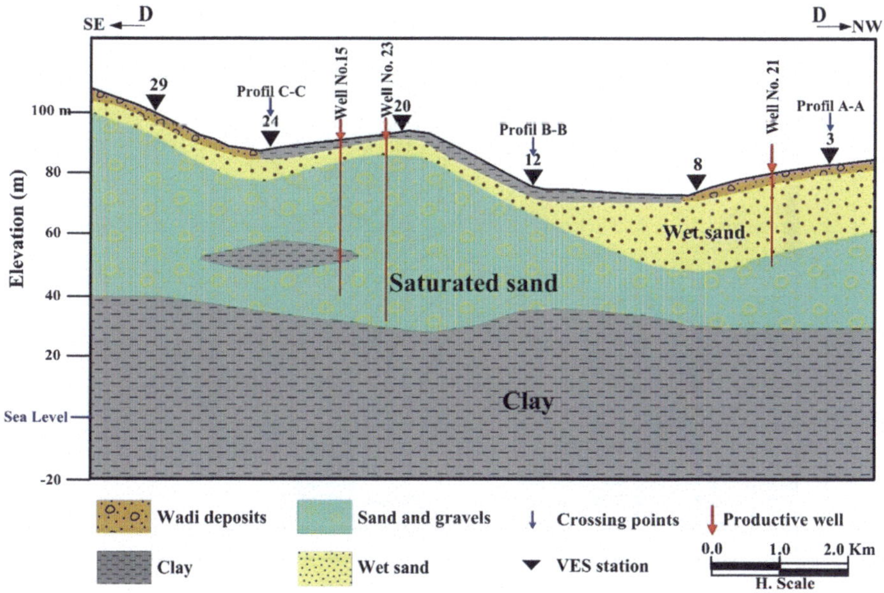

Fig. 8 Geoelectrical cross section D–D

the northeast, most likely due to the River Nile taking a different channel during the pre-Pleistocene era (Fig. 9).

The resistivity contour map of the Quaternary aquifer; the Pleistocene aquifer's resistivity shows that average values of around 40 Ω m on cultivated land (next to the Nile) increase to around 90 Ω m in the west. This could be due to a drop in saturation as you get closer to the scarp area. Due to the presence of a dry wadi deposit, the western part of the area has a high resistivity while the eastern part has a low resistivity on the resistivity contour map (Fig. 10).

Groundwater table above sea level; the contour map of the groundwater table above sea level (data from VES) reveals that high values are found in the southwest and

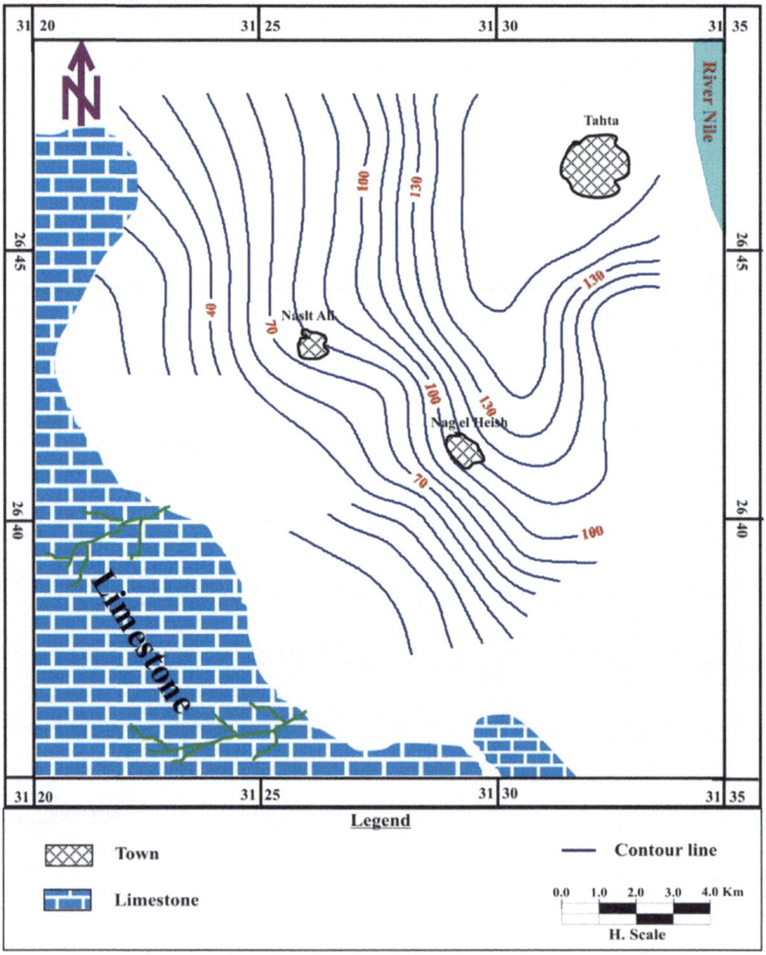

Fig. 9 Isopachyte map of the base of the saturated layer

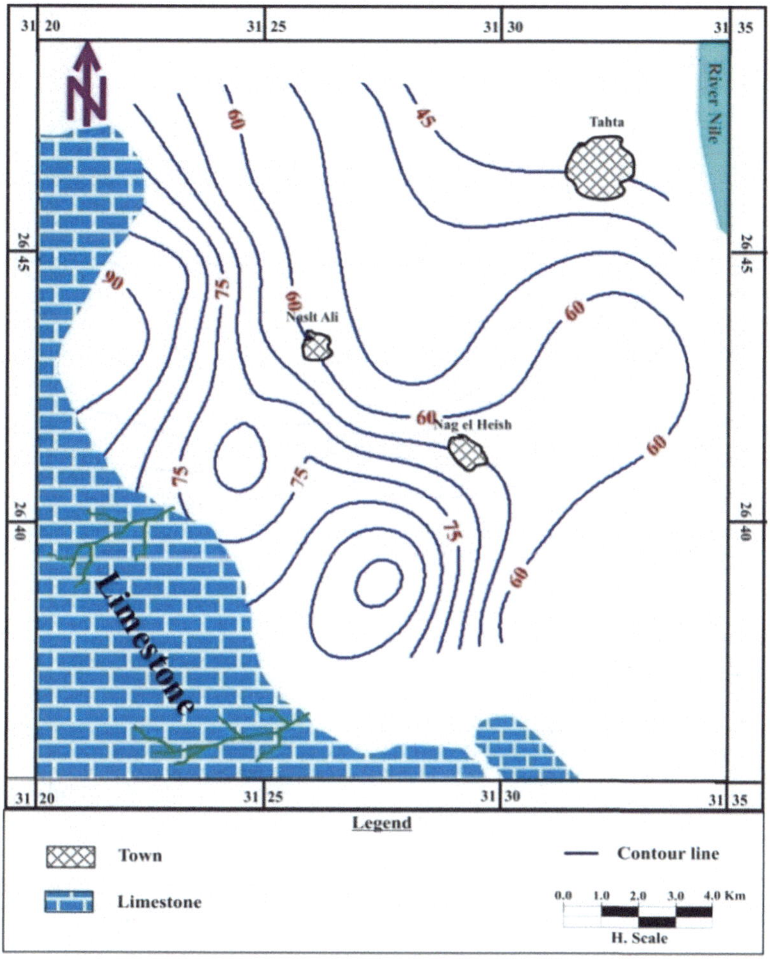

Fig. 10 True resistivity contour map of the Quaternary aquifer

low values are found in the northeast. We deduced that groundwater flows from the southwest toward the Nile River based on these data (Fig. 11).

4.2 *Hydrogeochemical Analysis*

The pH of surface water is between 7.1 and 7.6, while the pH of groundwater is between 6 and 7.7. It denotes a pH range of mildly acidic to slightly basic. According to BIS (1998), all of the water samples had pH levels between 6.5 and 8.5, which is within the allowed range. Surface water electrical conductivity (EC) ranges from

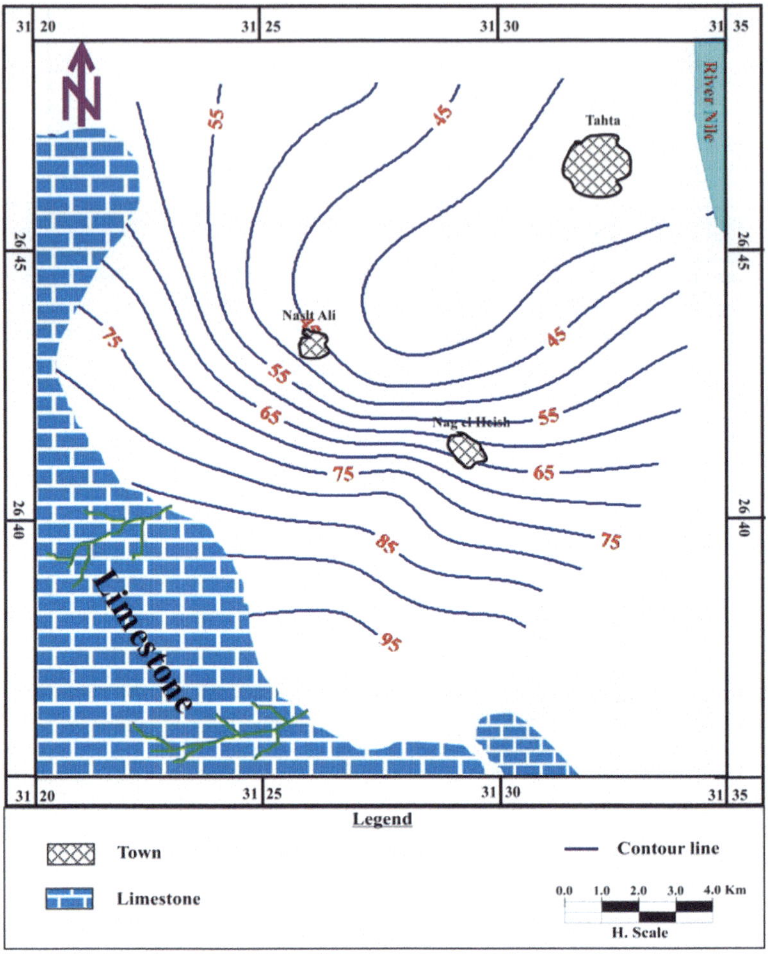

Fig. 11 Groundwater table above sea level (m)

230 to 2386 S/cm, while groundwater EC ranges from 356 to 7550 S/cm. Surface water and around 46% of the groundwater samples had conductivity levels below the 3000 S/cm BIS permissible limit (1998). Based on Hem (1985) classification, the salinity of the surface water is lower than that of freshwater, except for one sample, where the salinity of groundwater is between fresh and moderately saline. Increased salinity is caused by leaching processes and limestone dissolution during significant relief, as well as irrigation water evaporation.

Ion ratios are valuable for detecting prior hydrochemical activities such as groundwater contamination, mixing, leaching, and ion exchange that have impacted water quality. It offers a clear picture of the similarities and differences in hydrochemistry among various water types. The ratios are higher than 1.2 in the study area (rNa

+ rK/rCl) of surface water and 89% of groundwater samples, showing Ca and Mg replacement in the carbonates. In 4% of the groundwater samples, the ratio is less than one, indicating that Na has taken the position of Ca and Mg in the halogen.

7% of groundwater samples have values between 1 and 1.2, indicating alkaline replacement of Ca and Mg in their sulphates and minimal replacement of carbonates. The majority of surface water and groundwater samples contain an excess of sulphate over chloride, which can be linked to the widespread and growing usage of fertilisers. Most surface and groundwater samples have a greater Ca ion value than Mg ion value (rCa/rMg). The majority of these ratios are identical to those of Nile water, which appears to be the aquifer's primary source of recharge. The ion ratio shows how surface water and groundwater are related, indicating that surface water is the primary source of groundwater recharge.

4.3 Assessment of the Collected Groundwater for Different Purposes

4.3.1 Evaluation of Groundwater Quality for Drinking Purposes

Color, distinctive taste, turbidity, and excessive levels of dissolved salts should all be avoided in drinking water. When groundwater samples from the study area are compared to international standards for human drinking water and the Egyptian Committee for Water (2007), it is clear that the majority of surface water is suitable for drinking, but the majority of groundwater samples (71%) are unsuitable due to high salinity (1200–4453 ppm) and major ions above the permissible limits. The remaining samples (29%) are fit for consumption.

4.3.2 Evaluation of Water Quality for Domestic and Industrial Purposes

Hardness is a significant factor in the evaluation of groundwater for domestic and industrial purposes. 89% of the collected groundwater samples and 50% of surface water are in hard to very hard water, indicating their unsuitability for domestic and industrial purposes due to the high level of hardness, whereas the rest (16% groundwater and 50% surface water) are suitable. The high levels of total hardness of the groundwater samples reflect the high dissolution of the limestone in the western part of the study area.

4.3.3 Evaluation of Groundwater Quality for Irrigation Purposes

A appropriate irrigation practise requires consideration of water quality, soil types, and cropping practises. Excessive dissolved ions in irrigation water have a physical and chemical effect on plants and agricultural land, lowering output. The osmotic pressure in the plant structure cells is reduced as a result of these ions' physical action, preventing water from reaching the branches and leaves.

Plant metabolism is disrupted by the chemical effects. The following are the major chemical elements that influence the irrigation water's suitability:

- *Salinity index or total dissolved salts as computed by measured EC values*: Due to the plant's incapacity to complete ions in the soil solution for water, high salinity (electrical conductivity) in irrigation water has an impact on crop output. According to Bauder et al. (2007), half of the surface water samples are of outstanding quality, while the other half are of acceptable class. Although nearly half of the groundwater samples were deemed unsuitable for irrigation, the other half ranged between good qualities to doubtful for use in irrigation.
- *Sodium Adsorption Ratio (SAR) or sodicity index*: The salt concentration, which reflects reactions with the soil and reduces its permeability, is another essential component for water quality. Higher salt concentrations in water are generally undesirable for irrigation because they deteriorate soil properties. As a result, SAR is considered to be a better indicator of sodium (alkali) dangers in irrigation since SAR of water is directly related to sodium absorption by the soil. This is an important parameter for deciding if water is suitable for irrigation. When compared to calcium and magnesium, a high sodium level limits soil permeability, limiting the amount of water available to crops. The classification of water samples from the study area based on SAR values reveals that the study area's surface and groundwater are below excellent water class.
- *According to the U.S. Laboratory Staff's diagram* (Richards 1954): All of the surface water samples, in addition to 82% of the groundwater, are appropriate for irrigation. Under normal conditions, 18% of groundwater looks to be unsuitable for irrigation due to its high salinity (Fig. 12).
- *Percent sodium*: To understand the chemical features of groundwater, Richards (1954) and Wilcox (1955) methodologies are utilised, while the assessment of groundwater for irrigation depends on the mineralization of water and its effect on plants and soil. Clay particles absorbed sodium ions and replaced Mg^{2+} and Ca^{2+} ions in irrigation water with a high sodium concentration. The exchange of Na^{+} in water for Ca^{2+} and Mg^{2+} in soil lowers the permeability of the soil, resulting in poor internal drainage. When the soil is moist, air and water circulation is hindered, and the earth becomes hard when it dries (Saleh et al. 1999). Most surface water and 39% of groundwater are safe, according to Eaton's (1950), whereas 61% of groundwater samples are unsafe. Wilcox (1948) categorised groundwater for irrigation purposes by linking sodium content (i.e., sodium in irrigation water) with electrical conductivity. According to Wilcox's (1955) graphic (Fig. 13), the

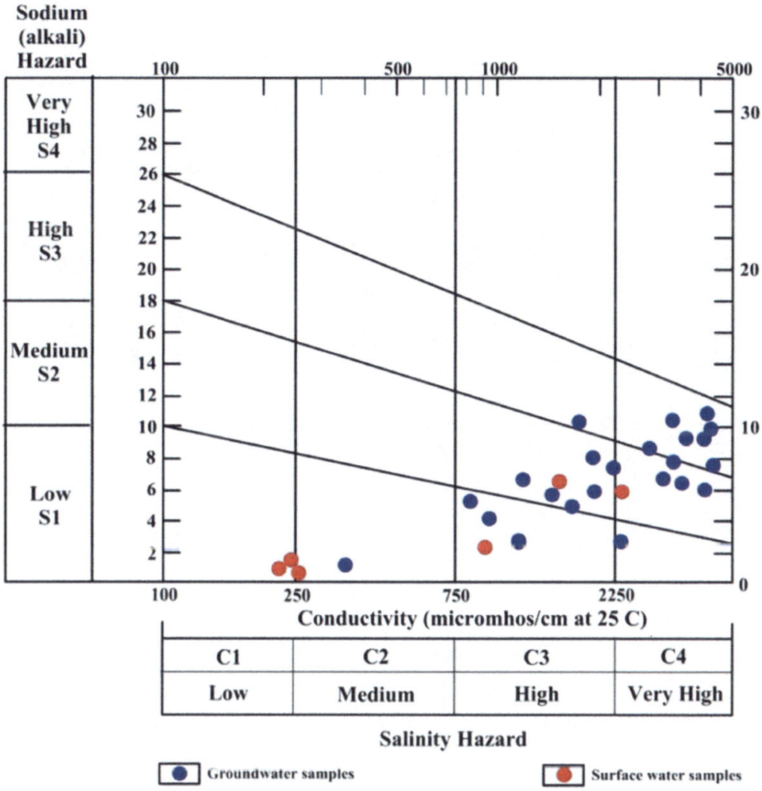

Fig. 12 U.S. Salinity Laboratory Staff diagram for the studied water samples

percentages of groundwater retrieved from western Tahta are as follows percentages: 54% unsuitable, 14% doubtful to unsuitable, 21% permissible to doubtful, 7% good to permissible, and 4% excellent to good.

- *Residual sodium carbonate*: The proportion of carbonate and bicarbonate in groundwater, in addition to the amount of calcium and magnesium, impacts the appropriateness of groundwater for irrigation. As the water in the soil becomes more concentrated, water with a high bicarbonate concentration has a tendency to precipitate calcium and magnesium. Excess sodium bicarbonate and carbonate are regarded deleterious to soil physical qualities because they allow the soil to absorb organic debris, resulting in a black stain on the soil surface as it dries. Wells with an RSC of less than 1.25 are safe and suited for irrigation, whereas those with a value greater than 2.5 are not (Eaton 1950; Ragunath 1987). By classifying the study area water samples in accordance with the RSC, it can be concluded that most of water samples (surface and groundwater) are suitable for irrigation.

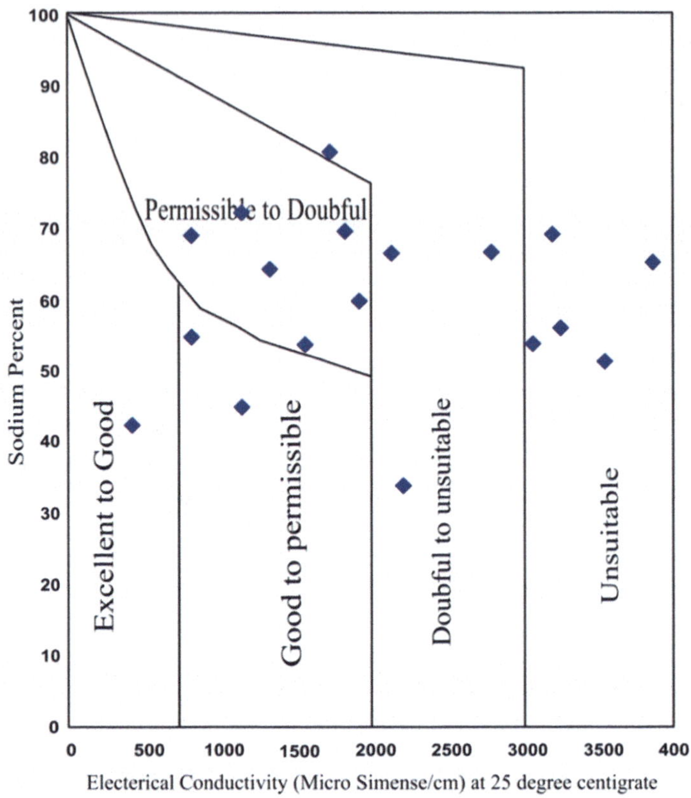

Fig. 13 Wilcox's diagram for the studied groundwater samples

4.4 Mechanisms Controlling Groundwater Chemistry

Finally, to understand the chemistry of groundwater and the interaction between the chemical components of the water and their respective aquifers, such as the chemistry of rock types, precipitated water chemistry, and evaporation rate. Gibbs (1970) proposed a graph that plots the ratio of dominating anions and cations versus the TDS value. The Gibbs diagram depicts the chemical data for the groundwater samples (Fig. 14). Chemical weathering of rock-forming minerals has an impact on groundwater quality through the breakdown of the host rock, according to the majority of groundwater samples taken from the research region.

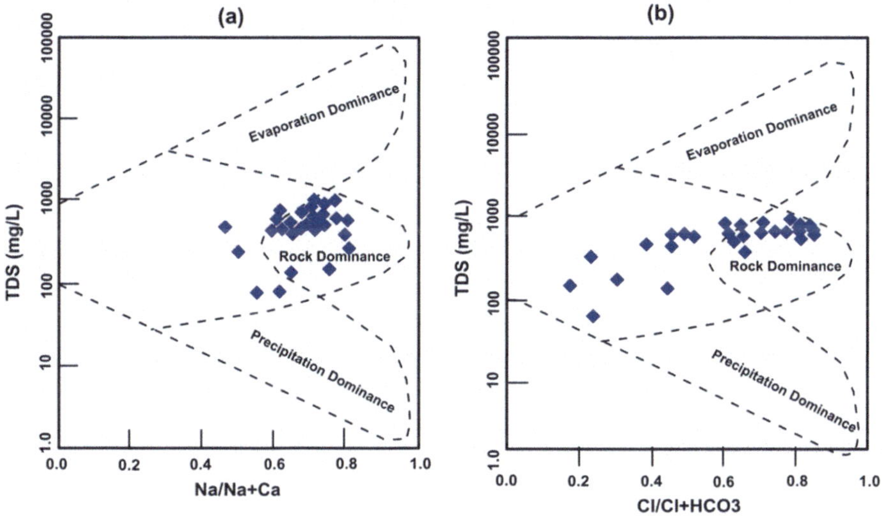

Fig. 14 Gibbs diagram of the studied groundwater samples

5 Conclusions

The study area is defined by a single aquifer system (Quaternary aquifer) that is mostly refilled by surface water, particularly irrigation canals, which play an important role in the water table's design. During evaporation, the aquifer was discharged.

Because of the presence and influence of wadi deposits in the west and clay in the east, the study area is a geophysical four-layer instance; the first layer (surface layer) has varying resistivity values from west to east. This layer is quite thin, ranging from 1 to 5 m thick. Wet sand makes up the second layer, which has resistivity values ranging from 111 to 325 Ω m and a thickness of 5–34 m. The main water-bearing formation in the research area is the third geoelectrical layer. Sand and gravel with resistivity values ranging from 28 to 99 Ω m make up the material. This layer's thickness decreases as it moves west, ranging from 18 to 144 m. Due to the limits of the device, the thickness of this layer is not defined in some areas. The fourth geoelectrical layer, which is represented by Pliocene clay with resistivity values ranging from 1.7 to 12.9 Ω m, is the base of the water-bearing layer. Due to the existence of wadi deposits apparent on the contour maps, the thickness of the Quaternary aquifer increases to the east, whereas the resistivity increases to the west. In the pre-Pliocene age, the highest thickness of the groundwater formation can be seen in the several courses of the Nile River.

The bulk of surface water looks to be drinkable, however the majority of groundwater samples (71%) are unfit for human consumption due to salinity levels that exceed the allowed limit. The groundwater samples from half of the wells are suitable for irrigation. Because of their excessive salinity, the other half are unusable. The chemical weathering of rock-forming minerals has an impact on groundwater quality by dissolving the host rock, according to the majority of groundwater samples taken from the research area.

6 Recommendations

In order to avoid the poor quality of the groundwater in the study area, it is highly recommended:

1. Continues monitoring quality of groundwater to assess any deterioration in water quality for different uses this reduce the risk of aquifers pollutions.
2. Reduce the using of pesticides, herbicides, and fertilizers, and expanding the use of organic fertilizers.
3. The Egyptian government must put a law that prevents the disposal of any waste water, and sewage treatment stations, into the drains of the study area without treatment.
4. The production wells must not worked at the same time and the total yield of these wells must be controlled.
5. Recommended detailed hydrogeological plan and reliable groundwater management studies to controlling the groundwater conditions of the study area.
6. The irrigation system must be changing from traditional method (flood) to new methods (drip and sprinkler) methods and improve the system of agriculture drainage over all the study area.

Acknowledgements The first author would like to thank Institute of Applied Geophysics, Montan Universität Leoben, Austria for providing support to make this work possible.

References

Abdel Moneim AA (1988) Hydrogeology of the Nile basin in Sohag province. M.Sc. thesis, Geology Department, Faculty of Science, Assiut University, Egypt, 131 p

Abdel Moneim AA (1992) Numerical simulation and groundwater management of Sohag aquifer, the Nile Valley, Egypt. Ph.D. thesis, Civil Engineering Department, Strathclyde University, Glasgow, Scotland, Great Britain

Abdel Moneim AA (1999a) Groundwater studies in and around Abydous temples, El-Balyana, Sohag, Egypt. Ann Geol Surv Egypt 22:357–368

Abdel Moneim AA (1999b) Geoelectrical and hydrogeological investigations of the groundwater resources on the area to the west of the cultivated land at Sohag, Upper Egypt. Egypt J Geol 43(2):253–268

Abotalib AZ, Heggy E, El Bastawesy M, Ismail E, Gad A, Attwa M (2021) Groundwater mounding: a diagnostic feature for mapping aquifer connectivity in hyper-arid deserts. Sci Total Environ 801:149760. https://doi.org/10.1016/j.scitotenv.2021.149760. Epub 2021 Aug 19. PMID: 34467900

Ahmed SM (1980) Geological studies on the west and southwest of Sohag. M.Sc. thesis, Geology Department, Faculty of Science, Assiut University, Sohag, Egypt

Ahmed AA (2007) Using lithologic modeling techniques for aquifer characterization and groundwater flow modeling of Sohag area, Egypt. In: Second international conference on geo-resources in the Middle East and North Africa, Cairo, 24–28 Feb 2007

Ahmed AA (2009) Using generic and pesticide DRASTIC GIS based models for vulnerability assessment of the Quaternary aquifer at Sohag, Egypt. Hydrogeol J 17:1203–1217

Ahmed AA, Mohamed H (2009) Hydrochemical evolution and variation of groundwater and its environmental impact at Sohag, Egypt. Arab J Geosci. https://doi.org/10.1007/s12517-009-0055-z

APHA "American Public Health Association" (2005) Standard methods for the examination of water and wastewater, 20th edn. In: Eaton AD, Clesceri LS, Greenberg AE (eds). APHA, AWWA, WPCF, New York

Attia FR (1985) Management of water systems in Upper Egypt. Ph.D. thesis, Faculty of Engineering, Cairo University, Egypt

Awad MA, Nada AA, Hamza MS, Froehlich K (1995) Chemical and isotopic investigation of groundwater in Tahta region, Sohag, Egypt. Environ Geochem Health 17:147–153

Awad A, Eldeeb H, El-Rawy M (2020) Assessment of surface water and groundwater interaction using field measurements: a case study of Dairut city, Assuit, Egypt. J Eng Sci Technol 1:406–425

Barber W, Carr DP (1981) Water management capabilities of the alluvial aquifer system of the Nile Valley, Upper Egypt. Technical report no. 11, water master planc. Ministry of Irrigation, Cairo, Egypt, 145 p

Bauder TA, Waskom RM, Davis JG (2007) Irrigation water quality criteria. Extension fact sheet no. 0.506. Colorado State University, 15 p

BIS (1998) Drinking water specifications, IS: 10500. Bureau of Indian Standards

Chumakove IS (1967) Pliocene and Pleistocene deposits of the Nile Valley in Nubia and Upper Egypt (in Russian). Trans Geol Inst Acad USSR 170:1–110

Diab MSh, El-Shayeb MH, Abdel Moneim AA, Said MM, Zaki SR (2002) Evaluation of water resources and land suitability for development in southern part of Sohag, Upper Egypt. Ann Geol Surv Egypt V XXV:487–512

Doneen ID (1964) Notes on water quality in agriculture. Published as a water science and engineering, paper 400. Department of Water Sciences and Engineering, University of California

EHCW Egyptian Higher Committee for Water (2007) Egyptian standards for drinking and domestic uses

Eaton FM (1950) Significance of carbonates in irrigation waters. Soil Sci 69:123–134. https://doi.org/10.1097/00010694-195002000-00004

El-Rawy M, Ismail E, Abdalla O (2019) Assessment of groundwater quality using GIS, hydrogeochemistry, and factor statistical analysis in Qena Governorate, Egypt. Desalin Water Treat 162:14–29

El-Rawy M, Abdalla F, El Alfy M (2020) Water resources in Egypt. In: The geology of Egypt. Springer, Cham, pp 687–711

El-Rawy M, Moghazy HE, Eltarabily MG (2021a) Impacts of decreasing Nile flow on the Nile Valley aquifer in El-Minia Governorate, Egypt. Alex Eng J 60(2):2179–2192

El-Rawy M, Makhloof AA, Hashem MD, Eltarabily MG (2021b) Groundwater management of quaternary aquifer of the Nile Valley under different recharge and discharge scenarios: a case study Assiut Governorate, Egypt. Ain Shams Eng J 12(3):2563–2574

El-Sayed SA, El-Shater AA (2010) Engineering aspects and associated problems of flood plain deposits in Sohag Governorate, Upper Egypt. J Am Sci 6(12):1614–1623

El-Sayed E, El Kashouty M, Kamel A (2010) The hydrochemical characteristics and evolution of groundwater and surface water in the western part of the River Nile, El Minia district, Upper Egypt. Arab J Geosci 5:637–652. https://doi.org/10.1007/s12517-010-0219-x

Gibbs RJ (1970) Mechanism controlling world water chemistry. Science 170:1088–1090

Gomaa AA (2006) Hydrogeologic and geophysical assessment of the reclaimed areas in Sohag, Nile Valley, Egypt. Ph.D. thesis, Geology Department, Faculty of Science, Ain Shams University, 272 p

Hassan AM (2005) Geochemical characteristics of the superficial Nile basin sediments and their environmental relevance, Sohag area, Egypt. M.Sc. thesis, Faculty of Science, South Valley University (Sogah Branch)

Heleika MA, Toney S, Ismail E (2021) Mapping of groundwater opportunities for multi-purposes use in Beni-Suef province, Egypt. Arab J Geosci 14:784. https://doi.org/10.1007/s12517-021-07123-1

Hem JD (1985) Study and interpretation of the chemical characteristics of natural water USGS. Water supply paper no. 2254, pp 117–120

IPI2Win (2003) Resistivity sounding interpretation. Moscow State University

Ismail E (2013) Studied the environmental investigation in the west Tahta region, Upper Egypt: hydrochemical, geophysical and remote sensing study. Ph.D. Department Angewandte Geowissenschaften und Geophysik, Montan Universität Leoben, Austria, pp 85–86

Ismail E, El-Rawy M (2018) Assessment of groundwater quality in West Sohag, Egypt. Desalin Water Treat 123:101–108

Ismail E, Zaki R, Rapantov N, Licbinska M, Sharawi H (2020) Hydrogeochemical characteristics of the groundwater in the quaternary aquifer of western fringes of El-Minia Governorate, Egypt using an integration of geochemical modeling and geo-statistical techniques. Desalin Water Treat 189:134–151. https://doi.org/10.5004/dwt.2020.25465

Ismail E, Abdelhalim A, Heleika M (2021) Hydrochemical characteristics and quality assessment of groundwater aquifers northwest of Assiut district, Egypt. J Afr Earth Sci 181:104260. https://doi.org/10.1016/j.jafrearsci.2021.104260

Joshi DM, Kumar A, Agrawal N (2009) Assessment of the irrigation water quality of river Ganga in Haridwar district India. J Chem 2(2):285–292

Kelly WP (1940) Permissible composition and concentration of irrigation waters. Am Soc Civ Eng 66:607–613

Korany EA, Omran AA, Abdel-Aal AA (2006) Hydrogeochemical approach for the assessment of the recharge routes of groundwater in the Quaternary aquifer, Nile Valley of Sohag area, Egypt. In: Proceedings of the 7th international conference on geochemistry, Alexandria, Sept 2006

Mahran TM, El-Haddad A (1992) Facies and depositional environments of Upper Pliocene-Pleistocene Nile sediments around Sohag area, Nile Valley, Egypt. J Saharian Stud 1(2):11–40

Mousa S, Attia FA, Abu El-Fotouh M (1994) Geoelectrical and hydrogeological study on the Quaternary aquifer in the Nile Valley between Asyut and Sohag governorates, Egypt. Geol J Egypt 38–2:1–20

Omer AAM (1996) Geological, mineralogical and geochemical studies on the Neogene and Quaternary Nile basin deposits, Qena-Assiut stretch, Egypt. Ph.D. thesis, Geology Department, Faculty of Science, South Valley University, 320 p

Omer AAM, Abdel Moneim AA (2001) Geochemical characteristics of the Pliocene and Pleistocene Nile basin deposits and their influence on the groundwater chemistry in Sohag area. Ann Geol Surv Egypt 24:567–584

Omran AA (2008) Integration of remote sensing, geophysics and GIS to evaluate groundwater potentiality—a case study in Sohag region, Egypt. In: The 3rd international conference on water resources and arid environments (2008) and the 1st Arab water forum

Omran AA, Korany EA, Abdel-Rahman AA (2006) An integrated approach to evaluate groundwater potentiality—a case study. In: Proceedings of the 3rd international conference of applied geophysical, ESAP, 18–20 Mar 2006. National Research Center, Cairo. J Appl Geophys 5, part 1

Paliwal KV (1972) Irrigation with saline water. Monogram no. 2 (new series). IARI, New Delhi, p 198

Ragunath HM (1987) Groundwater. Wiley Eastern, New Delhi, p 563

Richards LA (US Salinity Laboratory) (1954) Diagnosis and improvement of saline and alkaline soils. In: US department of agriculture hand book, p 60

RIGW, IWACO (1989) Groundwater development for irrigation and drainage in the Nile Valley: groundwater development in the area of West Tahta. Technical note 70.124-89-05 (internal report)

Said R (1960) Planktonic foraminifera from the Thebes Formation, Luxor. Micropaleontology 6:277–286

Said R (1981) The geologic evaluation of the river Nile. Springer-Verlag Inc., New York, p 151

Said R (1990) The geology of Egypt. Balkema Publ, Rotterdam, Netherlands, p 734

Saleh A, Al-Ruwaih F, Shehata M (1999) Hydrogeochemical processes operating within the main aquifers of Kuwait. J Arid Environ 42:195–209

Snousy MG, Li P, Ismail E (2021a) Trace elements speciation and sources characterization in the main watercourses, middle-upper Egypt. Hum Ecol Risk Assess Int J 27(7):1764–1785. https://doi.org/10.1080/10807039.2021.1902265

Snousy MG, Wu J, Su F, Ahmed A, Esam I (2021b) Groundwater quality and its regulating geochemical processes in Assiut province, Egypt. Expo Health. https://doi.org/10.1007/s12403-021-00445-1

Todd DK (1980) Groundwater hydrology, 2nd edn. Wiley, New York, p 315

Todd DK, Mays LW (2005) Groundwater hydrogeology. Wiley, New York

Velpen VDBA (1988) Resist, a computer program for the interpretation of resistivity sounding curves. An ITC M.Sc., research project. ITC, Delft, The Netherlands

Wilcox LV (1948) The quality of water for irrigation use. US Department of Agriculture, Technical Bulletin, Washington DC, 19 p

Wilcox LV (1955) Classification and use of irrigation waters, vol 969. U.S. Department of Agriculture Circular

World Health Organization (2011) Guidelines for drinking water quality, 4th edn. Recommendations. World Health Organization, Geneva, pp 1–4

World Health Organization (2017) Guidelines for drinking-water quality: incorporating first addendum. http://www.apps.who.int/iris/bitstream/10665/254637/1/

Zaki SR (2001) Hydrogeological studies and application of geographic information system for evaluation of water resources and land use projects in the southern part of Sohag Governorate. M.Sc. thesis, Geology Department, Faculty of Science, Monufiya University

Mapping Groundwater Recharge Potential in the Nile Basin Using Remotely Sensed Data and GIS Techniques

Mohamed Abdelkareem, Abbas Mansour, and Ahmed Akawy

Abstract Groundwater is a valuable source of fresh water in the arid/hyper-arid regions where the precipitation is limited. Thus, the demand for groundwater increased in many regions worldwide to sustain life. Following rainstorms, substantial amount of infiltrates the subsurface to replenish the groundwater aquifers. Here, we evaluate the groundwater recharge potential in Wadi Qena and Wadi Matula in the Eastern Desert of Egypt as a part of the Nile basin. Remotely sensed data were used followed by applying GIS techniques. Several evidential maps were derived from radar/optical remotely sensed data in addition to geologic and precipitation data, which control the occurrence, movements and infiltration capacities were combined after applying weight factors. The results revealed that about 3.56 and 10.36% of the surface areas of Wadi Qena, and Wadi Matula basins, respectively, were characterized by excellent recharging potential and represent the most promising zones for future groundwater extraction. Field data confirmed that the detected zones are cultivated and groundwater is suitable for irrigation of several strategic crops such as wheat, tomato and alfalfa, by pumping wells. It is concluded that storing flood water during rainy storms in man-made reservoirs would provide additional new water resources to help sustaining the life of the people in these remote areas.

Keywords Egypt · Groundwater · Nile basin · Recharge zones · Wadi Matula · Wadi Qena

---

M. Abdelkareem (✉) · A. Mansour · A. Akawy
Geology Department, Faculty of Science, South Valley University, Qena 83523, Egypt
e-mail: mohamed.abdelkareem@sci.svu.edu.eg

A. Mansour
e-mail: aman@svu.edu.eg

A. Akawy
e-mail: akawy@svu.edu.eg

A. M. Negm and M. El-Rawy (eds.), *Sustainability of Groundwater in the Nile Valley, Egypt*, Earth and Environmental Sciences Library,
https://doi.org/10.1007/978-3-031-12676-5_12

1 Introduction

The Great Sahara of North Africa is characterized by extremely arid/hyperarid conditions with very scarce rainfall. Most countries in this region experience water scarcities. The Nile River is the only dominant resource of surface water across the desert which is necessary to domestic uses, as well as industrial and agricultural activities. Its flow is essential to human life, this is mainly true in countries like Egypt, which is already below the water poverty line; Egypt's Nile flow is limited to 55.5 billion m^3 per annum. The Nile River has long been considered a corridor for humanity's welfare along the eastern Sahara, i.e., in Egypt and Sudan. The demand for fresh water increased because the increasing of overpopulation. In mean time, fresh water scarcity is common in several parts; it is expected that Egypt will have to rely some extent on groundwater resources for future national development projects. Population extension and the extremely need for food requirements threaten to worsen the circumstances in the future.

The Egyptian Government intends to invest in desert fringes to enlarge the cultivated area. This mainly requires additional water resources such as groundwater and water from flash flood. Therefore, it is necessary to find innovative techniques for exploring of new water resources (Abdelkareem et al. 2012a). The groundwater is invisible water resource below the land surface. Thus, detecting and expecting groundwater sites using indirect approaches represent a significant factor in revealing sites of the groundwater infiltration and thus, groundwater abstraction zones (Abdelkareem and El-Baz 2015a).

Groundwater is a vital source of freshwater utilized to sustainable development on the Earth, particularly in areas of arid/hyper arid zones. Importantly, the present defunct wadis have received substantial rainfall and fluvial deposits as well, primarily during past pluvial conditions and stored below strata that might have stored as fossil water (El-Baz 1998). Therefore, the fluvial deposits are the promising areas for future groundwater exploration as several wadis experienced subsidence and structural activities, that help in trapping groundwater in porous layers by developing walls of massive rocks that prevent flowing of water outside the basin (Abdelkkareem and El-Baz 2017). Crucially, the present defunct wadis of the Nile basin recharged by substantial surface water during the rainfall storms, i.e., the plausible annual recharge rates for Wadi Qena is about 14.7×10^6 m^3 per year (Gheith and Sultan 2002).

When precipitation accesses the land surface, some may get absorbed, infiltrate and penetrate the soil and the rest may move downslope in the kind of runoff (Lentswe and Molwalefhe 2020). The downward movement of surface water into the water table and reaching to groundwater aquifer is termed groundwater recharge (Freeze and Cherry 1979), where water flows from an unconfined aquifer through a confining condition to below aquifer (Healey 2010). Groundwater infiltration, accumulation and prospective zones were characterized by several techniques, such as geological, hydrogeological, geophysical and remotely sensed techniques. Remotely sensed data provides interesting pilot output in fast and inexpensive manner (Abdelkareem and El-Baz 2015a, b). Crucially, remotely sensed data is very important in characterizing the hydrogeological properties and conditions in many environments. However, even in well-mapped areas, satellite images reveal lithologic units, geomorphologic, and structural features that were not recognized by conventional approaches. These data enhance our information of the hydrogeological conditions (Abdelkareem 2017; Abdelkareem and El-Baz 2017).

Many studies (e.g., Murmu et al. 2019; Çelik 2019; Andualema and Demekeb 2019; Mukherjee and Singh 2020; Al-Djazouli et al. 2020; Ajay Kumar et al. 2020; Berhanua and Hatiyeb 2020; Ghosh et al. 2020; Benjmel et al. 2020; Lentswe and Molwalefhe 2020) have utilized remotely sensed images/data and GIS for groundwater infiltration, recharging and exploration and characterizing of artificial recharge zones in many area in all over the world. The remotely sensed images have been widely used to predict the suitable locations of groundwater infiltration by assessing and analyzing of geologic structures, lithology, geomorphic features, geophysical properties and their hydrologic characteristics. Abdalla 2012; Abdelkareem and El-Baz (2015a, b) utilized satellite images and GIS techniques to revealing groundwater prospective zones in Wadi Matula using various data (i.e. geologic/geomorphic, geologic structures, stream-networks, and topography).

Based on the aforementioned review, in order to sustaining development in arid regions it is necessary to enhance groundwater resources. This requires gathering scientific investigations to explore additional water resources. Thus, the aim of the present study is to combine several geologic, geomorphic, structural and hydrologic data using GIS techniques to predict the potential groundwater recharge zones.

2 Geologic and Hydrologic Setting of the Nile Basin

The Nile River covers nearly 6800 km from the Great Lakes of Equatorial Africa, whereas the rainfall belt, to its delta at the Mediterranean Sea at Egypt whereas the arid/hyperarid conditions (Fig. 1a). It crosses five different environments that distinguish in geomorphology, hydrogeomorphology, geologic structures and topography (Fig. 1b). They are from south (high elevation; ~5700 m) to north (lower elevation; 0): Lake Plateau, Sudd, Central Sudan, Cataract tract (from Khartoum to Aswan) and Egyptian Nile (Said 1981). The Egyptian Nile region (Fig. 2) as a part of the eastern Sahara extends for approximately 1530 km between the Egypt-Sudan borders at Aswan to the Mediterranean Sea (Said 1981). The Egypt's Nile section cutting the upper Cretaceous/Lower Tertiary plateau forming Nile gorge (Fig. 2a; Abdelkareem and El-Baz 2016) that separated Eastern Desert to east and Western Desert to west. It drains an area of approximately 326,751 km^2, about 10.5% of the total area of the Nile basin (Woodward et al. 2007). This section of the Nile basin collected surface water in the pluvial events (e.g., the late Eocene to early Miocene; Said 1990) through the present defunct wadis (e.g., Allaqi-Gabgaba, Qena, Matula, Shait, east Kom-Ombo, Tushka basin) that evidenced by paleorivers, lake-beds and fluvial deposits concealed by sand sheets (Paillou et al. 2009; Abdelkareem and El-Baz 2017). However, the Pleistocene epoch experienced arid conditions (Said 1990, p. 490) that intercalated with slight fluctuations as evidenced by the existence of paleochannels during this epoch (Haynes 1980; Said 1990). Said (1981) stated that the Nile River progressed in subsequent phases that given numerous terms; Eonile (Late Miocene), Paleonile (Pliocene), and Protonile, Prenile and Neonile of the Pleistocene.

The present arid-hyper-arid climates of Egypt, these wadis collect surface water sourced from the Red Sea Mountains during sporadic storms. Such precipitation infiltrated through fracture/fault zones recharging the groundwater aquifers. Noteworthy, several fault/fracture zones (Fig. 2b) around the Nile Valley on both flanks play conduits for recharging and discharging processes.

From a hydrological point of view, more rainfall showers (see Fig. 1a) occur on the Intertropical Convergence Zone; ITCZ), that representing the rain fall belt (see Fig. 1a) in the Equatorial Africa (~20 N to ~20 S). The Egypt's Nile region is situated in an arid-to-hyperarid conditions that collects <100 mm of precipitation per year on average (Gheith and Sultan 2002) as the area shows dryness and scarce rainfall (Abdelkareem and El-Baz 2015a).

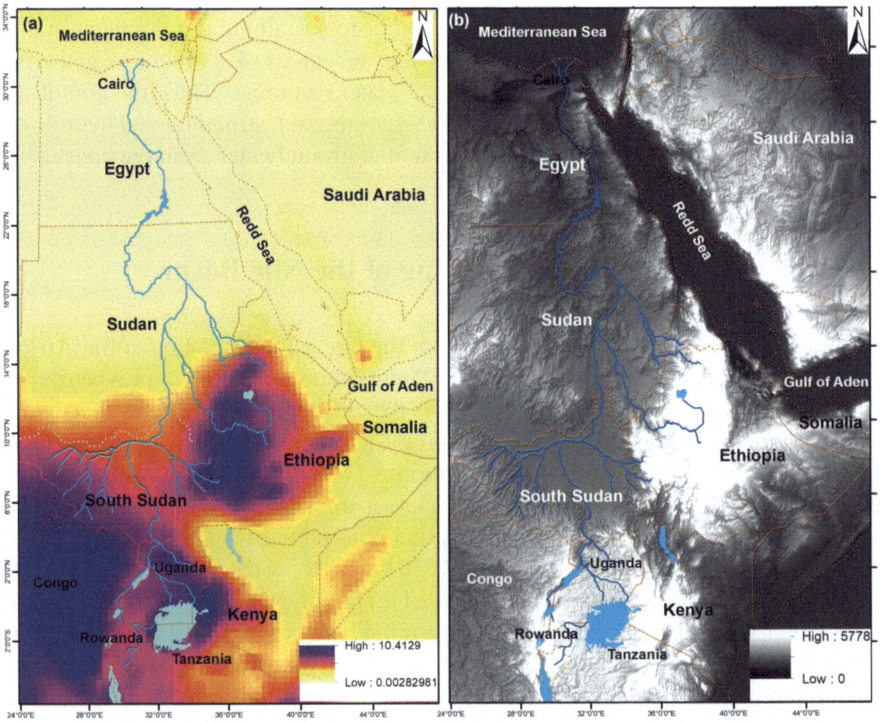

Fig. 1 Nile basin **a** Nile River streams on the precipitation data (range from 0.002 to 10.41 mm/day) derived from TRMM satellite; **b** Nile basin streams overlain the SRTM DEM; elevation ranges from 0 to 5778

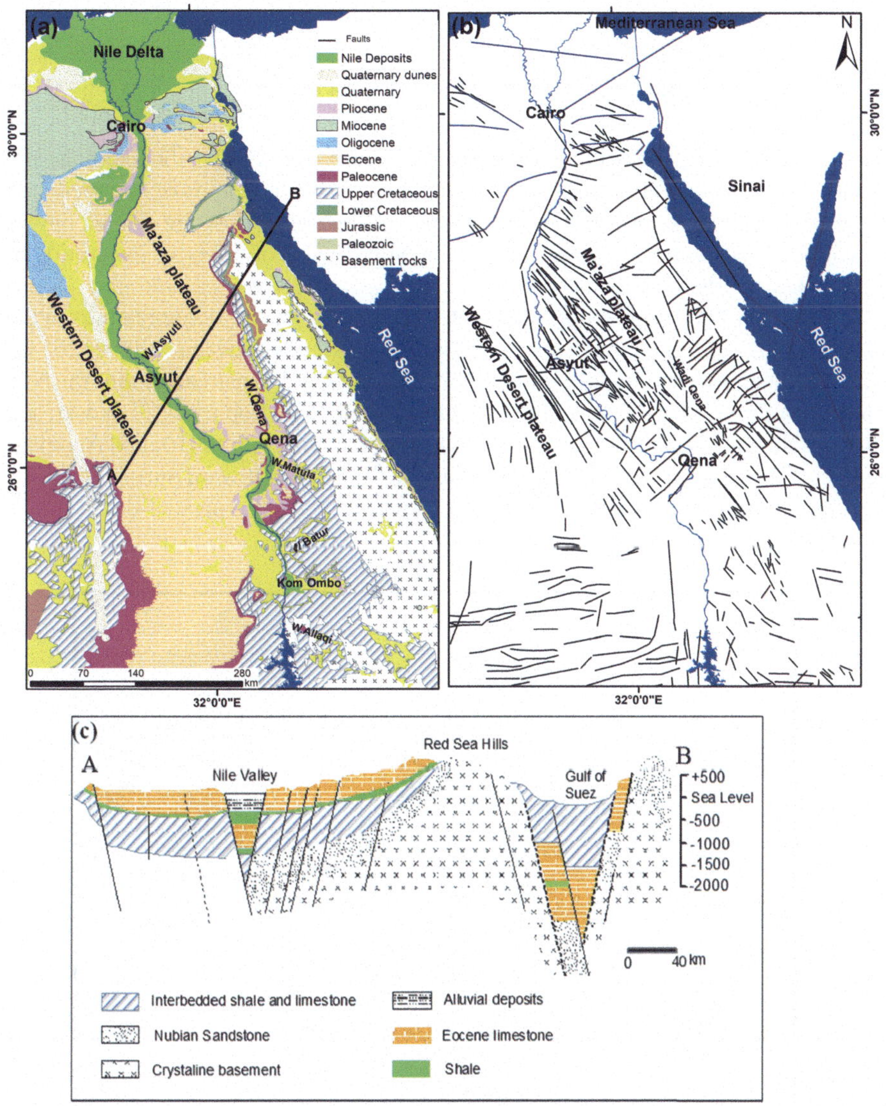

Fig. 2 Nile basin **a** geological map (part of geological map of Egypt by EGSMA 1981) showing the study area; **b** lineaments map of the study area derived from remotely sensed data and the geologic map of Egypt (EGSMA 1981); **c** the cross-section is modified from Sultan et al. (2007), after (RIGW 1993)

3 Case Studies

Several studies have been accomplished over the past few decades in order to delineate the prospective areas for water accumulation and infiltration in the Nile basin (Abdelkareem 2012; Abdelkareem et al. 2012a; Abdalla 2012; Abdelkareem and El-Baz 2015a). Such studies utilized satellite data and GIS approaches predicted the plausible areas for water infiltration and abstraction. Below are two examples including W. Qena and W. Matula east of the Nile (Fig. 3) of effective integration of multi-criteria in the Nile basin.

Wadi Qena (WQ) represents significant geomorphic and geologic features in the eastern Desert of Egypt, covering ~16,000 km^2 that extends between latitudes 26° 10′ to 28° 5′ N and longitudes 32° 31′ to 32° 45′ E (Fig. 4). It downstream drains the Nile at the Qena Bend. Based on results of integrating data from Radarsat-1, Landsat, and a fused Radarsat-1/Landsat image (Fig. 4a) Abdelkareem et al. (2010) and Abdelkareem and El-Baz (2015c) led to that Wadi Qena drained northward during its earlier stage, go along with the general slope of Africa; while at the present-day it drains

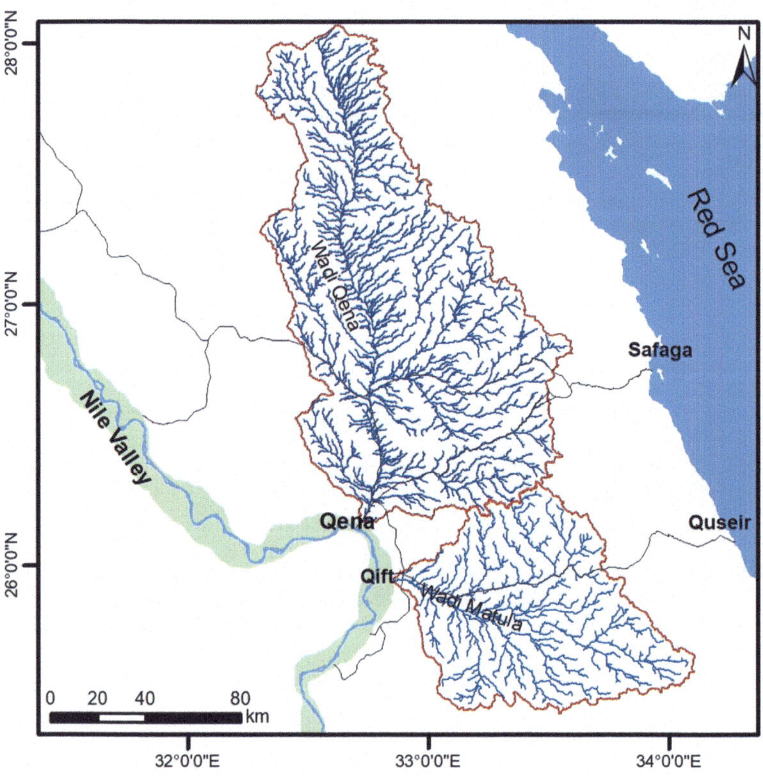

Fig. 3 Stream-networks and watersheds of Wadi Qena and Wadi Matula, east of Qena

southward opposite to the general drain of the Nile to the north. A fused Radarsat-1/Landsat image using PC Spectral Sharpening data fusion shows that the wadi initiated along the N–S fault trend that affected by the NE–SW right lateral Qena-Safaga shear zone in the southern part of the WQ basin (Fig. 4a). It also portrayed the wadi deposits in dark signatures along the main course, and showing extension beyond its northern boundary as it realistic image that combined the lithologic and structural characteristics (Abdelkareem 2012). It is built up of the rugged Precambrian basement rocks to east and Upper Cretaceous/Lower Tertiary succession topped by limestone, to the west, (Fig. 4b). The basement rocks represent the main catchments of the Red Sea highlands that reach to 1866 m (Fig. 5). The plateau extend westward till the border of the Nile Valley and is affected by major fault scarps that oriented N–S and NW–SE that form incised wadis. The main course of the wadi comprises thick succession of sand and gravel deposits revealing several cycles of flash floods.

On the eastern side of the Nile at the Qena Bend, Wadi Matula covers (7500 km^2) a large expanse that ranges from the Red Sea Mountains, to east, and the Nile Valley, to west (Fig. 6a, b). It mapped by Abdelkareem and El-Baz (2015c) using SRTM, Landsat, Radarsat-1, Advanced Land Observing Satellite (ALOS)/Phased Array type

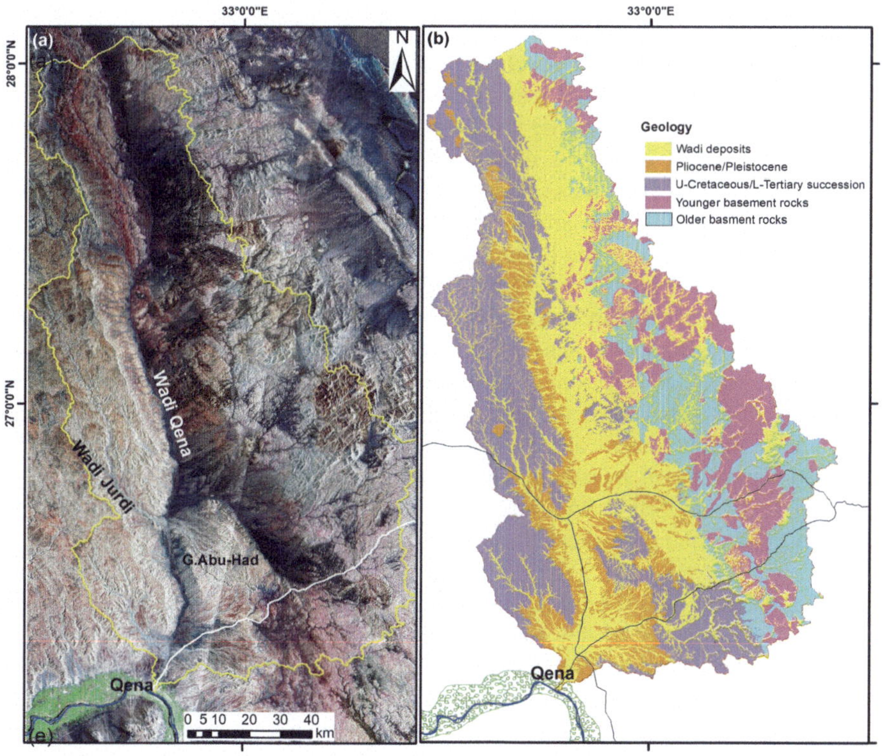

Fig. 4 Wadi Qena **a** a fused Radarsat-1/Landsat image of Wadi Qena; **b** geological map of Wadi Qena (Conoco 1987)

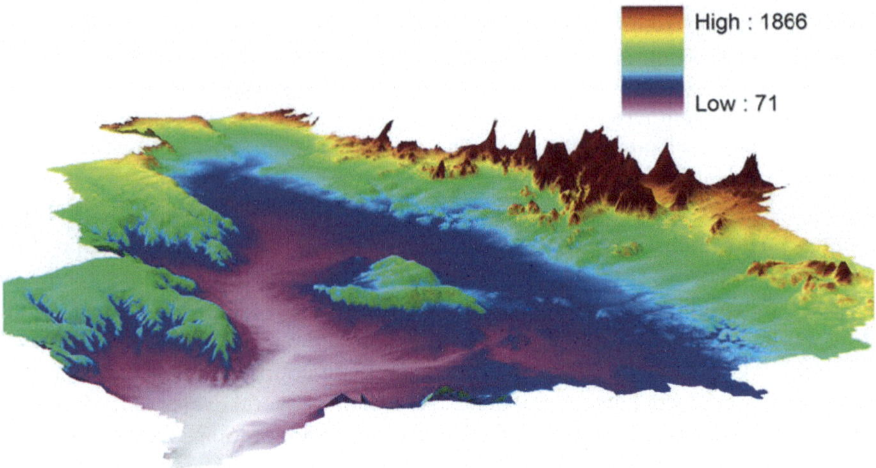

Fig. 5 Oblique, 3D perspective view of Wadi Qena showing the variations in elevation derived from SRTM DEM (the extension and orientation as the same as WQ watershed in Fig. 3)

L-band Synthetic Aperture Radar (PALSAR) data and TRMM data. Precipitation over the study catchment is relatively higher in the Red Sea highliands (Fig. 6c).

Wadi Matulla runs over the Nubian Sandstone plain between the limestone outliers southeast of Qena town, draining the Nile at Qift city. It build-up of Precambrian basement rocks to the east and sedimentary succession of Upper-Cretaceous/Lower Tertiary to west. The exposures of Taref Formation in Wadi Matulla covering about 1500 km^2 (20% of the total area of the studied basin) and stretched NNW-SSE. Unconformably, overlying the basement rocks and overlains by Quseir Formation (Fig. 7) and representing the main source of the wadi deposits.

In the study of Abdelkareem and El-Baz (2015c), Radarsat-1, ALOS/PALSAR, and fused optical/radar images (Fig. 8a) allowed revealing the near-surface fault/fracture systems and stream-networks. A fusion of Radarsat-1 and ETM + images clearly visualized the geomorphic and structural features unlike using the optical Landsat images alone in the downstream area. This image noticeably improved the views of incised stream-networks and fault/fracture lines. A fused subset of both ALOS/PALSAR and ETM + displayed high resolution and more diffusion through the sand sheet comparable to the fused Radarsat-1/ETM + image as the L-band of PALSAR data provided more penetration than the C-band and could characterize the hidden fault lines and structural geometry as well (Abdelkareem and El-Baz 2015c). They stated that the east–west lineaments cut the north–south and northwest trends. The overly the paleochannels and active stream-networks during rainfall storms derived from Landsat data and extracted streams from SRTM confirmed the stream capture process and allowed predicting the plausible areas of well abstraction (Fig. 8b).

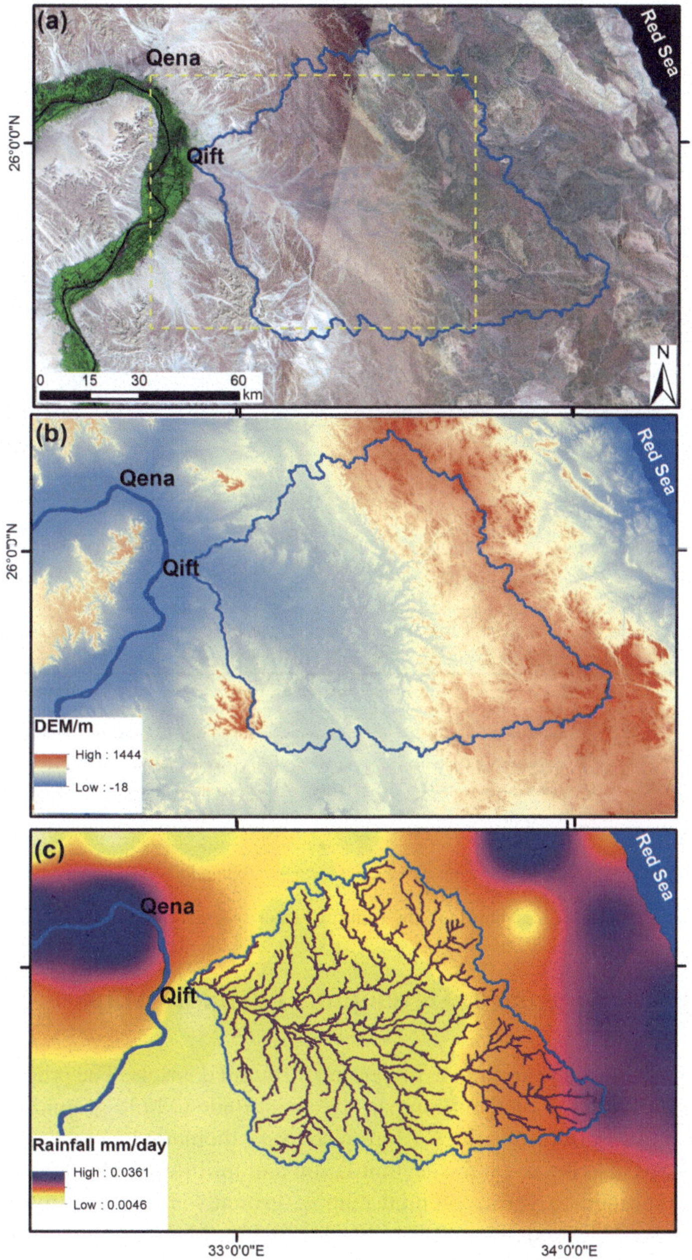

Fig. 6 Wadi Matula; **a** Landsat image overlain by Matula watershed; **b** SRTM DEM overlain by Matula watershed; **c** precipitation data overlain by watershed

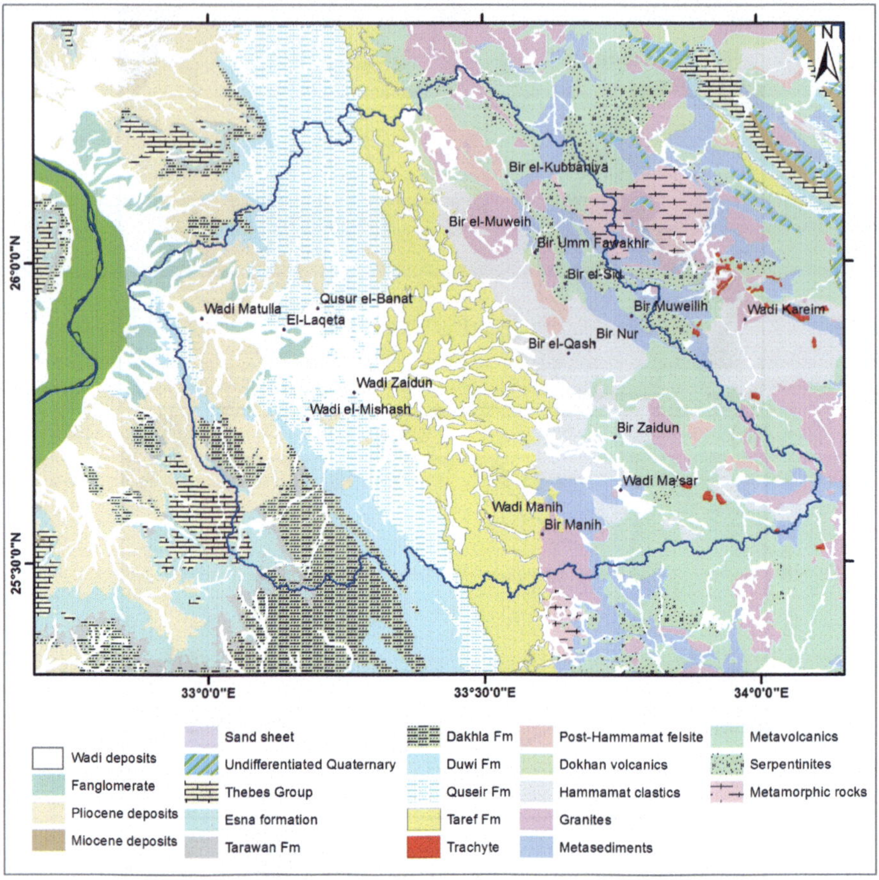

Fig. 7 Geologic map of Wadi Matulla in the Central Eastern Desert of Egypt

4 Analytical Approach: Developing Thematic Maps

GIS-based methods were applied to integrated different datasets. The prepared layers were ordered depending on their comparative magnitude to hold groundwater accumulation and infiltration capacities. Moreover, every thematic layer was categorized into sub-classes also based on the accumulation and infiltration characteristics, then the layers combined to get a potential map of groundwater infiltration capacities (Abdelkareem 2012; Abdelkareem and El-Baz 2015a; Mukherjee and Singh 2020; Ajay Kumar et al. 2020). The thematic layers illustrated below:

Lithologic characteristics are important geological factor manage the permeability, transmissivity and porosity of aquifer lithologic properties (Sreedhar et al. 2009; Abdelkareem et al. 2012a; Benjmel et al. 2020), therefore, surface water infiltration is definitely higher in zones of loose and more porous/permeable

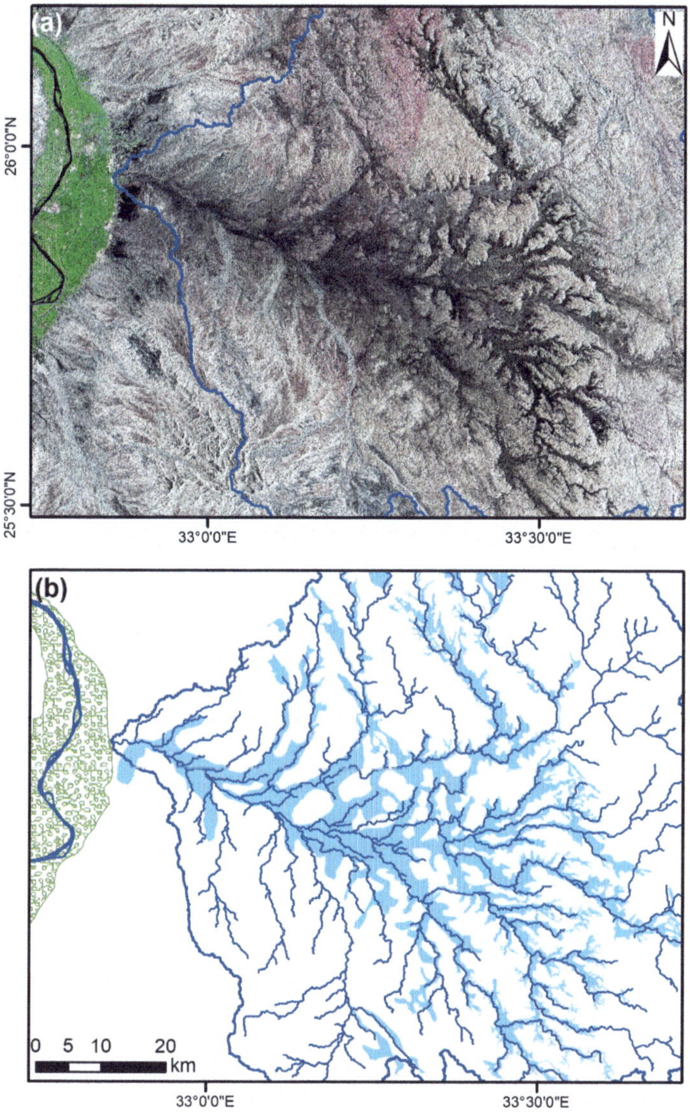

Fig. 8 **a** A fused Radarsat-1/Landsat image revealing incised wadi courses in the downstream area of Wadi Matula; **b** difitized paleochannels that overlain by predicted stream-networks

rocks/sediments rather than massive-dense impermeable rocks. Geological maps of W. Qena and W. Matula were digitized by Arc GIS using Conoco (1987) and EGSMA (1981) maps; it was divided into five categories in W. Qena as follows: wadi deposits, Pliocene/Pleistocene terraces, older basement rocks, sedimentary sequence and younger basement rocks (Fig. 9a) that assigned ranks of 5, 4, 3, 2, and 1, respectively. In Wadi Matula, lithologic map classified into wadi deposits,

Pliocene/Pleistocene, sedimentary sequence and basement rocks that given a rank range from 4 (wadi deposits) to 1 (basement rocks) based on their infiltration capabilities. The wadi bed (deposits) comprise mainly of substantial loose gravel, sand and silt, which characterized by pore spaces among the grains that simply let water to permeate within the subsurface layers, therefore, such conditions are very promising for groundwater recharging capacity, therefore, the permeable lithologic units promote the infiltration of water, through the underlying strata. Inversely, the impermeable rock units are rather promoting the surface runoff (Mukherjee and Singh 2020; Benjmel et al. 2020). The porous and permeable geologic deposits promote recharge and discharge processes; therefore, these areas represent groundwater abstraction wells. Therefore, extracting the loose sediments using Radarsat-1 data can help in characterizing potential areas of infiltration capacity.

Topographic characteristics allow governing the accumulation, direction movements and, therefore, it has an immense impact on groundwater existence. The topographic map (see Fig. 9b) of the areas under investigations area is taken from the SRTM ~1 arc second (30 m spatial resolution) data. The DEM data in W. Qena and W. Matula classified into four subclasses range from 4 (high topography) and 1 (low topography). Topographic density maps for the studied areas were classified the DEM into four groups based on their capacity of groundwater infiltration.

The variation between highest and lowest elevations yields land surface slopes that manage the movement of surface water in the direction of the lowermost topographic area in the south at W. Qena, and toward west at W. Matula as igneous and metamorphic rocks dominate the topmost topographic heights in both wadis. The steepness of land slope inversely related to potential recharge and infiltration process (Surabuddin et al. 2007). Overland flow is usually leisurely in regions with quiet slope, letting more time for infiltration through the soil and eventually recharge and penetrate to the underlying aquifers (Lentswe and Molwalefhe 2020). On the other hand, steeper slopes inversely related to infiltration capabilities (e.g., Simmers 1990; Abdelkareem et al. 2012a). Based on these notions, slope maps of the regions were prepared from the SRTM DEM 1 arc second. A slope density map, then was prepared for each area and grouped into 4 groups; low, moderate, high and very high, respectively that range from 4 (low slope; higher infiltration capacity) to 1 (very high slope; low infiltration capacity).

Precipitation during storms provides substantial amount of water resources that moved in form of surface water and infiltrate to replenish the groundwater aquifers. The TRMM satellite provides important information on the rainfall precipitation which allows monitoring storms, estimates the runoff of surface water, and the active drainage systems in space time to predict prone areas to flash flood hazards. The TRMM data was classified into 4 categories; low, moderate high and very high precipitation, the classes given weight factors range from 1 (low amount of precipitation) to 4 (high amount of precipitation). The higher rainfall intensities, the higher groundwater infiltration (e.g., Abdelkkareem 2012; Mukherjee and Singh 2020).

Lineaments present essential evidence on surface and sub-surface fracture/fault structures, which may play as conduits governing the movement and infiltration capability, which causing secondary porosity and allow for groundwater recharging

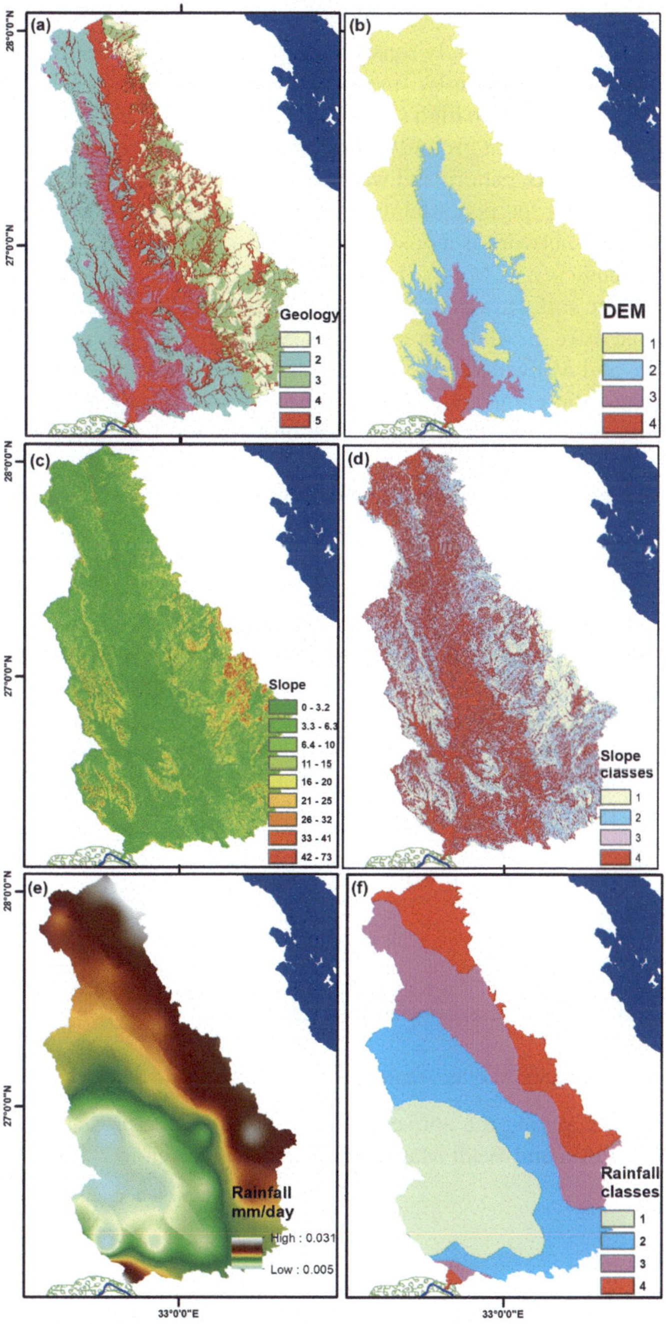

Fig. 9 Thematic maps of W. Qena **a** geology classes; **b** topographic classes derived from SRTM-DEM slope layer derived from SRTM; **c** slope; **d** slope classes; **e** TRMM data; **f** rainfall classes

(Abdelkkareem and El-Baz 2015a; Mukherjee and Singh 2020). Lineaments were manually digitized using remotely sensed data and geological maps through ArcGIS software packages. Subsequently, lineament density map was categorized into 4 classes; low (1); moderate (2); high (3), and very high (4); the area of high lineament density may have significant groundwater recharge and discharge capacity.

The drainage network pattern and watershed catchments that capturing excessive precipitation revealing the potential areas of infiltration and recharging zones. The density of drainage-network is a positively related to runoff and inversely related to the recharge processes (e.g., Mukherjee and Singh 2020) that largely depending on the permeability of the soil. However, zones with high drainage density signified highly infiltration capacity (e.g., Abdelkareem 2010; Abdelkareem et al. 2012a; Pande et al. 2018; Lentswe and Molwalefhe 2020). The drainage networks is realized by using the D8 (deterministic eight-node) flow routing algorithm (O'Callaghan and Mark 1984). Then, the drainage density map was classed into four categories; low (1), moderate (2), high (3) and very high (4).

The reclassified thematic maps (i.e., lithology, topography, slope, precipitation, lineaments, drainage networks, and soil derived from radar) were combined based on their given scores using the weighted overlay technique. Such method is also known as Multi-Criteria Evaluation (MCE; Voogd 1983). In this model, all the thematic maps are given the same weight; however, the inter-map classes in each input map are given different weights that mostly range from 1 to 4 where 4 is the most promising and 1 is the slightest based on the ability to hold and infiltrate water. The rank of every inter-map feature is divided by the total classes to generate the feature/sub-class weight (Sc). These weights are multiplied by the corresponding map weight in every thematic map to obtain the groundwater infiltration zones map (Eastman et al. 1995; Eastman 1996) using Eq. (1):

$$\mathrm{GWPF} = \sum \mathrm{Wi} * \mathrm{Sc} \tag{1}$$

where

GWP Groundwater Potential zone of infiltration
Wi map weight
Sc weight of inter-map class/feature.

The results of applying the above methodologies will be presented in two areas including Wadi Qena and Wadi Matula in the ongoing sections.

5 Wadis

5.1 Wadi Qena

Based on the composition and physical properties of the lithologic units at WQ, the lithologic density map categorized into five classes (Fig. 9a) range from 1 (low infiltration capacity) to 5 (high infiltration). The promise areas for water accumulation and infiltration are the main course of the wadi and areas of highly weathered and erosion deposits in the north east of the WQ catchment as they are characterized by high porosity and permeability. SRTM DEM showed that the elevation of WQ ranges from 67 to 1047 m (above sea level). The DEM density map (Fig. 9b) at the downstream area is the lowest part that would receive substantial amount of surface water. Moreover, the difference in elevation in Wadi Qena spans from 71 to 1866 m (above sea level) initiate the slope. The slope density map (Fig. 9c, d), revealed that the northeastern catchments and western parts of the area probably have poor groundwater recharge due to high slope gradient, that yields runoff, particularly in the Red Sea highlands. However, the majority of the wadi characterized by very gentle to gentle slopes (<5°) particularly along the main course. In this area, the wadi deposits probably are promise area for groundwater infiltration as a result of gentle slope. Rainfall precipitation data derived from TRMM revealed that most of the catchment areas (basement rocks) to east and north are receiving rainfall, particularly, during the rainy storms. Crucially, the higher precipitation is correspondence to high altitude, whereas the areas of lower altitudes receive low precipitation (Fig. 9e, f). The computed drainage network density map (Fig. 10a) displays that most of areas of high density consistent to the main courses of WQ that promoting infiltration capability, while, the majority of basement catchments areas to east are characterized by low-drainage density (Fig. 10b). This model has been implemented here, as high density of drainage networks will be given higher grade in judgement to low density or absent channel courses. The lineaments and lineament density map (Fig. 10c, d) showed that wide portion of the basement rocks of the Red Sea high lands are extensively deformed, fractured and faulted. This allows the more infiltration as a result of increasing the permeability and water conductivity through the fracture/fault zones rather than massive rocks. Radarsat-1 data allowed characterizing loose sediments opposite to bed rocks (Fig. 10e, f). Such deposits' mean vertical infiltration rate is approximately 2.6 m/day (Farrage et al. 1996). However, to east whereas the basement catchments are occurring, runoff is high.

Using the natural breaks/Jenks algorithm in the GIS platform, the prospective groundwater recharging map was categorized into seven categories; absent/barren, very low, low, moderate, high, very high and excellent zones, covering areas, 3.81, 18.20, 19.73, 22.27, 19.94, 12.49, and 3.56% of the study area (Fig. 11a). The potential areas of groundwater recharging are extends along the main wadi course at the northern and southern parts of the wadi. In order to verify the accuracy of the map, the high areas of recharging capacity compared to land-cover image derived from Landsat-8. The results revealed that the areas cover by vegetation as they cultivated

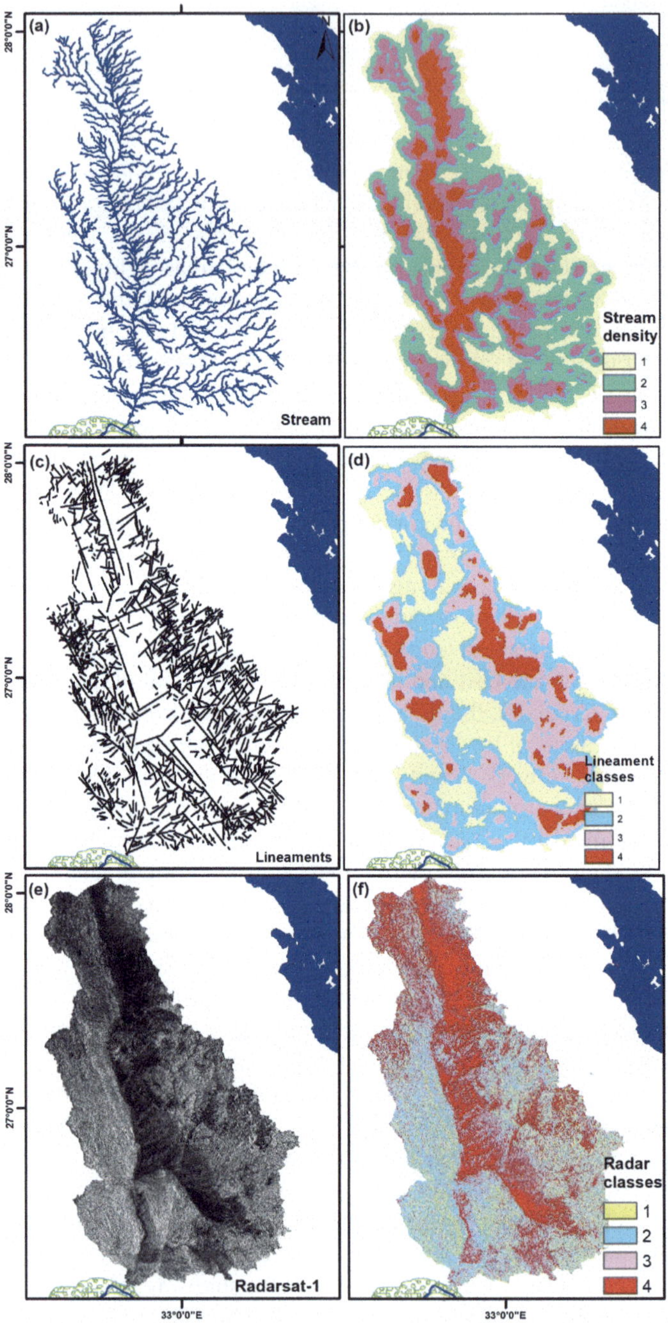

Fig. 10 **a** Drainage network; **b** drainage density; **c** lineament derived map; **d** lineament density map; **e** radar intensity; **f** radar classes

using groundwater. Moreover, the collected groundwater well data (Fig. 12) are consistent to the high areas of groundwater infiltration. Based on field observation, the areas of high groundwater recharging are also characterized by highly discharge process. Therefore, several farmers have been utilizing the groundwater from the quaternary aquifer to irrigate several crops such as wheat, alfalfa, Cantalob, grapes and tomatoes at the southern WQ (Fig. 12a–d).

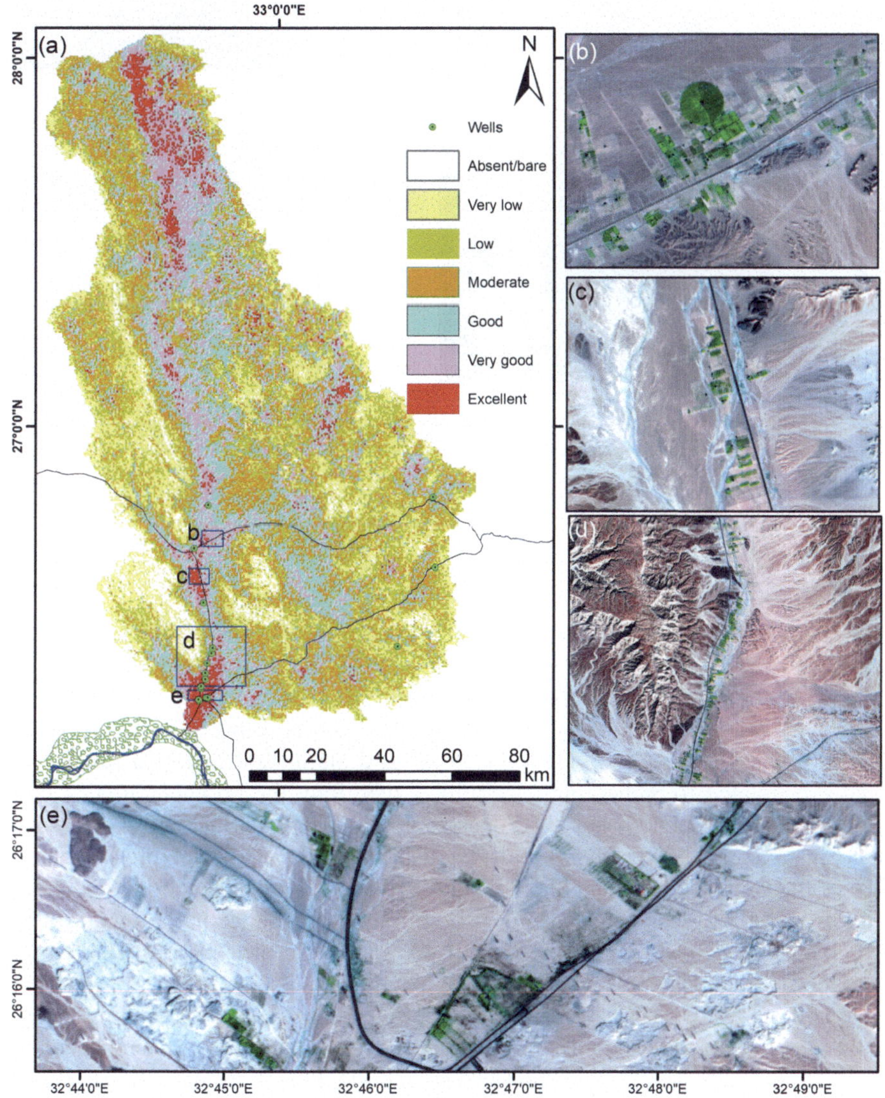

Fig. 11 **a** Groundwater prospective recharging zones map of Wadi Qena; **b**, **c**, **d**, and **e** subsets of Sentinel-2 revealing agricultural activities along WQ

Fig. 12 Field photographs of southern Wadi Qena; **a** a pumped productive water-well irrigating wheat crop; **b** excavated water trench through wadi deposits; **c** alfalfa crop at southern Wadi Qena; **d** cantalope crop in Wadi Qena

5.2 Wadi Matula

The results of SRTM DEM (30 m resolution) analyses, the variation in topography ranges from 73 to 1100 in W. Matula (Fig. 13a). In the topography density map, the downstream area is given the highest weight (4) as it represents the lowermost section that would collect large amount of water during rainy storms (Fig. 13a). The difference in elevation yields the slope as a significant factor in runoff and negatively related to recharging capabilities, therefore, the probable area of high recharging given weight 4 but low zones given 1 (Fig. 13b). Such areas represents the main course and downstream of the wadi. The main course that represents about 18% of the watershed area is filled by wadi deposits and representing the potential zone of recharging capacity that given high rank opposite to the basement rocks that cover about 36% of the area (Fig. 13c). Although, the water divides on the basement rocks that represent areas of high runoff, it received high precipitation. Areas of high rainfall (Fig. 13d) are representing a high recharging and given high rank (4) of infiltration capability rather than low precipitation (1). Lineaments map (Fig. 13d) of Wadi Matula also represents an important factor in characterizing areas of high infiltration. Therefore, the areas of the highest density (Fig. 13e) assigned higher weight factor (4) than lowermost (1). Furthermore, the delineated drainage network density map (Fig. 13f, g) shows that most of downstream areas are high density that facilitates the infiltration processes, while, the outline catchment areas are characterized by low-drainage density (Fig. 13g).

The results of the integration of many factors that control the movements and infiltration capacities of precipitation during the rainfall storms are influenced many hydrologic/geologic and geomorphic features such as topography, slope, geology, rainfall, lineaments, and stream-networks at the present chapter (Fig. 14). The best zones for groundwater accumulation and infiltration map, then, classified into six prospective zones; excellent, very good, good, moderate, low, and very low zones that covering approximately, 5.60, 16.39, 24.33, 23.79, 19.52 and 10.36% of the study area, respectively (Fig. 14a). Crucially, the zone of excellent groundwater infiltration/accumulation is consistent with the locations of the cultivated areas (Fig. 14b, c) and fitting to the collected groundwater wells (Fig. 14d, e). Therefore, the most appropriate zone for conducting of future abstraction/drilling is concordant with the zone of high recharging zone.

During present sporadic showers, Wadi Qena and Matula received substantial amount of surface water from several tributaries coming from the Red Sea highlands. The Red Sea Mountains represent the elevated part of the present study that includes crystalline rocks as the main catchment area for the basin. The rainy storms provide valuable water resources, the excessive water form flood of water that provisionally accumulated in lower lands as ponds and lakes (Fig. 15a) that infiltrate to the below aquifers. Storing such water in man-made reservoirs would provide new water resources plus the Nile River in Egypt that replenish the agriculture activities. Significantly, such source of water is very important for water security and sustainability. The sizable thickness of loose sediments that filling the wadi bed can allow for high infiltration capacity along channel beds in areas of high precipitation.

Importantly, these present defunct wadis recharged during the past pluvial conditions as the rest of the Sahara (Sanford 1929; Butzer and Hansen 1968). This allowed deposition of local sands and gravels in the forms of wadi deposits, fans, beds that distributed along the downstream (Butzer and Hansen 1968) and received substantial water amount that might have been stored below strata as fossil water (El-Baz 1998). Having in mind the sizable quantity of fluvial erosion that has performed, shattering of the plateau must be done before the deposition of fluvial deposits. Breaking apart of the plateau topped by limestone would be allowed accelerating the erosion as rain by itself will not erode a cliff (Sanford 1929) as the erosion has modified the structural elements (Fig. 16). This probably happened during Late Tertiary or later (Abdelkareem and El-Baz 2015a, b). During that time, the geometry of Equator was north (e.g., at the present-day latitudes of Sudan and Chad) than its present latitude, this would have promoted heavy rainfall conditions over Egypt (Abdelkareem et al. 2012b) that reflected the existence of fossil water.

The Red Sea rifting events activities caused disturbances in drainage orientations, topography, and fracture systems that provided evidences for structural piracy resulting of active movements Addelkareem and El-Baz (2015a). The formation of the E–W structural elements (fault/fracture systems) intersected and captured the N–S and NW–SE trends, provided further evidence for structural reactivation. The structural elements increase the porosity and permeability and play as conduits for

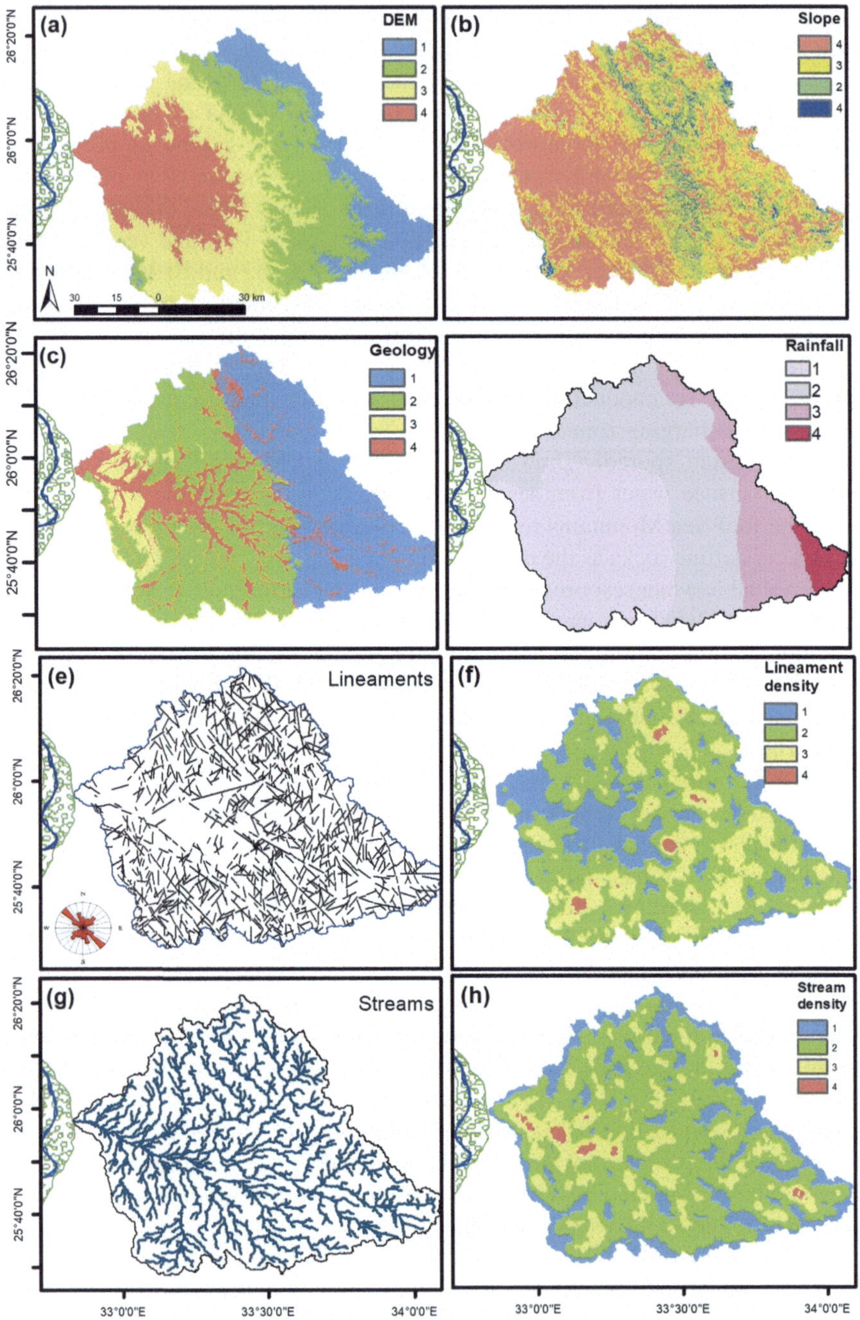

Fig. 13 **a** SRTM-DEM classes; **b** slope classes; **c** geology classes; **d** precipitation classes; **e** lineament derived map; **f** lineament density map; **g** drainage networks; **h** drainage density

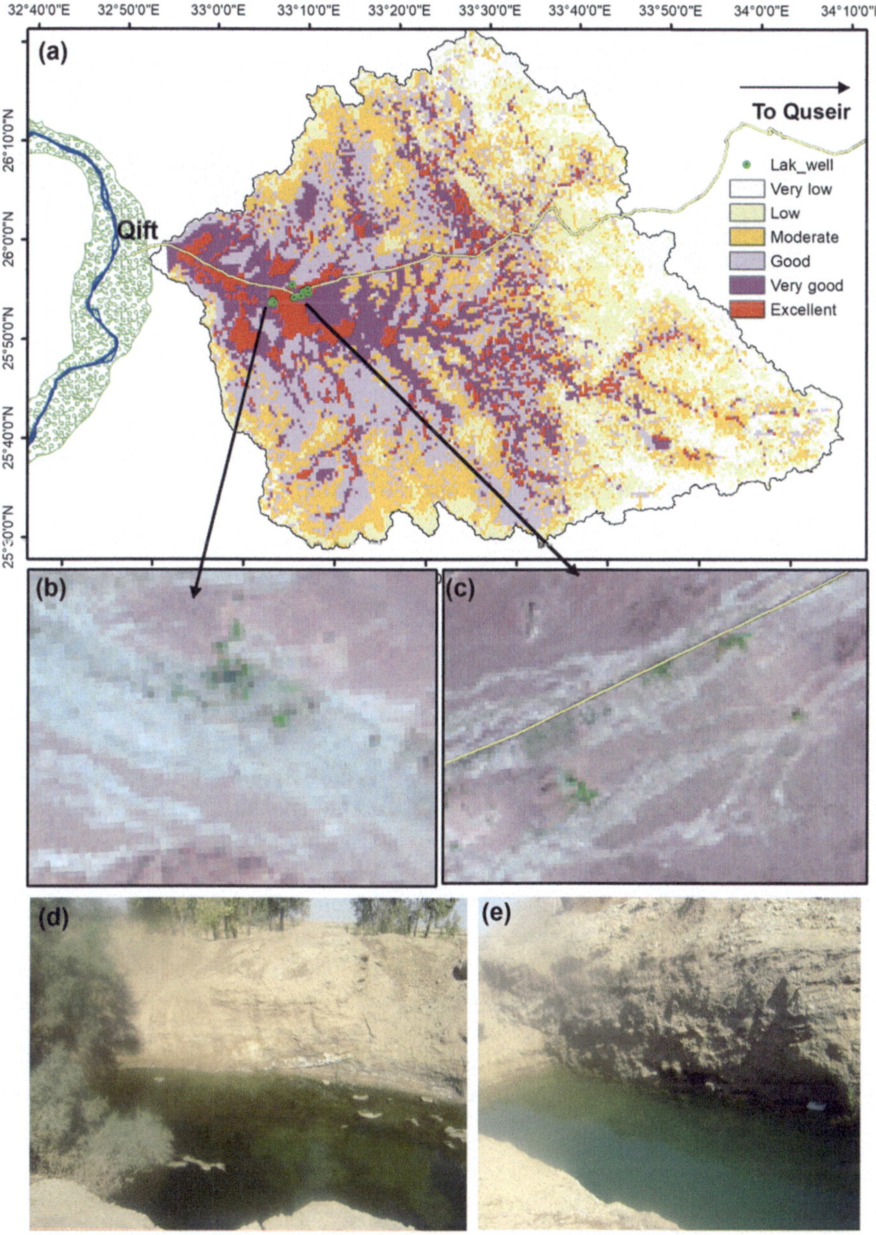

Fig. 14 Wadi Matula **a** groundwater prospective recharging zones map; **b, c** Landsat data revealing agricultural activities; **d, e** groundwater excavation trenches that consistent to excellent zone of recharging; **f, g** water accumulation during heavy rainfall, Qift-Quseir road

Fig. 14 (continued)

recharging and discharging processes. For example, the shear zone of Wadi Qena represents an important part for recharging and discharging. As a naturally flow water pipe in 1998 discharging quantities of water about 3450 m^3/day was (Abdel Moneim 2014, and references therein); however our estimation in November 2015 is about 2160 m^3/day.

6 Conclusions

In this study, a GIS-based knowledge driven approach was utilized to combine evidential maps derived from remotely sensed data and geologic maps for delineating groundwater potentiality via recognizing the recharge zones. Lithology, rainfall, topography, radar intensity, drainage density, lineament density, and slope evidential maps, were integrated for predicting groundwater recharging zones in W. Qena and W. Matula basins, in the Eastern Desert of Egypt. Validation of results using field data and satellite images showed that the knowledge-driven method performed reasonable prediction of the recharge potential of the groundwater aquifers. High recharge zones cover 3.56 and 10.36% of the areas of the catchments of Qena, and Matula basins. Such zones resemble to sites covered by thick fluvial deposits, highly fractures zones in low slope, and low topographic terrains. The groundwater recharge potentiality has been verified by matching with field data and agricultural areas derived from satellite images. The results obtained may be helpful for sustainable water resources in arid regions.

7 Recommendations

The main recommendations that esteemed from the present study are as follows:

- The use of remotely sensed data is necessary prior to any geophysical groundwater exploration and is an essential tool for decision-making.

Fig. 15 **a** Photo taken on 30–1–2010 reveals water accumulation after heavy rainy storm; **b** thick accumulation of sand and gravel deposits at Wadi Qena

- It is, advisable to combine geologic, structural and hydrologic data through numerical modeling and GIS techniques to provide fast and costly-effective information on the potential groundwater zones.
- Harvesting the precipitated data during rainy storms in man-made reservoirs would provide additional new water resources, and it is essential to maintain water security.

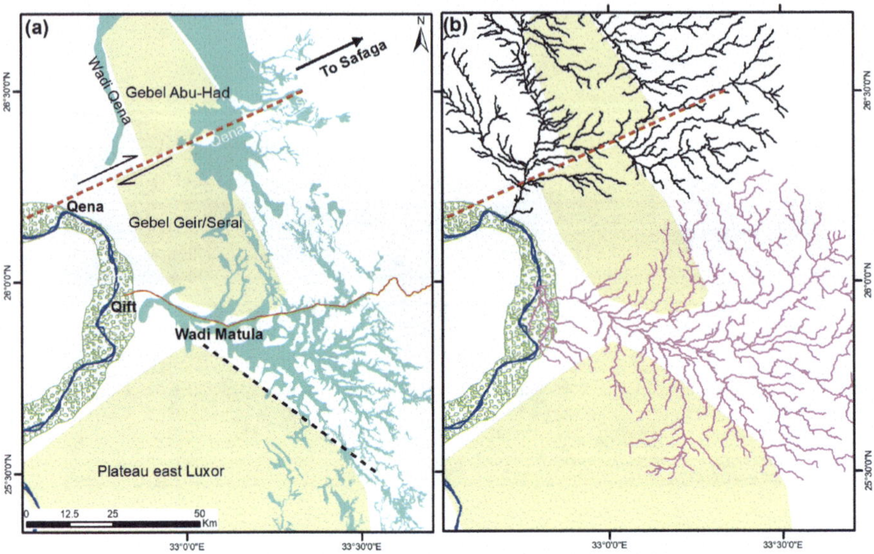

Fig. 16 Illustrations of the structural discontinuities between south Wadi Qena and Wadi Matula. **a** Digitized paleochannels derived from Radarsat-1 image and the right lateral movement of the blocks east of Qena: Qena-Safaga Shear Zone indicated by dashed red lines; **b** the relationship between the stream-networks of W. Qena and Matula, and structural blocks

References

Abdalla F (2012) Mapping of groundwater prospective zones using remote sensing and GIS techniques: a case study from the Central Eastern Desert Egypt. J Afr Earth Sci 70:8–17

Abdelkareem M (2010) Qena valley evolution, Eastern Desert Egypt. EGU 12:6869

Abdelkareem M (2012) Space data and GIS applications for arid region: Wadi Qena. Unpublished Ph.D, South Valley University, Egypt, p 251p

Abdelkareem M (2017) Targeting flash flood potential areas using remotely sensed data and GIS techniques. Nat Hazards J 85(1):19–37

Abdelkareem M, El-Baz F (2015a) Analyses of optical images and radar data reveal structural features and predict groundwater accumulations in the central Eastern Desert of Egypt. Arab J Geosci 8:2653–2666

Abdelkareem M, El-Baz F (2015b) Regional view of a trans-African drainage system. J Adv Res 6:433–439

Abdelkareem M, El-Baz F (2015c) Evidence of drainage reversal in the NE Sahara revealed by space-borne remote sensing data. J Afr Earth Sci 110:245–257

Abdelkareem M, El-Baz F (2016) Mode of formation of the Nile Gorge in northern Egypt by DEM-SRTM data and GIS analysis. Geol J 51(5):760–778

Abdelkareem M, El-Baz F (2017) Remote sensing of Paleodrainage systems west of the Nile River Egypt. J Geocarto Int 32(5):541–555

Abdelkareem M, El-Baz F, Askalany M, Akawy and Ghoneim E (2012a) Groundwater prospect map of Egypt's Qena Valley using data fusion. Int J Image Data Fus 3(2):169–189

Abdelkareem M, Ghoneim E, El-Baz F, Askalany M (2012b) New insight on paleoriver development in the Nile basin of the eastern Sahara. J Afr Earth Sci 62:35–40

Abdelkareem M, Ghoneim E, El-Baz F, Askalany M (2010) Did Egypt's Wadi Qena drain northward during its earlier stages? Geol Soc Am Abstr 42(5):645

Abdel Moneim AA (2014) Hydrogeological conditions and aquifers potentiality for sustainable development of the desert areas in Wadi Qena, Eastern Desert, Egypt. Arab J Geosci 7:4573–4591

Ajay Kumar V, Mondal NC, Ahmed S (2020) Identification of groundwater potential zones using RS, GIS and AHP Techniques: a case study in a part of deccan volcanic province (DVP), Maharashtra India. J Indian Soc Remote Sens 48:497–511. https://doi.org/10.1007/s12524-019-01086-3

Al-Djazouli MO, Elmorabiti K, Rahimi A, Amellah O, Fadil OAM (2020) Delineating of groundwater potential zones based on remote sensing, GIS and analytical hierarchical process: a case of Waddai, eastern Chad. Geo J. https://doi.org/10.1007/s10708-020-10160-0

Andualema TG, Demekeb GG (2019) Groundwater potential assessment using GIS and remote sensing: a case study of Guna tana landscape, upper blue Nile Basin Ethiopia. J Hydrol Reg Stud 24:100610

Benjmel K, Amraoui F, Boutaleb S, Ouchchen M, Tahiri A, Touab A (2020) Mapping of groundwater potential zones in crystalline terrain using remote sensing, GIS techniques, and multicriteria data analysis (case of the Ighrem Region, Western Anti-Atlas, Morocco). Water 12:471

Berhanua KG, Hatiyeb SD (2020) Identification of groundwater potential zones using proxy data: case study of Megech watershed Ethiopia. J Hydrol Reg Stud 28:100676

Butzer KW, Hansen CL (1968) Desert and river in Nubia: geomorphology and prehistoric environments at the Aswan Reservoir. University of Wisconsin, p 562

Çelik R (2019) Evaluation of groundwater potential by GIS-based Multicriteria decision making as a spatial prediction tool: case study in the Tigris River Batman-Hasan key Sub-Basin Turkey. Water 11:2630. https://doi.org/10.3390/w11122630

Conoco (1987) Geological map of Egypt scale 1:500,000. The Egyptian General Petroleum Corporation, Cairo

Eastman JR (1996) Multi-criteria evaluation and GIS. In: Longley PA et al (eds) Geographical information systems, vol 1, 2nd edn. Wiley, New York, pp 493–502

Eastman JR, Jin W, Kyem AKP, Toledano J (1995) Raster procedure for multi-criteria/multi-objective decisions. Photogramm Eng Remote Sens 61:539–547

EGSMA (1981) Geological map of Egypt scale 1: 2,000,000. The Egyptian Geological Survey and Mineral Authority, Cairo, Egypt

El-Baz F (1998) Sand accumulation and groundwater in the Eastern Sahara. Episodes 21:147–151

Farrage AA, Ibrahim HA, Attia MM (1996) Evaluation of the soil and groundwater resources in Wadi Attwani and Qusur El-Banat areas, Qift-Quseir road, Eastern Desert, Egypt. In: 1st international conference on basic sciences and advanced technology, Assuit University, Assuit, Egypt

Freeze RA, Cherry JA (1979) Groundwater. Prentice-Hall, New Jersey

Gheith H, Sultan M (2002) Construction of a hydrologic model for estimating wadi runoff and groundwater recharge in the Eastern Desert Egypt. J Hydrol 263(1–4):36–55

Ghosh D, Mandal M, Banerjee M, Karmakar M (2020) Impact of hydro-geological environment on availability of groundwater using analytical hierarchy process (AHP) and geospatial techniques: a study from the upper Kangsabati river basin. Groundw Sustain Dev 11:100419

Haynes CV (1980) Geological evidence of pluvial climates in the Nabta area of the Western Desert, Egypt. In: Wendorf F, Schild R (eds) Prehistory of the Eastern Sahara. Academic Press, New York (NY), pp 353–371

Healey RW (2010) Estimating groundwater recharge. Cambridge University Press, New York

Lentswe GB, Molwalefhe L (2020) Delineation of potential groundwater recharge zones using analytic hierarchy process-guided GIS in the semi-arid Motloutse watershed, eastern Botswana J Hydrol Reg Stud 28:100674

Mukherjee I, Singh UK (2020) Delineation of groundwater potential zones in a drought-prone semi-arid region of east India using GIS and analytical hierarchical process techniques. CATENA 194:104681

Murmu P, Kumar M, Lal D, Sonker I, Singh SK (2019) Delineation of groundwater potential zones using geospatial techniques and analytical hierarchy process in Dumka district, Jharkhand, India. Groundw Sustain Dev 9:100239. https://doi.org/10.1016/j.gsd.2019.100239

O'Callaghan JF, Mark DM (1984) The extraction of drainage networks from digital elevation data. Comput Vision Graph Image Proc 28:323–344

Paillou P, Schuster M, Tooth S, Farr T, Rosenqvist A, Lopez S, Malezieux JM (2009) Mapping of a major paleodrainage system in eastern Libya using orbital imaging radar: the Kufrah River. J Earth Planet Sci Lett 277:327–333

Pande CB, Khadri SFR, Moharir KN, Patode RS (2018) Assessment of groundwater potential zonation of Mahesh River basin Akola and Buldhana districts, Maharashtra, India using remote sensing and GIS techniques. Sustain Water Resour Manag 4:965–979

RIGW (1993) Hydrogeological map of Egypt, Asyuti Sheet. Research institute for groundwater, Academy of scientific research and technology: Cairo, Egypt

Said R (1981) The geological evolution of the River Nile. Springer, New York (NY), p 151

Said R (1990) The geology of Egypt. A.A. Balkema, Rotterdam, Brookfield, p 734

Sandford SK (1929) The Pliocene and Pleistocene deposits of Wadi Qena and of the Nile Valley between Luxor and Assiut (QAU). Q J Geol Soc 85:493–548

SimmersI (1990) Aridity, groundwater recharge and water resources management. Verlag Heinz Heise, Hanover

Sreedhar G, Kumar GTV, Krishna IVM, Kahya E, Demirel MC (2009) Mapping of groundwater potential zones in the Musi basin using remote sensing data and GIS. Adv Eng Softw 40:506–518

Sultan M, Yan E, Sturchio N, Wagdy N, Wagdy A, Gelil KA, Becker R, Manocha N, Milewski A (2007) Natural discharge a key to sustainable utilization of fossil groundwater. J Hydrol (amsterdam) 335(1–2):25–36

Surabuddin MM, Pandey AC, Garg RD (2007) Groundwater prospects evaluation based on hydrogeomorphological mapping using high resolution satellite images: a case study in Uttarakhand. J Indian Soc Remote Sens 36:69–76

Voogd J (1983) Multicriteria evaluation for urban and regional planning. Delftsche Uitgevers Maatschappij, Delft. https://doi.org/10.6100/IR102252

Woodward JC, Macklin MG, Krom MD, Williams MAJ (2007) The Nile: evolution, quaternary river environments and material fluxes. In: Gupta A (ed) Large rivers: geomorphology and management. Wiley, Chichester, 13:712

Sustainability of Groundwater in the Nile River Valley: Is It Possible After the Construction of GERD?

Mustafa El-Rawy and Abdelazim M. Negm

Abstract Because it is recharged from the Nile River by seepage from canals and excess agricultural water, the Nile Valley and Delta's shallow aquifer is considered to be a renewable water source extracted from shallow wells. The Grand Ethiopian Renaissance Dam (GERD) may cause a major problem for Egypt's groundwater resources by reducing the flow of the Nile to Egypt. The concept that improper GERD operation will negatively affect groundwater along the Nile valley aquifer and reduce or stop its sustainable nature is investigated in this chapter. Therefore, this chapter will summarize the key findings of the most current studies on the influence of GERD on the Nile River Valley's groundwater aquifer. Furthermore, it presents the results of a MODFLOW simulation to assess the impact of filling the GERD reservoir over a 10 year period using three different flow conditions: normal flow (NF), minimum of average flow (MAF), and minimum flow (MF) from the Blue Nile, as this is Egypt's preferred scenario of filling. The average groundwater levels decreased by 0.15 m, 0.573 m, and 1.25 m for NF, MAF, and MF, respectively, after 10 years of GERD filling with the NF, MAF, and MF conditions, compared to the base case. It can be stated that filling the GERD reservoir in 10 years may be the best solution, particularly in normal or lowest average flow conditions; otherwise, the aquifer's sustainability will be jeopardized.

Keywords Grand Ethiopian Renaissance Dam · Sustainability · Nile Valley aquifer · Groundwater level · Egypt

---

M. El-Rawy (✉)
Civil Engineering Department, Faculty of Engineering, Minia University, Minia 61111, Egypt
e-mail: mustafa.elrawy@mu.edu.eg

Civil Engineering Department, College of Engineering, Shaqra University, Dawadmi, Ar Riyadh 11911, Saudi Arabia

A. M. Negm
Water and Water Structures Department Engineering, Faculty of English, Zagazig University, Zagazig 44519, Egypt
e-mail: amnegm@zu.edu.eg

A. M. Negm and M. El-Rawy (eds.), *Sustainability of Groundwater in the Nile Valley, Egypt*, Earth and Environmental Sciences Library,
https://doi.org/10.1007/978-3-031-12676-5_13

1 Introduction

Egypt suffers from inadequate water supply and water quality deterioration, which represents a serious concern for the development of different sectors of industry, agriculture, municipality, tourism and the environment. Egypt has been considered one of the countries that are suffering from water scarcity by 2025 due to the rapidly growing in population (Diana 2000).

Egypt's water resources are restricted to the Nile River, which serves as the country's key source of freshwater. The Nile flow rate is determined by the amount of water collected in Lake Nasser to meet Egypt's annual water allocation, which was set at 55.5 billion m^3/year (BCM) in a 1959 agreement made with Sudan (El Arabi and Dawoud 2012; El-Rawy et al. 2020a). The second main source is the groundwater ranging from 7 to 8 BCM/year (El-Rawy et al. 2020a; El-Rawy and De Smedt 2020), where most of this amount is abstracting from the Nile River aquifer. The Nile River aquifer is considered a renewable source of water, extracted mainly from shallow wells along the Nile River, as it is supplied from the Nile River by seepage from canals and deep percolation as a result of flood irrigation. It is a highly productive, sustainable source (Attia and Lennaerts 1989; El-Rawy et al. 2020a). In many areas of Egypt, groundwater is the main source of agricultural and water supply. Over the last few years, rising abstraction to meet the growing demand for residential rations and the expansion of desert regeneration has raised questions about groundwater availability as a resource (Arabi and Dawoud 2012). In most Egyptian villages, groundwater is regarded as the principal source of drinking and domestic water. Egyptians are frequently confronted with a big dilemma in Ethiopia, where a variety of dams are being built that threaten Egypt's Nile water share and the long-term viability of groundwater in the Nile River Valley aquifer (NRVA). In April 2011, Ethiopia's government announced plans to construct the Grand Ethiopian Renaissance Dam (GERD), also known as the El-Nahda Dam (Tesfa 2013; GERD Fact Sheet 2014). The GERD is located 15 km from Sudan's border with Ethiopia on the Blue Nile. With a total and active storage volume of 74 BCM, the GERD will be Ethiopia's largest dam, measuring 1800 m long and 155 m high (Impregilo 2017; Sharaky 2014). Figure 1 illustrates a map depicting the location of GERD. The dammed water will drive the largest hydropower energy plant in Africa by 6000 MegaWatt (MW) (Tesfa 2013). Ethiopia has Downsized GERD's installed capacity to slash the installed capacity to 5150 MW (Heubl 2020). On the other hand, the building of the GERD is a major problem facing Egypt (Negm et al. 2018; Negm and Abdel-Fattah 2019; El-Rawy et al. 2021a).

The GERD's improper operation and certain failure might have a cascade of consequences for Sudan and Egypt (Soliman et al. 2017; Negm et al. 2018; Negm and Abdel-Fattah 2019). The effects of GERD on Egypt's water supplies and the fall in the water level in Nasser Lake have been investigated in several studies (Negm and Abdel-Fattah 2019). In conclusion, the findings of the previous analysis revealed that GERD has the following impact on Egypt:

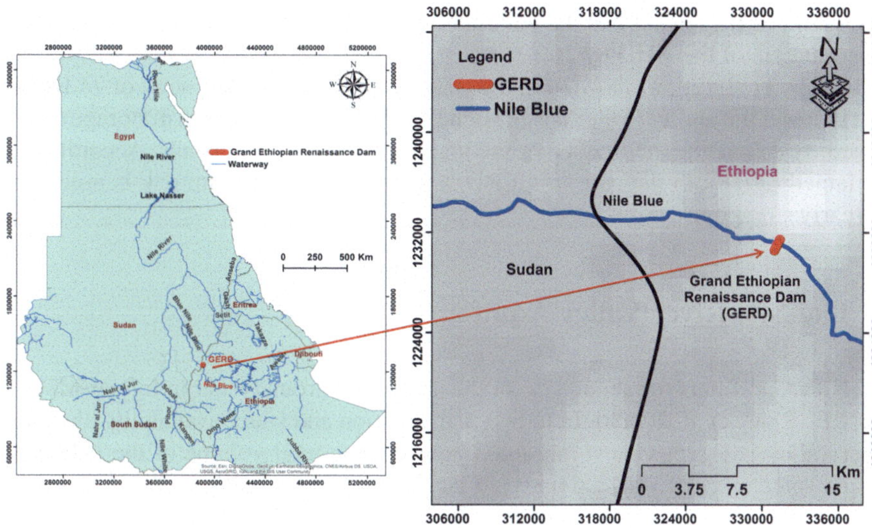

Fig. 1 Location of the grand ethiopian renaissance dam (GERD)

- According to Ismail (2013), the drop in AHD outflow will have an impact on irrigation pump stations along the Nile River.
- Mulat and Moges (2014) concluded that filling of GERD at normal flow case during 6 years has negative consequences on Egypt.
- According to Ramadan et al. (2015), GERD filling at normal Blue Nile flow for 2, 3, and 6 years reduces the water level of Lake Nasser by 37.263, 25.413, and 13.287 BCM/year. The Nasser Lake's water level is reduced by 45.105, 37.814, and 25.963 BCM/year by the GERD filling at an average minimum flow from the Blue Nile over 2, 3, and 6 years. Furthermore, the results showed that filling the GERD at lowest flow for 2, 3, and 6 years will reduce Nasser Lake storage by 55.138, 54.415, and 44.398 BCM/year, respectively.
- According to Abdelkader and Elsanabary (2015) and Elsanabary and Abdelkader (2018), the filling of the GERD reservoir causes a water shortage in Egypt and Sudan for an extended period of time, affecting agricultural lands and causing the relocation of millions of families.
- Kahsay et al. (2015) indicated that the GERD filling causes economic losses in Egypt, especially during the dry years. When the GERD becomes operational, the GERD's detrimental impacts on the Egyptian economy are reversed.
- According to Soliman et al. (2017), the failure of the GERD will be disastrous for Sudan, particularly Al Khartoum City. The AHD is also in jeopardy, but this is dependent on the water level in Lake Nasser at the time of the GERD collapse. If Lake Nasser reaches or approaches capacity, the dam body and all crossing structures from Aswan to Alexandria, as well as land and human lives, will be jeopardized.

- Ibrahim (2017) stated that the building of GERD would reduction the electricity production of the AHD by 20%.
- Sharaky et al. (2017) shown that the GERD, with a storage capacity of 74 BCM, will lower the ability of the Aswan High Dam to serve as a long-term storage reservoir by enhancing the frequency and magnitude of water shortfalls. According to Hamed (2018), in the case of GERD, a very high downstream risk is used as a yearly storage reservoir.

2 The Nile River Valley Aquifer

The NRVA accounts for about 87% of the abstracted groundwater in Egypt (MWRI 2005; El-Rawy et al. 2020a). In terms of abstraction and water availability, the Nile River valley is Egypt's most important aquifer. The total volume of the extracted groundwater in 2010 was estimated to be 6.2 BCM/year (MWRI 2012; El-Rawy et al. 2020a). This aquifer is mainly recharged by the Nile River, its branches and canals, and return irrigation water (El-Rawy et al. 2020a). The NRVAs located in two different parts in Egypt (Delta and upper Egypt). The Nile Delta is a 25,000 km^2 watershed in northern Egypt (Fig. 2). The Nile River runs through it on the south, the Mediterranean Sea on the north, the Suez and Ismailia canals on the east, and the El Nubaria canal on the west (MWRI 2014). The Nile Delta aquifer recharges about 6.78 BCM/year from the surface water (FAO 2013; El-Rawy et al. 2020a). The total outflow from the aquifer by pumping wells and drainage system is about 4.6 BCM/year (El Arabi 2012; El-Rawy et al. 2020a). The aquifer's saturated thickness ranges from 190 to 350 m (Al-Agha et al. 2015). The aquifer is categorized as a high permeable aquifer due to the high hydraulic parameters, where the transmissivity value is about 25,000 m^2/day (El Tahlawi et al. 2008; El-Rawy et al. 2020a). The Nile valley aquifer located in the Upper Egypt extended along the Nile River, from the Nasser Lake at the southern border of Egypt to the Nile River at Cairo (Fig. 2). The aquifer's saturated thickness ranges from 10 m at Luxor to 300 m at Sohag (Kamal 2004). The depth to groundwater in the Nile Valley aquifer varies from 3 to 20 m (Abdalla and Shamrukh 2016; El-Rawy et al. 2020a). The hydraulic conductivity ranges from 60 to100 m/day and the transmissivity values range from 2000 to 20,000 m^2/day (Abd El-Bassier 1997; El-Rawy et al. 2019a, b; El-Rawy et al. 2020a).

To assess the impact of GERD filling on groundwater aquifers, surface water-groundwater (SW–GW) interactions must be investigated. SW–GW interactions have been widely simulated by MODFLOW (Barlow and Harbaugh 2006; Furman 2008; El-Rawy et al. 2016, 2019a, b, 2020b, 2021a, b; Al-Maktoumi et al. 2016, 2018, 2020; Salem et al. 2018, 2020; El-Zehairy et al. 2018; Awad et al. 2020; Abdelrady et al. 2020). MODFLOW-2005 (Harbaugh 2005) was used in this chapter to assess the SW-GW interaction to study the impacts of lowering the water stage in the Nile River and its branches and canals on groundwater levels in the NRVA.

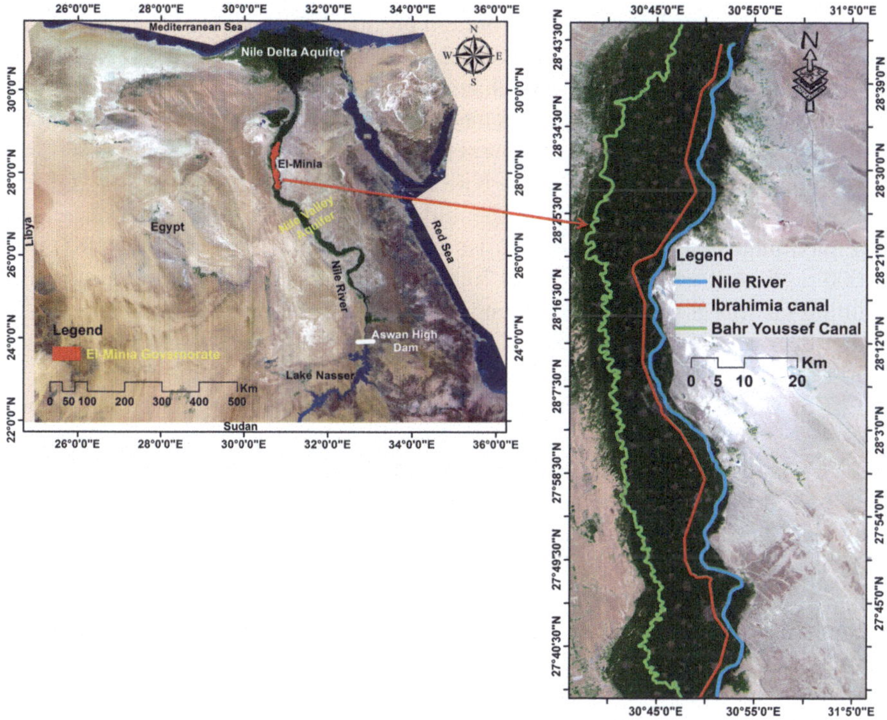

Fig. 2 Egypt map and the Nile River valley aquifer at El-Minia Governorate, Egypt

3 GERD Impacts on the Groundwater in the Nile River Valley

The impact of GERD filling on Egypt's groundwater aquifers has not been thoroughly investigated. Only three case studies have been published that look at the effects of the GERD predicted to fill on Egypt's groundwater aquifers due to lower Nile and main canal water levels. These case studies were done by Armanuos et al. (2017), Abd-Elhamid et al. (2019), and El-Rawy et al. (2021a).

Armanuos et al. (2017) investigated the impact of GERD and pumping scenarios on groundwater levels in the Nile Delta aquifer. They looked at three scenarios: reducing canal water depth by 25 and 50% owing to GERD, increasing pumping rates by 25 and 50%, and combining the first and second scenarios. As a result of a 25% fall in canal water depth, groundwater levels in the Nile Delta's western, middle, and eastern portions fell by 0.30 m, 0.5 m, and 0.2 m, respectively. Reducing the water depth by half resulted in groundwater levels dropping by 0.6 m, 0.7 m, and 0.4 m in the western, central, and eastern regions of the Nile Delta, respectively (Armanuos et al. 2017). They found that increasing pumping flows by 50% while decreasing water depth by 50% results in declines of 1.26 m, 1.7 m, and 1.35 m in the Nile Delta's western, central, and eastern portions, respectively (Armanuos et al. 2017). The possible impacts of the GERD on seawater intrusion in the Nile Delta aquifer were investigated by Abd-Elhamid et al. (2019). They develop numerical groundwater flow and solute transport models using SEAWAT (Langevin et al. 2008) for the Nile Delta aquifer in order to assess the detrimental consequences of the GERD and explore seawater intrusion in the Nile Delta aquifer. As result of filling GERD in 3 years, the groundwater level in the middle of the Nile Delta aquifer decreased to a maximum of 2.65 m. Also, the iso-concentration salinity lines of 1000 and 35,000 mg/L will advance 110.2 and 70.85 km inland, respectively. In the scenario of filling GERD in 6 years, the groundwater level in the middle of the Nile Delta aquifer decreased to a maximum of 1.4 m. Also, the iso-concentration salinity lines of 1000 and 35,000 mg/L will advance 108.25 and 67.3 km inland, respectively.

El-Rawy et al. (2021a) investigated the GERD's prospective impact on groundwater levels in the Nile Valley aquifer (along El-Minia Governorate, as shown in Fig. 2). Three GERD reservoir filling periods (6, 3, and 2 years) were simulated with three flow conditions from the Blue Nile to Lake Nasser (normal flow (NF), minimum of average flow (MAF), and minimum flow (MF). El-Rawy et al. (2021a) discovered that in the scenario of filling GERD in two years, the average groundwater level declined by 1.41, 1.76, and 2.21 m for NF, MAF, and MF, respectively, in comparison to the base case. In the scenario of filling GERD in three years, the average groundwater level declined by 0.93, 1.45, and 2.17 m, respectively, for NF, MAF, and MF, compared to the base case. In the 6-year scenario of filling GERD, the average groundwater level declined by 0.47, 0.96, and 1.72 m for NF, MAF, and MF, respectively, compared to the base case. Furthermore, they discovered that in the event of a GERD filling with NF, nearly 17 times the current extraction rate would have to be pumped to compensate for the loss of surface water. As a result, the groundwater level in the Nile Valley aquifer dropped by around 26 m compared to the base case, putting the aquifer in an unsustainable state. As a result, life in the Nile Valley may be jeopardized. As a result, in this section, the chapter will examine the impact of filling the GERD reservoir in 10 years on the decline of groundwater levels in the Nile Valley aquifer in order to assess its long-term viability.

4 Impact of Filling GERD Reservoir in 10 years on Groundwater Level

4.1 Study Area and Groundwater Model

The study area is the western plain of El-Minia governorate, which is situated on the Nile River's west bank (Fig. 2). The study area is approximately 1585 km^2 and is part of the El-Minia Governorate, which is situated 230 km south of Cairo and 125 km north of Assuit. The study area's width is about 12.5 km, and the North–South length of the governorate is about 128.5 km (Calculated by ArcGIS). The area is characterized by its geographical location because it serves as a gateway to many water sources, including the Nile River, Youssef, and Ibrahimia canal. About 700 canals in El-Minia government are used for irrigation purposes, as they flow northward and are conducted from Ibrahimia canal and Youssef canal (MWRI 2018). The groundwater in the study area represents a portion of the Nile aquifer system. The Nile River, its branches and channels, as well as excess irrigation water, recharge the aquifer (El-Rawy et al. 2021a).

A 3-D groundwater model for the research area constructed and calibrated (El-Rawy et al. (2021a)) used in this chapter to simulate the effects of decreasing water levels in surface water bodies in this study area throughout the 10-year filling of GERD. Scenarios NF, MAF, and MF will be simulated and studied in the situation of treating GERD in ten years. The Nile river levels declined by 0.02, 18, and 37% under the NF, MAF, and MF scenarios, respectively, while the canal water levels decreased by the same percentages. The outcomes of these simulated scenarios are based on changes in water balance components and groundwater levels compared to the baseline.

4.2 *Results and Discussions*

4.2.1 Water Balance Components for the Simulated Scenarios

The Nile River discharge, Ibrahimia canal, model boundaries, and ET are all part of the discharge components. The recharge components include inflow through the model boundaries as well as recharging to the aquifer from the Ibrahimia and Bahr Youssef canals.

Figure 3 depicts the change in water balance components for the simulated scenarios compared to the base case. In the scenario of filling GERD in 10 years under the normal flow condition (NF), the discharge from the aquifer to the Nile River and through the model boundaries increases by 6,860 m^3/day and 1744 m^3/day, respectively. The recharge to the aquifer from the Ibrahimia and Bahr Youssef canals is reduced by 232 m^3/day and 1923 m^3/day, respectively. Furthermore, ET decreases by 7070 m^3/day.

In the scenario of filling GERD in 10 years under the minimum of average flow condition (MAF), the discharge from the aquifer to the Nile River, model boundaries, the Ibrahimia canal increases by 53,090 m^3/day, 15.136 m^3/day, and 1012 m^3/day, respectively. The recharge to the aquifer from the Bahr Youssef canal is also reduced by 15,775 m^3/day. Moreover, the ET has decreased by 53,955 m^3/day compared to the base case.

In the scenario of filling GERD in 10 years under the minimum flow condition (MF), the discharge from the aquifer to the Nile River, model boundaries, the Ibrahimia canal are increased by 105,460 m^3/day 32,535 m^3/day, and 8308 m^3/day, respectively. The recharge from the Bahr Youssef canal to the aquifer are reduced by 25,212 m^3/day. Furthermore, the ET decreased by 89,274 m^3/day with respect

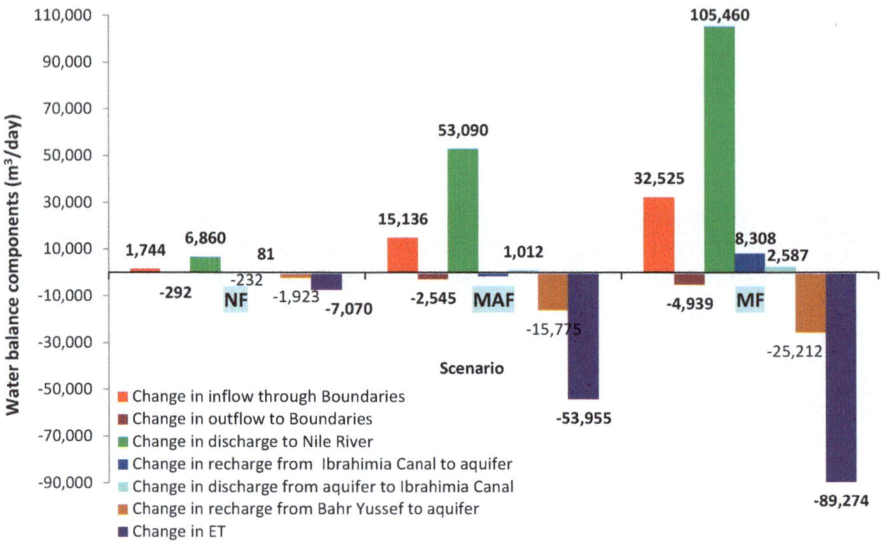

Fig. 3 Changes in water balance components for the scenario of filling GERD in 10 years with various flow conditions (NF, MAF, and MF) compared to the base case (BC)

to the base case. According to the water balance results, the Nile River gains water from the aquifer in all scenarios, whereas the Bahr Youssef and Ibrahimia canals lose water in all scenarios. When the Nile River's water depth decreases in the NF and MAF scenarios, the Nile River gets more water from groundwater and nearby canals, affecting the water balance of the closest adjacent canal (Ibrahimia canal).

4.2.2 Changes in Groundwater Levels for the Simulated Scenarios with Respect to the BC

The special distribution maps for the change in groundwater levels for the various scenarios compared to the BC are shown in Fig. 4. When filling GERD in ten years with NF, a 0.02% reduction in the Nile River and canals water depths resulted in a decline in groundwater levels ranging from 0 to 0.15 m, with an average of 0.062 m, when compared to BC (Fig. 4a). In the scenario of filling GERD in 10 years under the minimum of average flow condition (MAF) (a 18% reduction in the Nile River and canals water depths) leads in a drop in groundwater levels ranging from 0–1.05 m, with an average of 0.573 m with respect to the BC (Fig. 4b). In the scenario of filling GERD in 10 years under the minimum flow condition (MF) (a 37% reduction in Nile and canal water depths) results in a drop in groundwater levels ranging from 0–2.17 m, with an average of 1.25 m, compared to BC (Fig. 4c). The results of change in groundwater level show that the greatest drawdown occurs in the case of the minimum of average scenario (MF).

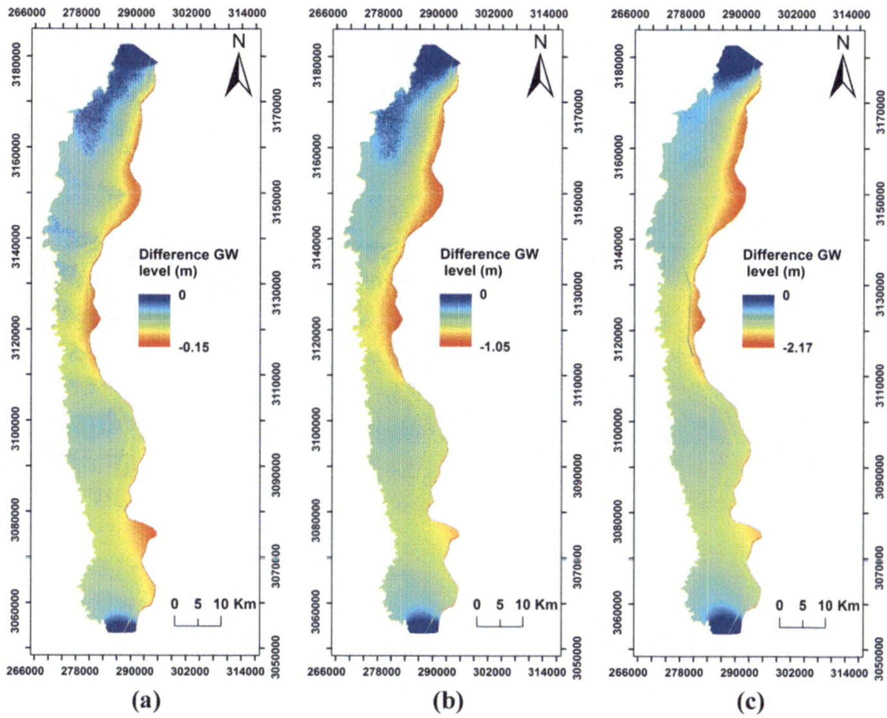

Fig. 4 Changes in groundwater level for the filling GERD scenarios in ten years with various flow conditions (NF, MAF, and MF) compared to the base case (BC): **a** NF–BC, **b** MAF–BC, **c** MF–BC

4.2.3 River-Aquifer Exchange Fluxes for the Simulated Scenarios

To assess the Nile River's losing and gaining portions, as well as the Ibrahimia and Bahr Youssef canals, the exchange fluxes between the river and the aquifer are analyzed and represented in Fig. 5. The fluxes of exchange between the aquifer and the river for the base case and the invistigated scenarios (NF, MAF, and MF) in Fig. 5a, b and c respectively. The color in the figures represents the direction and magnitude of the flux between the aquifer and the streams/rivers. Positive values indicate that the River is losing, while negative values show that the River is gaining.. The Nile River is a gaining stream at locations with an orange-red colour, while Ibrahimia and Bahr Youssef canals are a losing stream at locations with green–blue colour. In scenario MF, the magnitude of the gaining water of the Nile River was higher than the magnitude of gaining water from the aquifer of the BC and other scenarios (Fig. 5). These quantitative results are applicable only to the case study. However, qualtitively, we are expecting similar results for rest of the aquifer along the Nile River. We couldn't analyze the consequences of filling in 10 years on other parts of the aquifer or the full Nile River reach since we didn't have enough data.

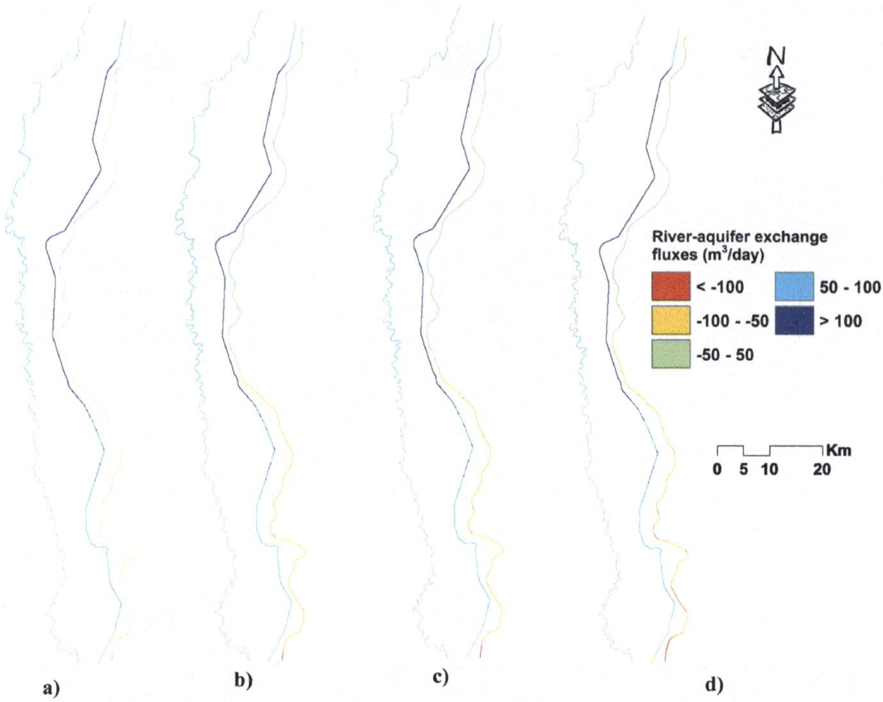

Fig. 5 Fluxes of river–aquifer exchange for various scenarios: **a** BC, **b** NF, **c** MAF, **d** MF. Positive values indicate that the River is losing sections, while negative values indicate that the River is gaining sections

5 Conclusions

The effects of the GERD filling on Egypt's Nile Delta and Nile Valley aquifers have been studied. There are three case studies available that look at the effects of the GERD predicted to fill on Egypt's groundwater aquifers due to lower Nile and main canal water levels.

The results showed that the groundwater level in the middle Delta dropped to a maximum of 2.65 m and 1.4 m, respectively, under scenarios 3 and 6. In addition, the iso-concentration salinity lines of 1000 and 35,000 mg/L will advance 110.2 and 70.85 km onshore, respectively, if GERD is filled in three years. The iso-concentration salinity lines of 1000 and 35,000 mg/L will advance 108.25 and 67.3 km inland, respectively, if GERD is filled in 6 years. According to this analysis, it was necessary to reduce groundwater extraction rates by 60% and 40%, respectively, in order to sustain freshwater supplies in the Nile Delta aquifer and with the same scenarios of GERD reservoir filling in 3 and 6 years.

The probable impcats of the GERD on groundwater level in the Nile valley aquifer have been studied based on three GERD reservoir filling periods (6, 3, and 2 years)

and three flow conditions from the Blue Nile to the Lake Nasser (NF, MAF, and MF). In the two-year filling GERD scenario, the average groundwater level declined by 1.41, 1.76, and 2.21 m for NF, MAF, and MF, respectively, compared to the base case. In the scenario of filling GERD in three years, the average groundwater level decreased by 0.93, 1.45, and 2.17 m for NF, MAF, and MF, respectively, compared to the BC. In the case of filling GERD in 6 years, the average groundwater level fell by 0.47, 0.96, and 1.72 m, respectively, for NF, MAF, and MF, compared to the BC. To compensate for the decline in surface water, almost 17 times the extraction rate of the base scenario would have to be pumped, and the groundwater level would drop by around 26 m, which is an unsustainable practice potential.

In the scenario of GERD filling in ten years, which comprises scenarios NF, MAF, and MF, average groundwater levels decreased by 0.15 m, 0.573 m, and 1.25 m, respectively, compared to the base case.

Based on the findings, it can be stated that GERD construction will have a negative influence on Egypt's groundwater resources, especially in the case of a longer filling period (GERD filling in 6–10 years). However, it may be determined that the GERD filling in ten years scenario is the greatest choice for keeping Egypt's groundwater resources in a sustainable state. The GERD construction will have an impact on Egypt's water resources, particularly if a single choice is made to fill the GERD reservoir from the Ethiopian side. As a result, the findings of this study should aid the Egyptian government in planning and managing water resources following the completion of the GERD.

6 Recommendations

The following suggestions can be made based on the above conclusions:

- It will be strongly advised to evaluate the impact of filling GERD in 10 years on the groundwater level of the Nile River full aquifer once data for the other reaches of the Nile River is available. This is necessary to ensure the long-term viability of groundwater-dependent activities in the Nile Valley.
- Because of the unique nature of the Nile River's natural flow time series, it is critical that the GERD's short- and long-term operations be shared with Egypt between Egypt, Sudan, and Ethiopia, in order to reduce the GERD's negative effects on water share and, as a result, the groundwater in the Nile River valley will be almost safe and sustainable.
- It is strongly recommended that the Nile Delta aquifer be re-investigated for saltwater intrusion, taking into consideration the various GERD reservoir filling scenarios, as well as climate change and sea level rise.
- An optimization modeling is needed to achieve best practices of pumping water from the Nile River valley aquifer to recompense the drop in surface water due to the filling GERD's reservoir.

- Because droughts are unpredictable, filling plans, operating policies, adaption options, protection policies, and emergency measures should all be in place prior to the GERD being filled and operated. Filling the GERD reservoir in 10 years may be the best choice under normal or minimum average flow conditions. During the filling of the GERD reservoir, Egypt's Ministry of Water Resources and Irrigation must have a workable plan for the High Aswan Dam discharge and water distribution in canal networks.
- Building modern irrigation system and improving traditional irrigation system could be a part of the solution to recompense for the drop in surface water caused bythe filling GERD's reservoir. Although, these types of management scenarios and strategies require time to be implemented.

References

Abd El-Bassier M (1997) Hydrogeological and hydrochemical studies of the Quaternary aquifer in Qena Governorate. Unpublished MSc thesis, faculty of science, Assiut University, Egypt

Abdalla F, Shamrukh M (2016) Quantification of River Nile/Quaternary Aquifer exchanges via riverbank filtration by hydrochemical and biological indicators. Assuit Egypt J Earth Syst Sci 125–8:1697–17711

Abd-Elhamid H, Abdelaty I, Sherif M (2019) Evaluation of potential impact of Grand Ethiopian Renaissance Dam on seawater intrusion in the Nile Delta aquifer. Inter J Environ Sc Technol 16(5):2321–2332

Abdelkader TA, Elsanabary M (2015) Hydrological and environmental impacts of grand Ethiopian renaissance dam on the Nile River. In: Proceeding of eighteenth international water technology conference, IWTC18. Sharm ElSheikh, pp 12–14 March, 336–347

Abdelrady A, Sharma S, Sefelnasr A, El-Rawy M, Kennedy M (2020) Analysis of the performance of bank filtration for water supply in arid climates: case study in Egypt. Water 12(6):1816. https://doi.org/10.3390/w12061816

Al-Agha D, Closas A, Molle F (2015) Survey of groundwater use in the central part of the Nile Delta. Activity Report, Australian Centre for International Agricultural Research

Al-Maktoumi A, El-Rawy M, Zekri S (2016) Management options for a multipurpose coastal aquifer in Oman. Arab J Geosci 9(14):1–14. https://doi.org/10.1007/s12517-016-2661-x

Al-Maktoumi A, Zekri S, El-Rawy M, Abdalla O, Al-Wardy M, Al-Rawas G, Charabi Y (2018) Assessment of the impact of climate change on coastal aquifers in Oman. Arab J Geosci 11(17):1–14. https://doi.org/10.1007/s12517-018-3858-y

Al-Maktoumi A, Zekri S, El-Rawy M, Abdalla O, Al-Abri R, Triki C, Bazargan-Lari MR (2020) Aquifer storage and recovery, and managed aquifer recharge of reclaimed water for management of coastal aquifers. Desalin Water Treat 176:67–77

Armanuos AM, Ibrahim MG, Mahmod WE, Negm A, Yoshimura C, Takemura J, Zidan BA (2017) Evaluation the potential impact of Grand Ethiopian Renaissance Dam and pumping scenarios on groundwater level in the Nile Delta Aquifer. Water Sci Techno Water Supply 2017037. https://doi.org/10.2166/ws.2017.037

Attia FA, Lennaerts ABM (1989) Economic aspects of groundwater development for irrigation and drainage in the Nile valley. Dev Water Sci 39:249–260

Awad A, Eldeeb H, El-Rawy M (2020) Assessment of surface and groundwater interaction using field measurements: a case study of Dairut City, assuit. Egypt J Eng Sci Technol 15(1):406–425

Barlow PM, Harbaugh A (2006) USGS directions in MODFLOW development. Ground Water 44(6):771–774

Diana RK (2000) Water sharing in the Nile River Valley project gnv011: using GIS/remote sensing for the sustainable use of natural resources, Geneva. Available at: http://www.grid.unep.ch/activities/sustainable/nile/nilereport.pdf

El Arabi N (2012) Environmental management of groundwater in Egypt via artificial recharge extending the practice to soil Aquifer treatment (SAT). Int J Environ Sustain 1(3):66–82

El Arabi NE, Dawoud MA (2012) Groundwater aquifer recharge with treated wastewater in Egypt: technical, environmental, economical and regulatory considerations. Desalin Water Treat 47(1–3):266–278

El Tahlawi M, Farrag A, Ahmed S (2008) Groundwater of Egypt: an environmental overview. Environ Geol 55(3):639–652

El-Rawy M, Vitaly Z, Marwan A, Al-Maktoumi A, Kacimov A, Abdalla O (2016) Conjunctive use of groundwater and surface water resources with aquifer recharge by treated wastewater: evaluation of management scenarios in the Zarqa River Basin. Jordan Environ Earth Sci 75:1146. https://doi.org/10.1007/s12665-016-5946-1

El-Rawy M, Al-Maktoumi A, Zekri S, AbdallaO A-A (2019a) Hydrological and economic feasibility of mitigating a stressed coastal aquifer using managed aquifer recharge: a case study of Jamma aquifer. Oman J Arid Land 11(1):148–159. https://doi.org/10.1007/s40333-019-0093-7

El-Rawy M, Ismail E, Abdalla O (2019b) Assessment of groundwater quality using GIS, hydrogeochemistry, and factor statistical analysis in Qena governorate. Desalination and water treatment. Egypt, p 14–29. https://doi.org/10.5004/dwt.2019b.24423

El-Rawy M, De Smedt F (2020) Estimation and mapping of the transmissivity of the nubian sandstone aquifer in the Kharga oasis. Egypt Water 12(2):604. https://doi.org/10.3390/w12020604

El-Rawy M, Abdalla F, El Alfy M (2020a) Water Resources in Egypt. In: Z. Hamimi et al. (eds), The geology of Egypt, regional geology reviews. Springer Nature Switzerland AG. https://doi.org/10.1007/978-3-030-15265-9_18

El-Rawy M, Batelaan O, Buis K, Anibas C, Mohammed G, Zijl W, Salem A (2020b) Analytical and numerical groundwater flow solutions for the FEMME-modeling environment. Hydrology 7(2):27. https://doi.org/10.3390/hydrology7020027

El-Rawy M, Makhloof AA, Hashem MD, Eltarabily MG (2021a) Groundwater management of quaternary Aquifer of the Nile Valley under different recharge and discharge scenarios: a case study assiut governorate. Egypt Ain Shams Eng Univ 12(3):2563–2574. https://doi.org/10.1016/j.asej.2021.02.023

El-Rawy M, Moghazy HE, Eltarabily MG (2021b) Impacts of decreasing Nile flow on the Nile Valley aquifer in El-Minia Governorate. Egypt Alexandria Eng J 60(2):2179–2192. https://doi.org/10.1016/j.aej.2020.12.037

Elsanabary MH, Ahmed AT (2018) Impacts of constructing the Grand Ethiopian Renaissance Dam on the Nile River. In: Negm AM, Abdel-Fattah S (eds), Grand Ethiopian Renaissance Dam Versus Aswan High Dam, Hdb Env Chem. Springer International Publishing AG. https://doi.org/10.1007/698_2017_228

El-Zehairy A, Lubczynski M, Gurwin J (2018) Interactions of artificial lakes with groundwater applying an integrated MODFLOW solution. Hydrogeo J 26(1):109–132

FAO (2013) Aquastat database. Food and agriculture organization of the United Nations (FAO)

Furman A (2008) Modeling coupled surface-subsurface flow processes: a review. Vado Zone J 7(2):741–756

GERD fact sheet (2014) The Grand Ethiopian Renaissance Dam. Available at: http://www.internationalrivers.org/resources/thegrand-ethiopianrenaissance-dam-fact-sheet-8213

Hamed KH (2018) Stochastic Investigation of the GERD-AHD Interaction through First Impoundment and Beyond. A.M. Negm and S. Abdel-Fattah (eds), Grand Ethiopian Renaissance Dam Versus Aswan High Dam. Hdb Env Chem. Springer International Publishing AG. https://doi.org/10.1007/698_2017_125

Harbaugh AW (2005) MODFLOW-2005, the U.S. Geological Survey modular ground-water model—the ground-water flow process. U.S. Geological Survey Techniques and Methods 6-A16. U.S. Geological Survey, Reston. Available at: http://pubs.usgs.gov/tm/2005/tm6A16/PDF.htm

Ben Heubl (2020) Why Ethiopia and Egypt can't agree over hydro dam. https://eandt.theiet.org/content/articles/2020/08/why-ethiopia-and-egypt-can-t-agree-over-the-grand-ethiopian-renaissance-dam-gerd-africa-s-largest-hydroelectric-power-project/

Ibrahim A (2017) Impact of Ethiopian Renaissance Dam and population on future Egypt water needs. Am J Eng Res (AJER) 6(5):160–171

Salini Impregilo (2017) Grand Ethiopian Renaissance Dam project. https://www.saliniimpregilo.com/en/projects/in-progress/dams-hydroelectric-plants-hydraulic-works/grand-ethiopian-renaissance-dam-project.html

Ismail S (2013) Flow reduction impacts along River Nile in Egypt. In: Proceedings of seventeenth international water technology conference (IWTC 17). Istanbul, Turkey, pp 5–7

Kahsay TN, Kuik O, Brouwer R, van der Zaag P (2015) Estimation of the transboundary economic impacts of the Grand Ethiopia Renaissance Dam: A computable general equilibrium analysis. Water Resour Econom. 10:14–30. https://doi.org/10.1016/j.wre.2015.02.003

Kamal E (2004) Geology of Luxor Area and its relationship to groundwater uprising under the Pharaoh Temples. Unpublished MSc thesis, Aswan Faculty of Science, South Valley University, Egypt.

Langevin CD, Thorne DT, Dausman AM, Sukop MC, Guo W (2008) SEAWAT Version 4: a computer program for simulation of multispecies solute and heat transport. Techniques and methods book 6, chapter A22. U.S. Department of the Interior, U.S. Geological Survey

Mulat AG, Moges SA (2014) Assessment of the impact of the Grand Ethiopian Renaissance Dam on the performance of the High Aswan Dam. J Water Resour Protec 6:583–598

(MWRI) Ministry of Water Resources and Irrigation (2005) National Water Resources Plan 2017. Ministry of Water Resources and Irrigation Planning Sector

Ministry of Water Resources and Irrigation (MWRI) (2012) Strategy of water resources of Egypt till 2050. Ministry of Water Resources and Irrigation

Ministry of Water Resources and Irrigation (MWRI) (2014) Climate change risk management in Egypt: proposed climate change adaptation strategy for the Ministry of Water Resources & Irrigation Egypt. Available at: http://www.eeaa.gov.eg/portals/0/eeaaReports/CCRMP/7.%20CC%20Water%20Strategy/CC%20Final%20Submitted%208-March%202013%20AdptStrtgy.pdf

Ministry of Water Resources and Irrigation (MWRI) (2018) Data bank in Excel sheets, Ministry of Water Resources and Irrigation

Negm AM, Abdel-Fattah S (2019) Grand Ethiopian Renaissance Dam Versus Aswan High Dam. Springer International Publishing, The handbook of environmental chemistry

Negm AM, Abdel-Fattah S, Omran ESE (2018) Update, conclusions, and recommendations for Grand Ethiopian Renaissance Dam Versus Aswan High Dam: a view from Egypt. In: Negm A, Abdel-Fattah S (eds) Grand Ethiopian Renaissance Dam Versus Aswan High Dam. The handbook of environmental chemistry, vol 79. Springer, Cham. https://doi.org/10.1007/698_2018_344

Ramadan SM, Negm AM, Smanny M, Helmy AHM (2015) Quantifying the impacts of impounding Grand Ethiopian Renaissance Dam Reservoir on Nasser Lake active storage. In: Proceedings of eighteenth international water technology Conference (IWTC 18). Sharm ElSheikh, Egypt, pp 12–14

Salem A, Dezső J, Lóczy D, El-Rawy M, Slowik M (2018) Modeling surface water-groundwater interaction in an Oxbow of the drava floodplain. In: Proceeding of 13th international conference on hydroinformatics (HIC 2018). Palermo, Italy, pp 1–6

Salem A, Dezső J, El-Rawy M, Lóczy D (2020) Hydrological modeling to assess the efficiency of groundwater replenishment through natural reservoirs in the Hungarian Drava river floodplain. Water 12:250. https://doi.org/10.3390/w12010250

Sharaky AM (2014) Geology of the Ethiopian Renaissance Dam and its effect on the dam safety. In: Conference of the "Nile Water Issue", Geography Department, Faculty of Arts, Cairo University (2014) pp 150–170

Sharaky AM, Hamed KH, Mohamed AB (2017) Model-based optimization for operating the ethiopian renaissance dam on the blue nile river. In: Negm AM, Abdel Fattah S (eds), Grand ethiopian renaissance dam versus aswan high dam, Hdb Env Chem. https://doi.org/10.1007/698_2017_188

Soliman AH, El Zawahry A, Bekhit H (2017) GERD failure analysis and the impacts on downstream countries. In: Negm AM, Abdel-Fattah S (eds), Grand ethiopian renaissance dam versus aswan high dam, Hdb Env Chem, Springer international publishing AG 2017. https://doi.org/10.1007/698_2017_135

Tesfa BC (2013) Benefit of grand ethiopian renaissance dam project (GERDP) for Sudan and Egypt. EIPSA Commun Article 1(1):1–12

Conclusions

Update, Conclusions and Recommendations for "Sustainability of Groundwater in the Nile Valley, Egypt"

Mustafa El-Rawy, El-Sayed E. Omran, and Abdelazim M. Negm

Abstract The current state of Nile Valley groundwater resources is framed by issues concerning groundwater evaluation, quality, management, and sustainability. Groundwater is Egypt's second most vital freshwater source after the Nile River, and it is used for irrigation, domestic, and industrial purposes. The Nile Valley aquifer in Egypt is the most important renewable aquifer, accounting for approximately 3.5 million cubic meters (MCM) per year, or roughly half of Egypt's total abstracted groundwater. Groundwater in the Nile Valley aquifer is critical to Egypt's long-term development and socioeconomic growth. This chapter summarizes groundwater resources in the Nile Valley (in terms of findings and recommendations) and offers ideas drawn from the volume cases. Furthermore, some updated findings from a few recently published research work on assessing groundwater resources and quality and groundwater sustainability covered themes are presented. This chapter introduces the current state of the groundwater state in the Nile Valley aquifer. It summarizes essential results found in the book to finally draw out some suggestions/recommendations to enhance, manage, and protect the groundwater resources in the Nile Valley for better sustainability uses.

M. El-Rawy (✉)
Civil Engineering Department, Faculty of Engineering, Minia University, Minia 61111, Egypt
e-mail: mustafa.elrawy@mu.edu.eg

Civil Engineering Department, College of Engineering, Shaqra University, Ar Riyadh 11911, Dawadmi, Saudi Arabia

E.-S. E. Omran
Soil and Water Department, Faculty of Agriculture, Suez Canal University, Ismailia 41522, Egypt

Institute of African Research and Studies and Nile Basin Countries, Aswan University, Aswan, Egypt

A. M. Negm
Water and Water Structures Department Engineering, Faculty of Eng, Zagazig University, Zagazig 44519, Egypt
e-mail: amnegm85@yahoo.com; amnegm@zu.edu.eg

A. M. Negm and M. El-Rawy (eds.), *Sustainability of Groundwater in the Nile Valley, Egypt*, Earth and Environmental Sciences Library,
https://doi.org/10.1007/978-3-031-12676-5_14

Keywords Nile valley · Groundwater resources · Groundwater quality · Water deficiency · Sustainability · Groundwater modeling · Agriculture · Wastewater · GIS · RS · Egypt

1 Introduction

Groundwater is critical to meeting the majority of the world's water needs in many regions (Suring 2020). Many ambitious water resource plans in developing countries are based on groundwater for achieving targeted agricultural expansion in response to population growth. Groundwater resources exceed 70% of total water consumption in many countries over the world (Nriagu 2019).

Egypt is currently facing a severe water shortage, with 13.5 billion cubic meters required annually, due to a 79.5 BCM annual water requirement and total available water of 66 BCM (MWRI 2014; Omar and Mousa 2016). Groundwater is considered the second most important source of freshwater after the Nile River, and it is used for irrigation, domestic, and industrial purposes (Abdel-Shafy and Kamel 2016; El-Rawy et al. 2020a). Groundwater is, therefore, a vital factor for economic growth and sustainable development. The Nile Delta and Nile Valley aquifers supply approximately 87% of the groundwater exploited (El-Rawy et al. 2020a). The Nile Valley aquifer represents about 3.5 million cubic meters (MCM) yearly, more or less 50% of the total abstracted groundwater in Egypt (El-Rawy et al. 2020a). The Nile Valley aquifer is considered as a shallow renewable aquifer. The Nile Valley aquifer is replenished by seepage from river canals and percolation from agricultural irrigation. Groundwater in the Nile Valley is easier to access and is invaluable as a source of drinking and irrigation water of excellent quality. Recently, two books on groundwater in the Nile Delta (Negm 2018) and groundwater in Egyptian deserts (Negm and Elkhouly 2021) were published. This book holds great promise and potential for groundwater resources in Egypt's Nile Valley, which is the first book on the groundwater resources in the Nile Valley in Egypt.

Many studies have been conducted to model groundwater resources in the Nile Valley to ensure better conditions for these resources, based on the wide variety of groundwater modeling advantages mentioned in Chap. 3. In various locations along the Nile Valley aquifer, more than ten groundwater flow and transport models have been developed (Shamrukh et al. 2001; Ahmed 2009; Campos 2009; Dawoud and Ismail 2013; El Fakharany and Fekry 2014; Sefelnasr et al. 2019; Abdelhalim et al. 2019; El-Rawy et al. 2021a,b). A 3-D groundwater flow model was developed along with the contamination transport in the Tahta region, Sohag, to assess the potential impact of chemical fertilizers on groundwater resources in the study area (Shamrukh et al. 2001). A regional numerical groundwater model was developed and used to assess the interaction between surface water forms and the Nile Valley's quaternary aquifer system (Dawoud and Ismail 2013). The model water balance calculations revealed that the current aquifer potentiality is around 1.32 BMC, which can be used in future development, but detailed potentiality maps are required to define where it can be used without aquifer deterioration. Recently, a hydrological model was developed to determine the best locations for the installation of bank filtration wells (Abdelrady et al. 2020). The findings revealed that decreasing the Nile level has a significant impact on the bank filtration parameters (e.g. travel time, bank filtrate share) in the wells' onset operation. For a better understanding of the interaction mechanism between surface water and groundwater in the Nile Valley aquifer, a numerical groundwater-surface water interaction model was developed in a section of the Nile Valley aquifer (El-Minia Governorate) (El-Rawy et al. 2021a). This model has also been used to investigate the effects of potential GERD incorrect filling scenarios. As a result, assessing the potential impacts on groundwater levels in the event of a future decrease in Nile water levels will be easy. A three-dimensional groundwater flow model was developed to simulate the behavior of the Assiut Quaternary aquifer under various recharge and discharge scenarios (El-Rawy et al. 2021b). The findings revealed that increasing the pumping discharge from wells or decreasing the Nile level has a significant impact on groundwater levels. Based on the foregoing, numerical groundwater modeling has been used successfully in the Nile Valley area to achieve various modeling objectives.

This book emphasized that there are different contamination sources, which included discharges of raw sewage, agricultural runoff, and industrial effluents (El Tahlawi et al. 2008; Abdalla and Shamrukh 2016; Environmental Affairs Agency 2017; El-Rawy et al. 2019a, b, 2020b). There are 124 point source of contaminations that discharged into Nile in the study area. Due to groundwater interaction with surface waters, aquifers' pollution is closely related to adjacent polluted surface water. Pollution is more severe on the Nile River's edges and desert fringes and in the shallow segments of the aquifers (Abdel-Shafy and Kamel 2016). The Nile Valley's intensive year-round agricultural practices, excessive use of chemical fertilizers and flash irrigation techniques to non-point sources of irrigation-return flow contamination (Abdalla and Khalil, 2018). However, the studies indicated that many Nile valley locations are highly polluted and need urgent measures to protect the Quaternary aquifer and mitigate further pollution. These studies reported that elevated heavy metals, COD, BOD, hardness, fecal coliform, and temperature. The present conditions indicated that groundwater in the study area would worsen unless urgent measures are taken. Accordingly, urgent mitigation measures have to be implemented to maintain. The groundwater quality has been assessed by many researchers in the Nile Valley aquifer (Hemdan and Abdel Rady 2011; Negm and Armanuos 2016; Fathi and El-Rawy 2018; Mamdouh et al. 2018; Megahed and Farrag 2019; El-Rawy et al. 2019a; Awad et al. 2020). Integrating remotely sensed data through the GIS technique is very important in characterizing the hydrogeological properties and assessing the groundwater quality and conditions in many environments (Elbeih et al. 2020; Shokr 2020; El-Rawy et al. 2020b). The integration of hydrochemical analysis and remote sensing and Geographic Information System (GIS) techniques with factor analysis is essential to understand groundwater chemistry factors and may aid decision-makers toward effective groundwater quality management.

This book also includes some chapters discussing the groundwater management and sustainability and how they can be achieved. In the Nile Valley aquifer, groundwater management for different scenarios of recharge and discharge was studied by many authors (Dawoud and Ismail 2013; Sefelnasr et al. 2019; Abdelhalim et al. 2019; El-Rawy et al. 2021a,b). The book also introduced the possibility of the correctness of the assumption that improper GERD operation will negatively affect groundwater along the Nile valley aquifer and reduce or stop its sustainable nature in these chapters. In addition, using multicriteria, mapping the groundwater recharge potential in two defunct Nile Basin areas. The book discusses how agricultural drainage strategies used in the Nile Valley can help save groundwater quality.

However, a few studies have looked into the current state of groundwater in the Nile Valley, and these findings have mostly been local, and do not cover the entire Nile Valley. Consequently, the intention of the book is to improve and address the following main theme.

- Groundwater Modeling.
- Assessment of Groundwater Quality.
- Sustainability of Groundwater.

The following section provides a review of the primary results of some of the most recent (updated) published studies on groundwater in the Nile River Valley, as well as the book chapters' main conclusions and recommendations for researchers and decision-makers. This chapter's update, findings, and recommendations are built on the data provided in the book.

2 Update

The following are the major update for the book project based on the main book theme.

2.1 Groundwater Modeling

In this part, one chapter was displayed related to groundwater modeling methods and a focused review on the groundwater models of the Nile Valley aquifer. The fundamentals of groundwater flow models are covered in this chapter, which includes boundary conditions, calibration methods, validation, sensitivity analysis, model uncertainties, and modeling technique limitations. It is also serves as a study of groundwater modeling applications in Egypt's Nile Valley aquifer. In some parts of the world, groundwater is critical for meeting much of the water needs (Suring 2020). Groundwater is constantly in contact with nearby bodies of water and the climate. Many shifts in natural processes, land uses, and field conditions (e.g. agricultural practices (Awad et al. 2021a,b), human activities, and urbanization) have an effect on it (Zhao et al. 2013)). These interactions have a big impact on the health of a specific groundwater resource. Groundwater models were created to aid in the analysis of the various impacts of climate, environmental, and human factors on groundwater supplies, thus improving and enhancing groundwater management (Kumar 2019). As a result, groundwater modeling is viewed as a visual medium for communicating key messages about groundwater supplies to the general public and decision-makers. Many researchers all over the world have used groundwater modeling to accomplish a variety of goals, including but not limited to:

- Assessment of potential impacts of climate change on aquifers (Al-Maktoumi et al. 2018).
- Assessment of the efficiency of groundwater replenishment through natural reservoirs (Salem et al. 2020).
- Investigate the role of chemical fertilizers and pesticides in groundwater contamination (Srivastav 2020; Eltarabily et al. 2017).
- Assessment of how aquifers' conditions can be impacted under different groundwater extraction scenarios (Al-Maktoumi et al. 2016, 2020; El-Rawy et al. 2019b; Eltarabily et al. 2018; Negm 2018).
- Understanding the mechanism of groundwater and surface water interactions (El-Rawy et al. 2016; Rawy et al. 2020b, 2021a,b).

2.2 Assessment of Groundwater Quality

Five potential approaches were identified for assessment of groundwater quality in Egypt. The first approach is related to the contamination sources along the Nile Valley, Egypt and its impact on groundwater. The main goals of this chapter were to look at how wastewater discharge from various sources interacts with surface and shallow groundwater resources in the Nile Valley, as well as to identify the main potential sources of pollution, especially from domestic, industrial, and agricultural sources. The Nile Valley is known for its year-round farming activities. Non-point sources of irrigation-return flow pollution result from excessive use of chemical fertilizers and flash irrigation techniques (two to three times per month) (Abdalla and Khalil 2018). The irrigation-return flows are usually mixing with various pollutants and agro-chemicals, fertilizers. The main sources of nitrate contamination are due to incorrect use of agro-chemicals, fertilizers and manure in agricultural areas.

The second potential way linked the role of environmental impacts of treated wastewater contaminates on groundwater quality in the Nile River Valley, Egypt. The potential health and environmental threat posed by industrial and domestic wastewater disposal sites into surface water bodies (the Nile, irrigation canals, and drains) either directly or indirectly is one of the major challenges facing sustainable management. The main objective of this chapter is to do reviews and summarize the most relevant studies and research that dealt with the impact of disposal of treated wastewater on groundwater and clarify this through a case study of the most extensive treatment plants in Egypt and the Middle East (Gabal EL Asfar WWTP with a capacity of 2.5 million m^3/day). A lot of research has been completed on the health risks of treated wastewater effluents in terms of water quality for human use and crop productivity (Custodio 2013). The Nile and its branches downstream of Cairo have high concentrations of coliform bacteria as a result of partly wastewater treatment (FAO 2016). Upper Egypt has 119 sewage treatment plants with a total capacity of 1.8 million m^3/day (Holding Company for Water and Wastewater 2020). Regions on the Nile's western bank transport their wastewater to the Giza Pump station, as well as the Zeinin and Abou-Rawash wastewater treatment plants (Osman et al. 2016).

The third approach is the hydrochemistry and hydrogeology aspects of alluvial aquifer in Aswan City, Egypt. The groundwater characteristics of the Aswan aquifer are highlighted in this chapter. The hydro-chemical and hydrological characteristics of the Aswan aquifer were examined in this study (i) to examine the quality of the pumped water, (ii) to define the suitability of the abstracted water for drinking and agricultural uses, and (iii) to propose different solutions to decline the groundwater table and avoid the consequences of increasing its level. The interaction between the Nile River and the aquifer is the primary input and output of the system. Water enters the aquifer from the Aswan Dam Lake in the south, flows north, and then exfiltrates back into the river. In addition to water quality issues, some areas saw an increase in groundwater table problems as a result of the shutdown of some extraction wells, which could cause infrastructure harm. A number of solutions have been suggested to address both issues. The bank infiltration technique will help ensure a steady supply of good-quality water when dealing with changes in the surface water. It does, however, only impact the water table near the wells, not the entire model domain. As the water table only dropped by 1.5 m after nine years at Kima area. Therefore, an increase in the extraction wells has to be combined with the bank filtration technique to help to lower the water table and serve good water quality.

The fourth approach is related to the hydrochemical analysis of groundwater in the area Northwest of El-Sadat City, West Nile Delta, Egypt. The importance of groundwater hydrochemistry in terms of groundwater quality cannot be overstated. It aids in determining the hydrochemical processes responsible for localized changes in groundwater chemistry (Şen 2015). The chemical composition of groundwater and its source is the subjects of hydrochemical research.

The final approach is "As a Water Resources Management Tool, Groundwater Quality Assessment for Irrigation in the Young Alluvial Plain of Western Nile Delta, Egypt." Sufficient quantity and quality of water are two of the most important human needs for survival, but maintaining desirable water quality is a difficult challenge for decision-makers. Water quality is determined by a variety of factors, including physical, chemical, and biological characteristics. Conclusions about water quality must be drawn from analyses of a wide number of variables, which are a challenging task (Mukatea et al. 2019). In place of conventional methods of water quality assessment, the water quality index for irrigation is an important tool for tracking and assessing the overall quality of water. Various researchers have produced various types of water quality index models (Khangembam and Kshetrimayum 2019; Mukatea et al. 2019; Solangi et al. 2019) to determine water quality. Several studies were carried out for groundwater quality evaluation in the Nile Delta, such as (Eltarabily and Negm 2018).

2.3 Sustainability of Groundwater

Five ways were identified for using biological control as a potential way for groundwater sustainability. The first potential practice is given an overview of agricultural drainage strategies in Egypt as a protection tool against groundwater contamination by fertilizers. Agricultural drainage is regarded as a crucial method for better groundwater management in both agricultural and urban areas (Hansen et al. 2019; Singh 2019). Agricultural drainage techniques can also influence the recharge capacity of aquifers in drained lands (Smith and Berg 2020). It was also reported that drainage systems can be employed to help in coping with adverse impacts of climate change on agriculture and the surrounding environment (Awad et al. 2021a). As a result, Egypt began to implement drainage techniques in agricultural land in the Nile River valley in order to ensure improved groundwater management, both quantity and quality, as well as increased agricultural productivity. Agricultural drainage strategies are needed in Egypt to ensure a higher quality of aquifers underneath agricultural land, as well as the conservation and long-term viability of agricultural productivity. Practicing agricultural drainage in agricultural land of the Nile River valley in Egypt led to increasing crop yield and enhancing the quality of aquifers that underlain these lands.

Another approach is the Pliocene aquifer characterization using TEM and VES geophysical techniques: Case study at the area to the East of Wadi El-Natrun City, West Nile Delta, Egypt. The main freshwater resources in Egypt are the River Nile, groundwater, and precipitation (El-Rawy et al. 2020a). Using an integrated analysis of vertical electrical sounding (VES) and transient electromagnetic (TEM) geophysical data, the current study aims to characterise the Pliocene groundwater aquifer east of Wadi El-Natrun city and northwest of El-Sadat city in the Western Nile Delta region. As a result, extensive geophysical studies should be carried out in order to detect groundwater aquifers, determine the best locations for drilling groundwater boreholes, and assess the groundwater quality in the most promising potential areas. This satisfies the Egyptian government's plan of the sustainable development strategy for the year 2030. Several geophysical and hydrogeochemical studies have been conducted in the region of the West Nile Delta. Ibraheem and El-Qady (2019) carried out a hydrogeophysical survey on the area of El-Nubariya–Wadi El-Natrun to the west of the study area. They conducted VES and TEM measurements and reported two main groundwater aquifers: Pleistocene aquifer, which is consisting of gavels and clayey sand deposits, and Pliocene aquifer, which constitutes of water-bearing sandy layers intercalated by clay faces.

The third potential conventional method is concerned with the evaluation of groundwater potential zones using electrical resistivity and hydrogeochemistry in

West Tahta Region, Upper Egypt. In this chapter, the geoelectrical and hydrogeochemistry studies were carried out in the western Tahta region to identify the subsurface hydrostratigraphic aquifer units and obtain a proposed potential groundwater development. The suitability of groundwater for domestic consumption and irrigation purposes has been examined by considering various indices. The results are assessed in accordance with the drinking water quality standards given by the World Health Organization (WHO 2017) and the Egyptian drinking water standards (2007).

The fourth potential practice provides an overview of the mapping groundwater recharge potential in the Nile Basin using remotely sensed data and GIS techniques. Groundwater recharge capacity in Wadi Qena and Wadi Matula, both in Egypt's Eastern Desert and part of the Nile basin, is assessed here. GIS methods were used after remote sensing data was used. Several evidential maps were derived from radar/optical remotely sensed data in addition to geologic and precipitation data, which control the occurrence, movements and infiltration capacities were combined after applying weight factors. The groundwater recharge potentiality has been verified by matching with field data and agricultural areas derived from satellite images. The results obtained may be useful for sustainable water resources in arid regions.

The fifth potential practice is the sustainability of groundwater in the Nile River Valley: is it possible after the construction of GERD? The chapter will investigate the impact of filling GERD reservoir in 10 years on the reduction of groundwater level of the Nile Valley aquifer to check its sustainability in this case. GERD construction would have an effect on Egypt's water supplies, especially if a single decision is made to fill the GERD reservoir from the Ethiopian side. As a result, the findings of this study should aid the Egyptian government in preparing and maintaining water supplies following the completion of the GERD. Once the data for the other reaches of the Nile River are available, it will be highly recommended to quantify the impact of filling GERD in 10 years on the groundwater level of the Nile River whole aquifer. This is vital to ensure the sustainability of the activities that depend on groundwater along the Nile Valley.

3 Conclusions

Several conclusions drawn from this book have been reached by the editorial teams throughout this book project. The chapter draws essential lessons from the book-cases on methodological concepts, specifically the promising aspects of groundwater modeling, groundwater quality assessment, and groundwater sustainability in the Nile Valley aquifer. This chapter sets out the current problems faced by managing the Nile Valley Aquifer's groundwater resources. These results are essential for the improvement of groundwater sustainability in the Nile Valley in Egypt. Based on the materials mentioned in all parts of this volume, the following results could be stated:

1. The authors relied on the information and data gathered from the published research and the official international reports over the last fifty years until

recent times that determined the groundwater state in the Nile Valley using the computational analysis to simulate the groundwater conditions. The main results include rising groundwater levels in the last 20 years in the middle and north of Egypt and the high affection of the groundwater in the Nile valley region by seawater intrusion towards the North while the impact decreases towards the south. The authors concluded that the groundwater levels increased in the middle of Egypt, the rising of salinity rates are getting higher on the north coast, and the groundwater solar desalination is suitable in the Northwestern Coast and the South towards Aswan.

2. The numerical groundwater modeling has been successfully used to achieve the various modeling objectives. Groundwater models have been developed and applied to different case studies along the Nile Valley aquifer for different model purposes. Results have shown that groundwater models have been instrumental in understanding the groundwater system's behavior in the Nile Valley and helped in the application of different management scenarios. The Nile Valley aquifer is exposed to deterioration due to the impacts of climate change and the construction and potential improper filling of various projects in the Nile River basin, such as the Grand Ethiopian Renaissance Dam (GERD), which causes a decrease in the water level in the Nile River as well as a decrease in the groundwater table.
3. There are 124 point sources of contaminations that discharged into the Nile in the study area. Seventy-two agricultural point sources discharge about 13.7 billion m^3/year. Industrial facilities, including seven sugar factories and two paper factories, discharge about 150 million m^3/year, where 14 power stations discharge 4.2 billion m^3/year. Amounts of domestic sewage are discharged directly into the River Nile or indirectly into main drains that discharge into the Nile, and its branches are about 1.06 BCM/year. These in addition to the discharge from dewatering effluents and cooling water. All these sources have elevated chemical and biological contaminants that influence and degrade the quality of Nile directly. Elevated heavy metals, COD, BOD, hardness, fecal coliform and temperature, are reported. Accordingly, urgent mitigation measures have to be implemented to maintain. The authors concluded the following: 124 point source of contaminations discharged into the Nile, these discharge about 13.7 billion m^3/year. Accordingly, high heavy metals, COD, BOD, hardness, fecal coliform, and temperature are reported. Accordingly, urgent mitigation measures have to be implemented to maintain the water resources from severe pollution.
4. Various solutions and scientific methods can be applied to develop the traditional groundwater assessment methodology to protect it from pollutants, especially partially treated wastewater. The integration of RS and GIS was used to simplify and accelerate groundwater quality assessment with acceptable accuracy. The use of mediators for spatial data principle can be constructive by rapidly providing systematical information. The most common form of digital mediator technology currently is the Geographical Information System (GIS). This technology makes it simple to combine layers of digital data from different

sources and drive the different layers related to each other. Also, GIS technology enables to link with the satellite imagery system and serves as geodata management by spectral, temporal, and spatial information and ground measurements. The groundwater quality assessment and study of many characteristics in the Nile River Valley are becoming more convenient with the growing satellite imagery. Also, it gets to be cost-effective than conventional methods. The authors concluded that the detection of pathogenic bacteria and heavy metals in Nile Valley villages provides an indication of biological and chemical contamination from the sewage system. Contaminants in the Delta region are relatively less than in the Nile Valley, but still, the problem of affected groundwater by the disposal of treated wastewater and agricultural drainage. Besides, the area around the major two conventional wastewater treatment plants El Gabal EL Asfar and El Berka in Greater Cairo, findings that studied drains were highly polluted due to the discharge of the effluents from wastewater treatment plants.

5. The water quality and quantity problems of Aswan's aquifer and present solutions with numerical simulations have been investigated. Hence, to provide the people of the area with adequate water for domestic, agricultural, and industrial uses. The methods used for the study are summarized in chemical analysis for water samples collected from the area to determine its suitability for various usages, in addition to hydrological data acquisition from authorities. These data have been used in constructing a model for simulating the groundwater flow as a helping tool for decision making. The main outcomes included a clear picture of the aquifer's situation and the water's suitability for different purposes. They were, besides, spotting the light on possible future degradation of the water quality. Additionally, a discussion of the available solutions for groundwater rise problems in the Kima industry area is supported by the aquifer's numerical visualization. The authors concluded that the water quality was found to be within WHO guidelines for drinking water. However, an increase in the organic matter concentration of the abstracted water was observed compared to the surface water systems, and the potential risk of forming carcinogenic trihalomethane compounds during the treatment process is still needed to be investigated. The groundwater table rise in the Kima industry area can be solved with bank infiltration techniques and moderating water extraction rates.
6. The hydrochemical settings of groundwater for both its quality assurance and characterization have been studied in the area northwest of El-Sadat city, hence, deciding if this water is suitable for different utilities or not. Chemical analyses for 57 groundwater samples have been tested for the major cations, major anions, electrical conductivity, total dissolved solids, and some trace elements. Depending on TDS values, water samples were categorized into freshwater-brackish water. Distribution maps of TDS, EC, and TH show a high EC and TDS level zone at the northwestern part, while there are moderate concentration zones in the central and eastern parts, but the lowest values are in the southwestern sector. Moreover, EC exhibits a strong relation with TDS, TH, Ca^{2+}, Cl^-, Na^+ and intermediate linkage with HCO_3^-, SO_4^{2-}, Mg^{2+} indicating that such cations and anions have an identical source and are implicated in the

reactions of ion exchange. Hardness levels reveal that groundwater could be grouped into very hard, hard, moderately hard, and soft waters. Groundwater quality for agricultural purposes was performed depending on several numerical indicators such SC, SAR, PI, RSC, MH, Kelly's ratio, and chloride classification. USSL diagram reveals that 93% of groundwater samples show a wide range of both alkalinity and salinity. The authors concluded that hydrochemical outcomes reveal a clear harmony between EC, TDS, and TH distribution maps. The EC and TDS values suggest fresh- to brackish water, while hardness indicates hard to very hard water in most of the boreholes. This analysis suggests groundwater is rich with Ca^{2+} and Mg^{2+} salts in the study area. Groundwater is characterized by the dominance of Alkaline earth and strong acids. Different water quality indices such as SAR, SC, PI, MH, and RSC suggest safe and suitable groundwater for agricultural activities.

7. The study evaluated groundwater applicability for irrigation in the region between the Rosetta Nile Branch and the El Nubariya Canal, west Nile Delta Egypt. Seventy-five groundwater samples were collected and chemically analyzed. The collected samples' chemical results were used to determine individual parameters such as TDS, Na%, SAR, RSC, Cl, KI, PI, MH, Cl, CAI, and CR. Also, a water quality index for irrigation was performed. GIS software was used for the spatial distribution of different parameters. The main results included that TDS of 74.4% of wells are within a suitable range. According to Na%, SAR, and RSC, 53.3%, 82.6%, and 90.7%, respectively, of the groundwater samples fall within the permissible limit. All samples are within the desirable limit based on MH, while KI values for most samples are unsuitable. 82.3% of the collected samples have corrosive characters. 30.3% of groundwater samples belong to class I of PI. The southern parts have fresh characters based on chloride content. According to the water quality index for irrigation, the study area is classified into three zones; high, medium, and low water quality in the south, central, and north respectively. The authors concluded that the groundwater in the central and southern parts is suitable for irrigation needs according to the measured and calculated parameters and indices. Infiltration from irrigation canals represents a source of groundwater recharge within the studied area.
8. The authors presented the implementation and application of agricultural drainage systems in the Nile Valley as a brilliant tool that allows better control of groundwater resources in terms of quantity and quality. It was concluded that the role of agricultural drainage systems in saving the groundwater quality also referred to the official published data from the Egyptian Ministry of Water Resources and Irrigation to cover the progress in the implementation and rehabilitation of agricultural drainage projects in Egypt. The main results included a conclusion of practicing agricultural drainage strategies in Egypt and how these strategies can save groundwater quality. Some brief recommendations were also listed to ensure that Egypt's agricultural drainage systems continue operations with high efficiency, ensuring better groundwater resource conditions

shortly. Egypt relies mainly on groundwater as a first alternative to compensate for any shortage in water quantity delivered to irrigated-agriculture lands. Also, groundwater is the leading water supplier in agricultural lands that are not accessible by water supply networks. Therefore, agricultural drainage strategies became indispensable in most irrigated-agriculture lands in Egypt to mitigate the excessive move of fertilizers down to aquifers, thus ensuring a better quality of groundwater resources.

9. The hydrogeological regime and information about the water-bearing formations in the investigated area regarding the groundwater potentialities have been evaluated. The vertical electrical sounding and time-domain electromagnetic data have been jointly interpreted to characterize the Pliocene water-bearing formation. The study included a detailed description of the Pliocene aquifer's hydrogeological properties in the Nile Valley area. The authors conclude that the Pliocene groundwater aquifer is the main aquifer in the investigated area. It is divided into two water-bearing horizons; the upper one has brackish water at a depth range of 30–70 m and a thickness of 20–60 m, whereas the lower one is much fresher with a depth range of 95–135 m. The northwest portion is the most suitable location for future boreholes. Results can provide valuable information to professionals, planners, decision-makers, and stakeholders. Therefore, it can be considered as an essential document for further advanced geophysical and hydrogeological investigations.
10. The Nile Valley aquifer's geometry has been evaluated at West Tahta Region, Upper Egypt, and the hydrogeochemical characteristics to assess the water quality for drinking and irrigation purposes. Vertical electrical soundings in the Schlumberger array were performed in the area under investigation. The result of chemical analysis of thirty-four water samples has been used to detect the hydrochemical characteristics and evaluate the collected water samples for different uses. The study area is characterized by a single aquifer system (Quaternary aquifer) which is mainly recharged from surface water and discharged during evaporation. The study area is a geophysical four-layer case; the first layer (surface layer) has various resistivity values from west to east due to the presence and influence of wadi deposits in the west and clay in the east. The second layer shows resistivity values between 111 and 325 Ω m and a variation in thickness between 5 and 34 m. This layer is composed of wet sand. The third geoelectrical layer is the main water-bearing formation in the study area. It consists of sand and gravel with resistivity values ranging from 28 to 99 Ω m. The thickness of this layer decreases towards the west and ranges from 18 to 144 m. The fourth geoelectrical layer represents the base of the water-bearing layer and is represented by Pliocene clay. The majority of the groundwater samples (71%) are unsuitable for human drinking because salinity is higher than the permissible limit. Half of the groundwater samples are suitable for irrigation purposes. The other half are unsuitable for their high salinity. The majority of groundwater samples collected from the study area suggested that rock-forming minerals' chemical weathering impacts groundwater quality by dissolving the host rock.

11. This chapter aims to map the groundwater recharge potential in two defunct areas of the Nile Basin using multicriteria. The methodology utilized to integrate ranked evidential maps derived from optical and radar remotely sensed data, geologic map, and structural features. The combined evidential maps through Geographic Information System techniques (GIS) include geology, topography, slope, lineaments, stream-networks, rainfall, and radar intensity. The groundwater recharge potential classified to 7 potential zones at Wadi Qena range from excellent to absent/bare and six zones at Wadi Matula. The predicted zones were validated using well data and evidence of agricultural activities derived from optical remote sensing data. Integrating remotely sensed data through the GIS technique is very important in characterizing the hydrogeological properties and conditions in many environments. Moreover, storing floodwater during rainstorms in human-made reservoirs would provide additional new water resources to sustain the people's lives in these remote areas.
12. The possibility of the assumption correctness that GERD's improper operation will negatively affect the groundwater along the Nile valley aquifer and reduce or stop its sustainable nature has been investigated. The authors introduced the most important findings of the previous recent studies connected to GERD's impact on the Nile River valley's groundwater aquifer. Also, it aims to investigate the impact of filling of GERD reservoir in 10 years based on three different flow conditions. A groundwater flow model using MODFLOW was developed with helping of the Geographic Information System techniques (GIS) for a part of the Nile Valley aquifer (El-Minia Aquifer). The study of the potential impacts of the GERD on groundwater level in the Nile valley aquifer has been done based on available data of three filling periods of the GERD reservoir (10, 6, 3, and 2 years) and three flow conditions from the Blue Nile to the Lake Nasser. The authors concluded that the GERD construction would negatively impact Egypt's groundwater resources even in the more extended filling period (filling GERD in 6–10 years). The filling of GERD's reservoir in 10 years could be the best option, especially in cases of normal flow or minimum of average flow conditions; otherwise, the aquifer's sustainability will not be maintained.

4 Recommendations

The capability to adapt to future issues is a crucial component of managing groundwater resources in the Nile Valley. We confirm that water resources need integrated flexibility in order to achieve this objective. Editorial teams have identified specific aspects that could be explored for further improvement throughout this book project. Based on the authors' results and findings, this chapter offers some recommendations that provide suggestions for future researchers to go beyond this book's scope.

1. The rising of the groundwater levels in the last 20 years in the middle and north of Nile Valley and the high affection of the groundwater in the Nile valley region by seawater intrusion towards the North while decreasing south. Therefore, the

Egyptian government should consider a substantial water conservation policy to face the current challenges and reduce contamination by drilling extraction wells. Future boreholes should be at least 300 m away from any adjacent well and 1 km from any nearby canal.

2. Data collection represents a significant issue in the groundwater modeling process. Parameters with a high effect on model results should get the most attention in the data collection and the calibration process. Future research should be performed based on a continuous recording of groundwater level, water level changes in canals, an actual case of groundwater abstraction, actual representations of heterogeneity of the Nile Valley aquifer system. More field or laboratory works or designed well-pumping tests to measure/estimate the Nile Valley aquifer parameters is essential.
3. In order to reduce and minimize the adverse effects of discharging wastewater into the Nile and its branches, it is highly recommended that (i) More intensive investigation for the whole area, especially that dealing with the environmental pollution as a result of pollutants accumulation and detailed distribution maps should be constructed for the most common pollutants along the Nile valley and delta. (ii) Emphasizing compliance with environmental protection regulations, especially the Environment's Laws regarding wastewater disposal, minimizes these pollutants' risk. (iii) Effective coordination among geoscientists, environmentalists, and planners/policy makers is essential to resolve these problems and throw light on the adverse effect of excessive use of organic (manures) and chemical fertilizers. (iv) Development and completion of the sewage system network in the area and construct drain water treatment stations at the downstream sites of these drains before mixing with the Nile water.
4. Continuous monitoring of groundwater quality in the Nile valley regions is needed to evaluate any degradation of water quality for various uses, which decreases the possibility of contamination in aquifers. Continue and consider the assessment of water quality by GIS and RS techniques since they are instrumental in establishing a time–cost effective method for the routine monitoring of groundwater to protect it from pollutants, especially treated wastewater disposal.
5. An increase in the organic matter concentration of the abstracted water in the Kima industry area (Aswan) was observed compared to the surface water systems. The potential risk of forming carcinogenic trihalomethane compounds during the treatment process is still needed to be investigated. It is recommended that continuous monitoring of groundwater quality is essential. Also, it is a required remedy to construct a sewage network to minimize the risks of the septic tanks. Also, continuous operation of existing groundwater wells with a minimum capacity of 1440 cubic meters per hour and drilling number of wells must provide for irrigation purposes. Moreover, further studies are necessary to be conducted by coupling the soil subsidence model with the groundwater model to regulate the proposed solutions and ensure that the water table's lowering has no significant effects on the soil and will not lead to future instabilities of the infrastructure.

6. For future water supply boreholes, it is highly recommended to choose places in the southwestern, central, and eastern parts of the area. However, before any groundwater exploitation process in this region, more studies must be carried out to manage the basis for groundwater abstraction to overcome as many possible problems associated with exaggerated groundwater abstraction.
7. The authors recommended that decision-makers should stop digging any wells in the northern portion of the study area. The drilled wells' pumping rate in the central and southern regions should also be changed to keep groundwater quality from degradation.
8. Widespread use of the so-called "controlled drainage practices" in Egypt, as such practices contribute much in mitigating the excessive move of fertilizers down to aquifers. Also, the authors recommend ensuring that each component of the agricultural drainage network, whether above ground (like drains) or underground (like drainpipes), works with its designed capacity. That will help drainage systems to perform well and thus ensuring better groundwater quality.
9. More studies are necessary to establish a sustainable level of exploitation from the Pliocene aquifer and assess the vulnerability to avoid any potential pollution associated with excessive groundwater abstraction. Public awareness programmers concerning water conservation's significance should be introduced through different media to build up a new water culture based on the conservation principle.
10. The authors presented some recommendations for managing and protecting the groundwater in the Nile Valley aquifer: Continues monitoring the groundwater quality to assess any deterioration in water quality. Reduce the using of pesticides, herbicides, and fertilizers and expanding the use of organic fertilizers. The irrigation system must be changing from traditional methods (flood) to new methods (drip and sprinkler). The production wells must not be worked simultaneously, and the total yield of these wells must be controlled.
11. Integrating remotely sensed data using GIS technology is very important in characterizing hydrogeological properties and conditions in many environments. Therefore, it is highly recommended that the use of remotely sensed data is necessary before any exploration of geophysical groundwater and is an essential decision-making tool. It is advisable to combine geological, structural, and hydrological data with numerical modeling and GIS techniques to provide rapid and cost-effective information on potential groundwater zones. Harvesting precipitated data during rainstorms in man-made reservoirs would provide additional new water resources and are essential for maintaining water security.
12. In order to mitigate the negative impact of the GERD on the deterioration of the Nile Valley aquifer in Egypt, it is highly recommended that the impact of the 10-year filling of the GERD on the groundwater level of the Nile River Nile Valley aquifer be quantified to ensure the sustainability of groundwater-dependent activities along the Nile Valley. Also, re-investigate the saltwater intrusion in the Nile Delta aquifer, taking into account the different filling scenarios of the GERD reservoir, climate change, and seawater rise.

References

Abdalla F, Shamrukh M (2016) Quantification of River Nile/Quaternary aquifer exchanges via riverbank filtration by hydrochemical and biological indicators. Assuit Egypt J Earth Syst Sci 125–8:1697–1711

Abdalla F, Khalil R (2018) Potential effects of groundwater and surface water contamination in an urban area, Qus City, Upper Egypt. J Afr Earth Sc 141C:164–178

Abdelhalim A, Sefelnasr A, Ismail E (2019) Numerical modeling technique for groundwater management in Samalut city, Minia governorate. Egypt Arab J Geo 12:124

Abdelrady A, Sharma S, Sefelnasr A, El-Rawy M, Kennedy M (2020) Analysis of the performance of bank filtration for water supply in arid climates: case study in Egypt. Water 12:1816

Abdel-Shafy HI, Kamel AH (2016) Groundwater in Egypt issue: resources, location, amount, contamination, protection, renewal, future overview. Egypt J Chem 59(3):321–362. https://doi.org/10.21608/EJCHEM.2016.1085

Ahmed AA (2009) Using lithologic modeling techniques for aquifer characterization and groundwater flow modeling of the Sohag area. Egypt Hydrogeol J 17:1189–1201

Al-Maktoumi A, El-Rawy M, Zekri S (2016) Management options for a multipurpose coastal aquifer in Oman. Arab J Geosci 9:1–14

Al-Maktoumi A, Zekri S, El-Rawy M, Abdalla O, Al-Wardy M, Al-Rawas G, Charabi Y (2018) Assessment of the impact of climate change on coastal aquifers in Oman. Arab J Geosci 11:1–14

Al-Maktoumi A, Zekri S, El-Rawy M, Abdalla O, Al-Abri R, Triki C, Bazargan-Lari MR (2020) Aquifer storage and recovery, and managed aquifer recharge of reclaimed water for management of coastal aquifers. Desalin Water Treat 176:67–77

Awad A, Eldeeb H, El-Rawy M (2020) Assessment of surface water and groundwater interaction using field measurements: a case study of Dairut City, Assuit. Egypt J Eng Sci Technol 15(1):406–425

Awad A, Luo W, Zou J (2021a) DRAINMOD simulation of paddy field drainage strategies and adaptation to future climate change in lower reaches of the Yangtze River basin. Irrig Drain 1–13. https://doi.org/10.1002/ird.2564

Awad A, Wan L, El-Rawy M, Eltarabily MG (2021b) Proper predictions of the water fate in agricultural lands: indispensable condition for better crop water requirements estimates. Ain Shams Eng J. https://doi.org/10.1016/j.asej.2021.02.003

Campos E (2009) A groundwater flow model for water related damages on historic monuments—case study West Luxor Egypt. Vatten 65:247–254

Custodio E (2013) Trends in groundwater pollution: loss of groundwater quality & related services. Groundwater governance: a global framework for action, GEF-World Bank-UNESCO. http://www.groundwatergovernance.org/fileadmin/user_upload/groundwatergovernance/docs/Themat

Dawoud MA, Ismail SS (2013) Saturated and unsaturated River Nile/groundwater aquifer interaction systems in the Nile Valley. Egypt Arab J Geosci 6:2119–2130. https://doi.org/10.1007/s12517-011-0483-4

El Tahlawi M, Farrag A, Ahmed S (2008) Groundwater of Egypt: "an environmental overview." Environ Geol 55:639–652

Elbeih SF, Negm AM, Kostianoi AG (eds) (2020) Environmental remote sensing in Egypt. Springer Nature

El-Fakharany Z, Fekry A (2014) Assessment of New Esna barrage impacts on groundwater and proposed measures. Water Sci 28:65–73

El-Rawy M, Al-Maktoumi A, Zekri S, Abdalla O, Al-Abri R (2019a) Hydrological and economic feasibility of mitigating a stressed coastal aquifer using managed aquifer recharge: a case study of Jamma Aquifer. Oman J Arid Land 11(1):148–159

El-Rawy M, Ismail E, Abdalla O (2019b) Assessment of groundwater quality using GIS, hydrogeochemistry, and factor statistical analysis in Qena Governorate. Egypt Desalin Water Treat 162:14–29

El-Rawy M, Abdalla F, El Alfy M (2020a) Water resources in Egypt. In: The geology of Egypt. Springer, Cham, pp 687–711

El-Rawy M, Batelaan O, Buis K, Anibas C, Mohammed G, Zijl W, Salem A (2020b) Analytical and numerical groundwater flow solutions for the FEMME-modeling environment. Hydrology 7(2):27

El-Rawy M, Fathi H, Abdalla F (2020c) Integration of remote sensing data and in situ measurements to monitor the water quality of the Ismailia Canal, Nile Delta. Egypt Environ Geochem Health 42:2101–2120. https://doi.org/10.1007/s10653-019-00466-5

El-Rawy M, Moghazy HE, Eltarabily MG (2021a) Impacts of decreasing Nile flow on the Nile Valley aquifer in El-Minia Governorate, Egypt. Alexandria Eng J 60:2179–2192

El-Rawy M, Makhloof AA, Hashem MD, Eltarabily MG (2021b) Groundwater management of Quaternary aquifer of the Nile Valley under different recharge and discharge scenarios: a case study Assiut Governorate, Egypt. Ain Shams Engineering University 12(3):2563–2574. https://doi.org/10.1016/j.asej.2021b.02.023.

Eltarabily MG, Negm AM, Yoshimura C, Saavedra OC (2017) Modeling the impact of nitrate fertilizers on groundwater quality in the southern part of the Nile Delta. Egypt Water Sci Technol Water Supply 17:561–570

Eltarabily MG, Negm AM, Yoshimura C, Takemura J (2018) Groundwater modeling in agricultural watershed under different recharge and discharge scenarios fort Quaternary aquifer eastern Nile Delta. Egypt Environ Model Assess 23:289–308

Environmental Affairs Agency (2017) Achievements report of the Egyptian Ministry of Environment for the Year 2016 (in Arabic)

FAO (2016). AQUASTAT country profile—Egypt. http://www.fao.org/3/i9729en/I9729EN.pdf

Fathi H, El-Rawy M (2018) GIS-based evaluation of water quality index for groundwater resources nearby wastewater treatment plants. Egypt. Pollut Res 37:105–116

Hamdan A, Abdel Rady A (2011) Vulnerability of the groundwater in the quaternary aquifer at El Shalal-Kema area, Aswan. Egypt Arab J Geosci 6:337–358. https://doi.org/10.1007/s12517-011-0363-y

Hansen A, Jakobsen R, Refsgaard J, Højberg A, Iversen B, Kjaergaard C (2019) Groundwater dynamics and effect of tile drainage on water flow across the redox interface in a Danish Weichsel till area. Adv Water Resour123:23–39

Holding Company for Water and Wastewater (2020) Treatment plants in upper Egypt. Avilable on. https://www.hcww.com.eg

Ibraheem IM, El-Qady G (2019). Hydrogeophysical investigations at El-Nubariya–Wadi El-Natrun area, West Nile Delta, Egypt. In: Negm A (ed), Groundwater in the Nile Delta. The handbook of environmental chemistry. vol 73, Springer, Cham, pp 235–271

Khangembam S, Kshetrimayum KS (2019) Evaluation of hydrogeochemical controlling factors and water quality index of water resources of the Barak valley of Assam, Northeast India. Groundwater for sustainable development

Kumar C (2019) An overview of commonly used groundwater modelling software. Int J Adv Res Sci Eng Technol 6:7854–7865

Megahed HA, Farrag AE-HA (2019) Groundwater potentiality and evaluation in the Egyptian Nile Valley: case study from Assiut governorate using hydrochemical, bacteriological approach, and GIS techniques. Bull Nat Res Centre 43:48

Mamdouh SM, Farrag AA, El Sayed EEA (2018) Assessment of groundwater resources for irrigation purposes in Assiut governorate, upper Egypt. Int J Res-Granthaalayah 6:248–273. https://doi.org/10.5281/zenodo.1248172

MukateaS ,Wagha V, Panaskara D, Jacobsb JA, Sawantc A (2019) Development of new integrated water quality index (IWQI) model to evaluate the drinking suitability of water. Ecol Ind 101:348–354

MWRI (2014) Arab Republic of Egypt, Ministry of water resources and irrigation. Water scarcity in Egypt. mfa.gov.eg.

Negm AM (Ed) (2018) Groundwater in the Nile Delta (73). Springer International Publishing. https://doi.org/10.1007/978-3-319-94283-4

Negm AM, Armanuos AM (2016) GIS-based spatial distribution of groundwater quality in the western Nile Delta, Egypt. In: The Nile delta. Springer, Cham, pp 89–119

Negm AM and Elkhouly A (2021) Groundwater in Egyptian deserts. Springer International Publishing (accepted for publication)

Nriagu JO (2019) Encyclopaedia of environmental health. Elsevier

Omar ME, Mousa AMA (2016) Water management in Egypt for facing the future challenges. J Adv Res 7(3):403–412. https://doi.org/10.1016/j.jare.2016.02.005

Osman R, Ferrari E, Mcdonald S (2016) Water scarcity and irrigation efficiency in Egypt. Water Econ Policy 2:1650009

Salem A, Dezső J, Lóczy D, El-Rawy M, Loczy D (2020) Hydrological modelling to assess the efficiency of groundwater replenishment through natural reservoirs in the Hungarian Drava River floodplain. Water 12:250

Sefelnasr AM, Omran AA-K, Abdel-Hak HA, El Tahawy WS (2019) GIS-based numerical modelling for the groundwater assessment: a case study in the Quaternary aquifer, assiut governorate. Egypt Arab J Geosci 12:1–13

Şen Z (2015). Practical and applied hydrogeology. Elsevier, 407

Shamrukh M, Corapcioglu MY, Hassona FA (2001) Modeling the effect of chemical fertilizers on ground water quality in the Nile Valley aquifer. Egypt Groundwater 39:59–67

Shokr ME (2020) Environmental applications of remote sensing in Egypt: a review and an outlook. Environ Remote Sens Egypt 95–126

Singh A (2019) Poor-drainage-induced salinization of agricultural lands: management through structural measures. Land Use Policy 82:457–463

Smith EA, Berg AM (2020). Potential groundwater recharge rates for two subsurface-drained agricultural fields, southeastern Minnesota, 2016–18. US geological survey

Solangi GS, Siyal AA, Babar MM, Siyal P (2019). Groundwater quality evaluation using the water quality index (WQI), the synthetic pollution index (SPI), and Geospatial tools: a case study of Sujawal district, Pakistan. Human Ecol Risk Assess. https://doi.org/10.1080/10807039.2019.1588099

Srivastav AL (2020) Chemical fertilizers and pesticides: role in groundwater contamination, in agrochemicals detection. Treatment and remediation, pp 143–159

Suring LH (2020) Freshwater: oasis of life—an overview

World Health Organization (2017). Guidelines for drinking-water quality: incorporating first addendum. http://www.apps.who.int/iris/bitstream/10665/254637/1/

Yasmin G, Islam D, Islam MT, Shariot-Ullah M, Adham AKM (2019) Evaluation of groundwater quality for irrigation and drinking purposes in Barishal district of Bangladesh. Fundam Appl Agric 4(1):632–641. https://doi.org/10.5455/faa.301258

Zhao W, Lin J, Wang S.-F, Liu J-L, Chen Z-R, Kou W-J (2013) Influence of human activities on groundwater environment based on coefficient variation method. Huan jing ke xue= Huanjing kexue (34):1277–1283

The manufacturer's authorised representative in the EU is Springer Nature Customer Service Centre GmbH, Europaplatz 3, 69115 Heidelberg, Germany. If you have any concerns regarding our products, please contact ProductSafety@springernature.com

Printed and bound by CPI Group (UK) Ltd, Croydon, CR0 4YY
12/08/2026
02194532-0004